LIVERPOOL JMU LIBRARY

The Normalized Difference Vegetation Index

The Normalized Difference Vegetation Index

Nathalie Pettorelli
Institute of Zoology,
Zoological Society of London

The Normalized Difference Vegetation Index. First Edition. Nathalie Pettorelli.
© Nathalie Pettorelli 2013. Published 2013 by Oxford University Press.

Great Clarendon Street, Oxford, OX2 6DP,
United Kingdom

Oxford University Press is a department of the University of Oxford.
It furthers the University's objective of excellence in research, scholarship,
and education by publishing worldwide. Oxford is a registered trade mark of
Oxford University Press in the UK and in certain other countries

© Nathalie Pettorelli 2013

The moral rights of the author have been asserted

First Edition published in 2013

Impression: 1

All rights reserved. No part of this publication may be reproduced, stored in
a retrieval system, or transmitted, in any form or by any means, without the
prior permission in writing of Oxford University Press, or as expressly permitted
by law, by licence or under terms agreed with the appropriate reprographics
rights organization. Enquiries concerning reproduction outside the scope of the
above should be sent to the Rights Department, Oxford University Press, at the
address above

You must not circulate this work in any other form
and you must impose this same condition on any acquirer

Published in the United States of America by Oxford University Press
198 Madison Avenue, New York, NY 10016, United States of America

British Library Cataloguing in Publication Data
Data available

Library of Congress Control Number: 2013937138

ISBN 978–0–19–969316–0

Printed and bound by
CPI Group (UK) Ltd, Croydon, CR0 4YY

Links to third party websites are provided by Oxford in good faith and
for information only. Oxford disclaims any responsibility for the materials
contained in any third party website referenced in this work.

Preface

As the human population and its resource requirements increase exponentially, major environmental changes are occurring at increasingly faster rates. Climate change is only part of a suite of environmental changes characterizing the last hundred years, as human expansion has translated into land-use change, habitat degradation, and fragmentation. Our ability to anticipate the effects of such changes on biodiversity, ecosystem services, and ultimately human wellbeing is fundamental to designing appropriate adaptation and mitigation strategies.

Over the last few decades, theoretical and applied ecology have increasingly relied on the wealth of opportunities and information provided by new technological developments, as exemplified by animal tracking devices such as Global Positioning Systems (GPS), camera traps, or increased computational speed for sophisticated analyses such as Bayesian analyses. The Normalized Difference Vegetation Index (NDVI) is among those technological advances useful for ecology: the NDVI is a satellite-based index that has been shown to be highly correlated with plant canopy-absorbed photosynthetically active radiation, photosynthetic capacity, net primary production, leaf area index, fraction of absorbed photosynthetically active radiation, carbon assimilation, and evapotranspiration. The NDVI opens the possibility of addressing questions on scales inaccessible to ground-based methods alone as it is mostly freely available, with global coverage across several decades, and high temporal resolution.

There has been a surge of interest in remote sensing and its use in ecology over the past few years. While there are related books in the area of remote sensing and ecology, surprisingly, none focuses explicitly on the NDVI. With this book I therefore aim to fill this gap and supply an overview of the principles and possible applications of the NDVI in ecology, biodiversity research, environmental management, and conservation. NDVI data can provide valuable information about temporal and spatial changes in vegetation distribution, productivity, and dynamics; allowing monitoring of habitat degradation and fragmentation, or assessment of the ecological effects of climatic disasters such as drought or fire. The NDVI has also provided ecologists with a promising way to couple vegetation with animal distribution, abundance, movement, survival, reproductive parameters, or population dynamics. Over the last few decades, numerous studies have highlighted the potential key role of satellite data and the NDVI in macroecology, plant ecology, animal population dynamics, environmental monitoring, habitat selection and habitat use studies, movement ecology, and paleoecology. In short, this book aims to become the 'ABC' of the NDVI for anyone interested in terrestrial ecology, environmental and wildlife management, and conservation.

The book begins by detailing what vegetation indices and the NDVI are, as well as the principles behind the NDVI, its correlation with climate, the available NDVI datasets, and the possible complications and errors associated with the use of this satellite-based index. Later chapters go on to discuss the possible applications of NDVI in ecology, biodiversity research, environmental and wildlife management, and conservation. The intended audience comprises primarily terrestrial ecologists and conservation biologists, yet the book is targeted as a handbook and will be suitable for advanced undergraduate and postgraduate students in the biological and ecological sciences, as well as specialists in the fields of conservation biology, biodiversity monitoring, remote sensing, and natural resource management.

Acknowledgements

I would like first to thank all my colleagues for their support during the past two years. Special thanks go to Clare Duncan, James Duffy, Michele Walters, Jonathan Baillie, Ben Collen, Sam Turvey, Muki Haklay, Ioan Fazey, Jean-Michel Gaillard, Patrick Duncan, Nils Christian Stenseth, Linda daVolls, Katy Gandon, Woody Turner, Compton Tucker, and Molly Brown: thank you for stimulating discussions; for providing me with opportunities that led to me being able to write such a book; for helping me design a strong book proposal; for helping me think about figures and tables; for helping me compile relevant information; and for your continuous encouragement and understanding during the whole process. Special thanks go to Guy Cowlishaw, Gregoire Dubois, Martin Wegmann, Iain Gordon, Paul de Ornellas, Jakob Bro-Jørgensen, Atle Mysterud, Kamran Safi, Tammy Davies, Christopher Chandler, Aliénor Chauvenet, and William Cornforth for willingness to review the various chapters and making some excellent suggestions for improvements.

I would also like to thank my family and friends for their backing during the writing process. I am particularly grateful to my dad, Jean-Rémy Pettorelli, without whom my work and this book would not have been possible. Thank you for being proud of me through my successes and through my failures. Knowledge is power, said Francis Bacon, and this quote has always been one of your mottos: thank you for showing me the way to access the world of possibilities that a lifetime has to offer. Writing a book is a process that requires, among many things, perseverance: there have been days when my enthusiasm waned—even days when I felt I'd never finish this project. And then there have been days when my mind isolated itself for hours; when I wouldn't move away from the screen for a whole weekend; when the topics of my conversation were limited to just this book. Through the low and high points of this adventure, I had someone who was there for me: Paul George, thank you for your support, love, understanding, and patience. Family and friends is what makes my world go round, being an endless source of inspiration, happiness, and comfort. Thank you Micheline and Angélina Le Somptier, Ken and Beryl George, Dave and Emma Abrey, Myriam and Patrick Martinet, Stéphane Paris, Eve Zeyl, Leif Christian and Ashild Stige, Jeanne-Sophie Aas, Anne Hilborn, Sarah Durant, Seirian Sumner, and Valeri and Dorothee Belov.

Finally I wish to extend my thanks to Oxford University Press staff, especially Ian Sherman, Helen Eaton, and Lucy Nash for their patience and continual support. My experience of book publication is probably among the best one can get! It has been a real pleasure working with you on this project.

Contents

Abbreviations	ix
1 Proxy-based approaches in ecology and the importance of remote sensing	**1**
1.1 A century of unprecedented global changes	1
1.2 Scope for remote sensing	6
1.3 Getting started: the remote sensing user toolkit	12
1.4 Conclusions	17
1.5 Organization of the book	17
2 Vegetation indices	**18**
2.1 Vegetation indices: definition and value in ecological studies	18
2.2 A brief history of vegetation indices	21
2.3 Characteristics of a successful vegetation index	27
2.4 Conclusions	29
3 NDVI from A to Z	**30**
3.1 How does it work?	30
3.2 Available datasets	36
3.3 Known caveats and limitations	37
3.4 Complementing NDVI with other datasets	41
3.5 Conclusions	43
4 Climate and the NDVI: a complex story	**44**
4.1 Climatic variability and the NDVI: global trends	44
4.2 Climatic variability and the NDVI: from global to regional and national trends	48
4.3 Conclusions	55
5 NDVI and environmental monitoring	**56**
5.1 Mapping ecosystem distribution	56
5.2 Predicting disturbances and assessing their impacts on ecosystems	58
5.3 Monitoring changes in the functional attributes of ecosystems	63
5.4 Monitoring habitat loss and degradation	67
5.5 Conclusions	69

6 NDVI and plant ecology — 70

- 6.1 NDVI and plant richness — 70
- 6.2 NDVI and plant distribution — 72
- 6.3 NDVI as a predictor of plant attributes — 74
- 6.4 NDVI and plant physiological status — 75
- 6.5 Informing agriculture — 78
- 6.6 Conclusions — 79

7 NDVI and wildlife management — 81

- 7.1 NDVI, habitat use, and animal movement — 83
- 7.2 NDVI and life histories — 90
- 7.3 NDVI and animal abundance — 95
- 7.4 NDVI and species richness — 96
- 7.5 Conclusions — 98

8 NDVI for informing conservation biology — 100

- 8.1 Supporting the management and expansion of protected areas — 101
- 8.2 NDVI for support of habitat assessments for reintroductions — 104
- 8.3 Identifying corridors — 107
- 8.4 Predicting climate change effects on wildlife — 109
- 8.5 NDVI and invasive species — 110
- 8.6 Conclusions — 113

9 NDVI falls down: exploring situations where it does not work — 115

- 9.1 Diet matters — 115
- 9.2 Scale matters — 119
- 9.3 Location matters — 124
- 9.4 Confounding effects — 125
- 9.5 Conclusions — 126

10 Ecosystem services, NDVI, and national reporting: matches and mismatches — 127

- 10.1 International conventions and platforms on the preservation of biodiversity and ecosystem services — 128
- 10.2 NDVI and the legal sphere — 135
- 10.3 Conclusions — 141

11 Future directions and challenges — 142

- 11.1 Availability of NDVI data for conservation work — 142
- 11.2 NDVI datasets in the future — 146
- 11.3 NDVI in the freshwater and marine environments — 148
- 11.4 Alternatives to the NDVI — 150
- 11.5 Constraints on the use of remote sensing information — 154
- 11.6 General conclusions — 155

Bibliography — 157
Index — 193

Abbreviations

AE or AET	Actual EvapoTranspiration	FPAR	Fraction of Photosynthetically Active Radiation
ALOS	Advanced Land Observing Satellite		
ANPP	Above-ground Net Primary Production	GBIF	Global Biodiversity Information Facility
AO	Arctic Oscillation	GEF	Global Environment Facility
ASAR	Advanced Synthetic Aperture Radar	GEO-BON	Group on Earth Observations Biodiversity Observation Network
ASTER	Advanced Spaceborne Thermal Emission and Reflection Radiometer	GIMMS	Global Inventory Modelling and Mapping Studies
AVHRR	Advanced Very High Resolution Radiometer	GIS	Geographic Information Systems
BRDF	Bidirectional Reflectance Distribution Function	GPP	Gross Primary Production
		GPS	Global Positioning System
CBD	Convention on Biological Diversity	GVI	Global Vegetation Index
CITES	Convention on International Trade in Endangered Species of Wild Fauna and Flora	HRG	High Resolution Geometrical instrument
CMS	Convention on Migratory Species	HRVIR	Haute Résolution dans le Visible et l'Infra-Rouge
CNES	Centre National d'Etudes Spatiales		
COP	Conference of the Parties	INDVI	Integrated NDVI
DEM	Digital Elevation Model	IOD	Indian Ocean Dipole
DOPA	Digital Observatory of Protected Areas	IPBES	Intergovernmental Platform on Biodiversity and Ecosystem Services
EBV	Essential Biodiversity Variable	IPCC	Intergovernmental Panel on Climate Change
ENSO	El Niño Southern Oscillation		
ENVISAT	Environmental Satellite	IRS	Indian Remote Sensing Satellite
ESA	European Space Agency	IUCN	International Union for Conservation of Nature
ERS	Earth Remote Sensing Satellite		
ERTS	Earth Resources Technology Satellite		
ET	EvapoTranspiration	JAXA	Japan Aerospace Exploration Agency
ETM	Enhanced Thematic Mapper		
ETM+	Enhanced Thematic Mapper Plus sensor	LACIE	Large Area Crop Inventory Experiement
EOS	Earth Observing System	LAI	Leaf Area Index
ERS-2	European Remote Sensing satellite 2	LiDAR	Light Detection And Ranging
EVI	Enhanced Vegetation Index	LISS	Linear Imaging Self-scanning Sensor
		LTDR	Land Long Term Data Record
FAO	Food and Agricultural Organization		
fAPAR	fraction of Absorbed Photosynthetically Active Radiation	MEA	Multilateral Environmental Agreements
		MEI	Multivariate ENSO Index
FASIR	Fourier-Adjusted, Sensor and Solar zenith angle-corrected, Interpolated, Reconstructed	MERIS	MEdium Resolution Imaging Spectrometer
		MODIS	MODerate resolution Imaging Spectroradiometer

MSS	MultiSpectral Scanner	RGB	Red, Green, Blue
MVC	Maximum Value Composite	RVI	Ratio Vegetation Index
NASA	National Aeronautics and Space Administration	SAR	Synthetic-Aperture Radar
		SAVI	Soil Adjusted Vegetation Index
NAO	North Atlantic Oscillation	SDM	Species Distribution Model
NDVI	Normalized Vegetation Difference Index	SeaWiFS	Sea-viewing Wide Field-of-view Sensor
NIR	Near InfraRed	SIR-C	Shuttle Imaging Radar band C
NOAA	National Oceanic and Atmospheric Administration	SOI	Southern Oscillation Index
		SPOT	Satellite Pour l'Observation de la Terre
NPP	Net Primary Productivity	SRTM	Shuttle Radar Topography Mission
		SST	Sea Surface Temperature
PAL	Pathfinder AVHRR Land	SWIR	Short-Wave InfraRed
PALSAR	Phased Array type L-band Synthetic Aperture Radar	TM	Thematic Mapper
PAR	Photosynthetically Active Radiation	TRMM	Tropical Rainfall Measurement Mission
PET	Potential EvapoTranspiration	TVI	Transformed Vegetation Index
PDO	Pacific Decadal Oscillation		
		UNEP	United Nations Environment Programme
QSCAT	Quick SCATterometer	UNFCCC	United Nations Framework Convention on Climate Change
RADAR	RAdio Detection And Ranging	USGS	United States Geological Survey
REDD	Reducing Emissions from Deforestation and forest Degradation	VIS	VIsible Spectrum

CHAPTER 1

Proxy-based approaches in ecology and the importance of remote sensing

Difficulties strengthen the mind, as labor does the body.   **Lucius Annaeus Seneca**

Whenever I open a book on remote sensing, or read a review article on the subject, there is always a section, generally at the start of the document, about why ecologists should become familiar with, and making use of, remote sensing techniques and information. Rarely, however, do I find myself reading about why remote sensing experts should be interested in ecological research. This is to be deplored, as getting the remote sensing community engaged with ecology and conservation is, I believe, as important as getting the ecologists engaged with remote sensing.

With this first chapter, and more broadly with this book, rather than hoping merely to enthuse ecologists about remote sensing, I aim to enthuse the current and future scientific community about carrying out research at the interface between ecology, conservation, and remote sensing. To this end, Section 1.1 begins with a brief overview of the main challenges posed by environmental change to scientists, wildlife managers, and human societies as a whole, as setting such a context is helpful for understanding the full potential of remote sensing to support biodiversity monitoring, as well as the assessment of the consequences of environmental change on biological diversity and the functioning of ecosystems. Section 1.2 discusses the scope for remote sensing to inform ecological research and conservation efforts, and introduces some of the sensors found widely in the ecological literature. Section 1.3 sheds light on some of the technical aspects inherent to remote sensing data manipulation and interpretation, such as access to relevant hardware and software facilities. Section 1.4 explains the structure of the book.

1.1 A century of unprecedented global changes

Many would agree that the twentieth century has been quite unique in terms of human expansion and global changes: between 1911 and 2011, the global human population has experienced its highest ever growth rate—doubling, and doubling again from an estimated 1.6 billion <http://www.un.org> to roughly seven billion (a number thought to have been reached in October 2011 <http://www.census.gov>). With more than five billion extra people to feed, the agricultural sector has had (and must continue) to increase its production: failure to secure access to an affordable food supply can trigger political instability (see, e.g., Brinkman and Hendrix 2011), leading to food security being currently placed high on the global political agenda. Increasing food requirements are expected to put a greater pressure on land, water, and biological resources. It is anticipated that there will need to be a 43% increase in crop production and a 124% increase in meat production to achieve the world's food requirements by 2030 (Food and Agriculture Organization 2009). Growing food demands are expected to lead to an increasing amount of land being managed for agricultural production. Production is likely to intensify on land currently used for agriculture (Smith et al. 2010). A growing demand for food coupled with increasing food prices is then likely to increase the pressure on non-agricultural sources of protein: this could mean increasing bushmeat trade and increasing pressure on targeted wildlife species (Gordon et al. 2012).

Increased pressure on arable land has led to, and is still leading to, disruptions in the natural nitrogen cycle (Stark and Richards 2008). These disruptions originate from fertilizers being used to boost crop production; planting of legumes, including soybeans (legumes are attractive hosts for nitrogen-fixing microbes and therefore enrich the soil where they grow); and formation of nitrogen oxide during fuel combustion. Between 1900 and 1950, global soil nitrogen surplus was estimated to have almost doubled to 36 trillion grams per year, whereas between 1950 and 2000, this global surplus increased to 138 trillion grams per year (Bouwman et al. 2011). Among the consequences of disruptions in the natural nitrogen cycle are the reduction of the protective ozone layer; the contribution towards increased frequency of acid rain; the eutrophication of aquatic systems; and the contamination of drinking water.

Parallel to the increase in human population, changes in the spatial distribution of humans have occurred, and within a hundred years the proportion of people living in urban areas has increased from 13% to an estimated 50% (<http://www.un.org>; Thomas 2008). Urbanization is an important driver of land use change, greenhouse gas emissions, and pollution (Grimm et al. 2008): for example, deforestation in the tropics has been reported to be positively correlated with local urban population growth (DeFries et al. 2010). Urbanization has also been reported to raise consumption levels, leading to an increase in demand for agricultural products (DeFries et al. 2010). Yet the processing and transport sectors, as well as the animal production sector, are important contributors to carbon dioxide emissions, probably explaining why most of the increase in carbon dioxide in the atmosphere is attributable to energy consumption in the world's cities (Thomas 2008).

The twentieth century has moreover seen our world become a significantly smaller place. The development of the commercial aviation and car industries has dramatically increased our ability to travel, both in terms of travelling distance and frequency of travel, increasing immigration, emigration, as well as technology and goods exchange worldwide. However, greater travel and exchange opportunities have facilitated intentional and unintentional introductions of new species and diseases into new environments. Such introductions have sometimes led to these species becoming established and dispersing, negatively impacting on local ecosystems and species. The increased emergence of the so-called invasive alien species is having dramatic effects on biodiversity (Blackburn et al. 2010), and this influence is expected to continue. In 2005, the International Union for Conservation of Nature Red List database described the causes of extinction for 170 species: invasive alien species contributed to the loss of 54% of them, and were the sole cause of extinction for 20% (Clavero and García-Berthou 2005). By comparison, habitat destruction contributed to the demise of 48% of these 170 extinct species (Blackburn et al. 2010).

With the advent of cars and aeroplanes during the twentieth century, oil has rapidly become the dominant source of energy: in 2005, total worldwide energy consumption was ~500 EJ, with 80–90% derived from the combustion of fossil fuels <http://www.eia.doe.gov>. Fossil fuel combustion leads to the emission of greenhouse gases such as carbon dioxide, methane, and nitrous oxide, and greater fossil fuel consumption leads to greater greenhouse gas emissions (Intergovernmental Panel on Climate Change (IPCC) 2001; 2007). The energy sector has not been solely responsible for greenhouse gas emissions: the loss of terrestrial ecosystem carbon stores through land use change, primarily deforestation, currently accounts for >20% of anthropogenic emissions (equivalent to ~5.8 Gt of carbon dioxide per year; IPCC 2007; Figure 1.1).

Greater greenhouse gas emissions have led to observable increases in global atmospheric gas concentrations: carbon dioxide rose from a pre-industrial value of ~280 ppm to 379 ppm in 2005, methane from ~715 to 1774 ppb, and nitrous oxide from ~270 to 319 ppb. These changes began to impact climate and ecosystem functioning during the last century, and in 2007 the IPCC (see Box 1.1

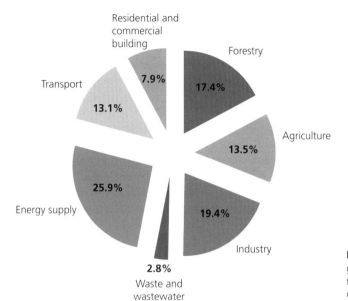

Figure 1.1 Source of greenhouse gas emissions, as assessed by the Intergovernmental Panel for Climate Change (2007).

Box 1.1 What is the IPCC?

The Intergovernmental Panel on Climate Change, or IPCC, is the leading international body for the assessment of climate change. The panel was established by the United Nations Environment Programme (see also Box 10.1) and the World Meteorological Organization in 1988 in order to provide nations with a clear scientific view on the current state of knowledge in climate change and its potential environmental and socio-economic impacts. The IPCC is an intergovernmental scientific body whose task is to review and assess the most recent information produced worldwide relevant to the understanding of climate change. The Panel's work is shared among three Working Groups, a Task Force and a Task Group: Working Group I deals with 'The Physical Science Basis of Climate Change'; Working Group II with 'Climate Change Impacts, Adaptation and Vulnerability'; Working Group III with 'Mitigation of Climate Change'. The main objective of the Task Force is to develop and refine a methodology for the calculation and reporting of national greenhouse gas emissions and removals, while the Task Group on Data and Scenario Support for Impacts and Climate Analysis was established to facilitate co-operation between the climate modelling and climate impacts assessment communities. More information on the IPCC can be found at <http://www.ipcc.ch>.

for more information) published a now-famous report clearly linking human activity, greenhouse gas emissions, and changes in climatic conditions (IPCC 2007). That year, the world learned that within one hundred years, the global surface temperature had increased by $0.74 \pm 0.18°C$. This increase translated into glacier reduction, sea surface ice reduction, and sea level increase (Meehl et al. 2005; IPCC 2007); for example, since 1900 the maximum areal extent of seasonally frozen ground has decreased by about 7% in the Northern Hemisphere, with decreases in spring of up to 15% (IPCC 2007). Global average sea level rose at an average rate of 1.8 (95% CI: 1.3–2.3) mm per year over 1961 to 2003 and at an average rate of about 3.1 (95% CI: 2.4–3.8) mm per year from 1993 to 2003 (IPCC 2007); eleven of the twelve years preceding 2007 (1995–2006) rank among the twelve warmest years in the instrumental record of global surface temperature since 1850 (IPCC 2007).

Typically, climate change is considered in terms of rising average temperature, but shifts in seasonal patterns also occur. For example, the average start of the growing season shifted by eight days from 1989 to 1998 in Europe (Chmielewski and Rötzer 2002) and by five to six days from 1959 to 1993 in North America (Schwartz and Reiter 2000). Springtime in

Europe was also shown to have advanced by eight days on average over the last 80 years and even faster over the last 40 years (Ahas 1999; Ahas et al. 2002). Analysis of climatological variables, such as last frost date of spring or first frost date of fall, confirmed such trends, with an estimated lengthening of the growing season of 1.1–4.9 days per decade since 1951 (Menzel et al. 2003). In North America, significant delays in the beginning, peak, and closing stages of the monsoon have also been reported in recent decades (Grantz et al. 2007). There is then growing evidence at global, regional, and local scales that climatic disasters (such as cyclones, droughts, wildfires, or floods) are becoming more frequent (Easterling et al. 2000; IPCC 2012). Munich Re, one of the world's largest insurance companies, has been compiling data on the occurrence of natural disasters around the world: the company recently reported that the collated information indicates a significant increase in the number of climate disasters over the period 1950–2010 <http://www.munichre.com>. Since these data come from an insurance company, it could be argued that this upward trend is linked to more clients living in areas likely to be hit by climatic disasters. Yet other studies report trends that are in line with this general assessment: for example, an overall increase in the number of warm days and nights, as well as in the number of heat waves, has been reported at the global scale over the same period (IPCC 2012). Droughts were shown to be on the increase over the second half of the last century (Dai et al. 2004; Barnett et al. 2005), and they have recently been shown to occur more frequently in the Mediterranean over the last 100 years (Hoerling et al. 2011). An upward trend in both the frequency and intensity of heavy precipitation events has been found over the USA during the twentieth century (Karl and Knight 1998), and a subsequent study of global precipitation gauge records confirmed these positive trends in wet extremes over much of the world (Groisman et al. 2005). In Canada, forest fire occurrence is expected to increase by 25% by 2030 and by 75% by the end of the twenty-first century (Wotton et al. 2010). Finally, several studies suggest that increased concentrations of greenhouse gases have already produced a substantial rise in Atlantic tropical cyclone activity: recent modelling work projects a near-doubling of the frequency of category 4 and 5 storms by the end of the twenty-first century, despite a decrease in the overall frequency of tropical cyclones (Mann and Emmanuel 2006; Holland and Webster 2007; Bender et al. 2010).

Plants and animals are directly and/or indirectly responding to such changes in climatic patterns (Parmesan 2006; Foden et al. 2008; Steltzer and Post 2009): species distribution ranges have been reported to shift; migrant birds have been observed arriving earlier and starting to breed earlier (Walther et al. 2002; Webb et al. 2011). In Western Europe, recent climate warming has resulted in a significant upward shift in plant species optimum elevation averaging 29 m per decade (Lenoir et al. 2008). In the UK, 23 of the 24 resident temperate odonates (dragonflies and damselflies) have been reported to have expanded their northern range limit, with a mean northward shift of 88 km over the period 1960–1995 (Hickling et al. 2005). In central California, 70% of the 23 butterfly species monitored advanced their first flight date by an average of 24 days during a period of just over 30 years (Forister and Shapiro 2003). Extreme climatic events can generate high mortalities in wildlife populations: in Svalbard, an extreme icing event ('locked pastures') led to a reindeer *Rangifer tarandus* population being reduced by 80% over a single winter (Chan et al. 2005); in Belize, a population of black howler monkeys *Alouatta pigra* inhabiting a tropical forest was severely deprived of its top food items by hurricane Iris, resulting in a loss of 42% of the population (Pavelka and Behie 2005). With extreme climatic events expected to occur more frequently and with greater intensity in the coming decades (IPCC 2012), there is increasing awareness of the potential threat that they could represent to biodiversity (Ameca et al. 2012). This might be especially true for those populations that are already being weakened by increased habitat loss, fragmentation, and degradation.

Based on these observations, it is evident that we have entered a time of unprecedented global change due to the increasing impacts of humans (Millennium Ecosystem Assessment Board 2005), with biodiversity being substantially altered from its historic state. Current rates of species extinction are unparalleled. Most indicators of the state of biodiversity show declines, with no significant recent reductions in rate, whereas indicators of pressures on biodiversity (including resource consumption, invasive alien species, nitrogen pollution, overexploitation, and climate change impacts) show increases (Butchart et al. 2010). Driven mainly by human activities, species are currently being lost one hundred to one thousand times faster than the natural rate: many therefore believe that we are currently witnessing the sixth mass extinction in the history of life on Earth (Ceballos et al. 2010).

Why should we be bothered about this? Biological diversity lies at the very heart of the maintenance and future sustainability of the functioning of the ecosystems upon which we depend for food and fresh water, health and recreation, as well as protection from natural disasters. For example, grasslands with more plant species have been reported to be associated with greater productivity and stability (Tilman et al. 2012), while increased pollinator diversity has been shown to result in greater yield of some agricultural crops (Hoehn et al. 2008). Current trends in biodiversity loss are bringing us closer to a number of potential tipping points that would catastrophically reduce the capacity of ecosystems to provide these essential services (see Box 1.2 for a detailed definition of ecosystem services): marine biodiversity loss, for example, is increasingly impairing the ocean's capacity to provide food, maintain water quality, and recover from perturbations (Worm et al. 2006). As well as being detrimental to our wellbeing, biodiversity loss is costly for our society as a whole, particularly for economic actors in sectors that depend directly on ecosystem services. To illustrate this statement, insect pollination in the EU has been recently reported to have an estimated economic value of €15 billion per year, meaning that the continued decline in bees and other pollinators could have serious consequences for Europe's farmers and agri-business sector (Gallai et al. 2009; <http://www.teebweb.org>). The conservation of biodiversity also makes an important contribution to moderating the scale of climate change and reducing its negative impacts on the functioning of ecosystems. This makes biodiversity loss the most important global environmental threat alongside climate change – and the two are inextricably linked (Collen et al. 2013).

How can we reduce and mitigate the impact of global environmental change on the delivery of ecosystem services and tackle the challenges

Box 1.2 Some definitions

Biodiversity: This is a composite term embracing the variety of types, forms, spatial arrangements, processes, and interactions of biological systems at all scales and levels of organization, from genes to species and ecosystems, along with the evolutionary history that led to their existence (CBD 1992; Scholes et al. 2008).

Ecosystem functioning: This refers to the collective effect of multiple ecosystem processes that ultimately determine the rates of matter and energy fluxes (Hooper et al. 2005). Such ecosystem processes include primary production, gas ecosystem exchange, energy balance, evapotranspiration, nitrogen mineralization, decomposition, and nutrient losses.

Ecosystem services: The term 'ecosystem service' was first used by Ehrlich and Ehrlich (1981), and broadly refers to the benefits that people obtain from ecosystems (Millennium Ecosystem Assessment Board 2005). As numerous definitions of ecosystem services exist, several scholars consider the term as being an evolving concept (Carpenter et al. 2006; Sachs and Reid 2006). Costanza et al. (1997), for example, defined ecosystem services as the representation of goods and services derived from ecosystem functions, while Daily (1997) considered ecosystem services as the conditions and processes of natural ecosystem fulfilling human life. Ecosystem services can be divided into four groups: provisioning services (representing the products obtained from ecosystems); regulating services (representing the benefits obtained from regulation of ecosystem processes); cultural services (representing the non-material benefits obtained from ecosystems); and supporting services (representing the services necessary for the production of all other ecosystem services). Examples of provisioning services include food, fresh water, fuel wood, fibre, biochemical and genetic resources. Examples of regulating services include climate regulation, pest regulation, run-off regulation, water purification, pollination, and erosion regulation. Examples of cultural services include recreation and tourism, aesthetic and inspirational values, educational value, and cultural heritage. Finally, examples of provisioning services include soil formation, nutrient cycling, primary production, provision of habitat, and oxygen production (Millennium Ecosystem Assessment Board 2005).

of maintaining healthy, functioning ecosystems (see Box 1.2 for a definition of ecosystem functioning), for the benefits of human wellbeing? Many would agree that addressing such challenges requires, among other things, the ability to detect, understand and predict the impact of environmental change on biodiversity and ecosystem functioning (Naeem et al. 1999). To do so ideally requires being able to monitor biodiversity, ecosystem distribution, and ecosystem functioning, as well as the effectiveness of management activities using standardized, comparable methodologies across the globe. Field data are generally difficult to use in assessing and predicting how environmental change affects ecosystems and wildlife, as such data are traditionally collected at small spatial and temporal scales, and vary in their type and reliability. Remote sensing data, on the other hand, offer a relatively inexpensive and verifiable means of recurrently deriving consistent and spatially complete environmental information (Duro et al. 2007).

1.2 Scope for remote sensing

1.2.1 Remote sensing: definition

There are many definitions of remote sensing: Cowell defined it as the act of examining photographic images for the purpose of identifying objects and judging their significance (Cowell 1983); Clarke viewed remote sensing as a tool to extract information from objects or land areas, using sensors to detect electromagnetic energy (Clarke 2007); Jones and Vaughan recently defined it as the acquisition of information by the use of a sensing device separated from the target (Jones and Vaughan 2010). The term is thought to have been first used in the USA in the 1950s, and remote sensing is now commonly used to describe the science (some people might even speak of art) of identifying, observing, and measuring an object without coming into direct contact with it. So, taking a family picture is, for example, a form of remote sensing, and remote sensing data can be collected using a range of observing platforms, from hand-held (e.g. cameras) to aircraft and satellites.

Over the past decades, remote sensing has been very successful in ecology, as scientists and wildlife managers have increasingly been drawn towards new technological developments for non-invasive, remote-based biodiversity monitoring. Examples include camera traps (O'Connell et al. 2011) or the increased use of microphone arrays for bio-acoustic monitoring (Blumstein et al. 2011). Throughout this book, however, I will use the term remote sensing primarily to refer to the detection of electromagnetic radiation (or energy) from the Earth's surface by sensors on board satellites (Turner et al. 2003).

1.2.2 Remote sensing: how does it work?

Remote sensing relies on the principle that all objects emit radiation, albeit in various amounts and at differing wavelengths. Radiation travels in a wave-like manner and the distance between wave peaks is known as the wavelength, also referred to as band or layer. The collective organization of these emissions by wavelength forms the electromagnetic spectrum (Jensen 2007; see also Box 1.3 for a definition of radiation and electromagnetic spectrum).

All matter, whether gaseous particles or rock, under specific conditions will have a specific and repeatable reaction to incident radiation. There are four ways in which a material can respond; absorption, reflection, transmittance, and scattering. All objects reflect different wavelengths of light to different extents (Figure 1.2), and for most surfaces it can be assumed that the remaining portion is absorbed. For example, a red carpet appears red because the carpet absorbs red light poorly. Likewise, chlorophyll *a* absorbs in the red band but poorly in the green band, meaning that objects which contain a lot of chlorophyll *a*, such as healthy leaves, tend to be green (Jensen 2007). The amount of light reflected from a surface for a specific wavelength is determined by the amount of irradiance at that wavelength that impinges the surface, as well as the reflectance properties of the given surface (Jackson and Huete 1991).

Humans can only see a very small part of the electromagnetic spectrum, a part referred to as the visible spectrum. Electromagnetic radiation in this range of wavelengths is called visible light. A typical human eye will respond to wavelengths from about 400 to 750 nm. However, unlike bees or butterflies, we cannot see what is happening in the ultraviolet or near-infrared part of the spectrum. Sensors (see Box 1.3 for definitions), on the other hand, are able to capture information across a much wider range of

Box 1.3 Some more definitions

Radiation: Energy transferred as particles or waves through space or other media. In remote sensing radiation often comes from the sun although it can also come from the sensor (i.e. in the case of active sensors; see definition below).

Electromagnetic spectrum: This represents the range of wavelengths of electromagnetic radiation. Remote sensing applications typically use wavelengths that include the visible (blue to red), the infrared, and microwave regions of the electromagnetic spectrum. The shorter wavelength ultraviolet, X-ray, and gamma-rays, as well as the long wavelength radio waves, are not typically used.

Infrared: This refers to the portion of the electromagnetic spectrum that lies between the visible and microwave wavelengths (0.7 nm to 100 μm; see Figure 1.3).

Sensor: A device that is capable of recording the intensity of electromagnetic radiation.

Radiometer: An instrument that measures the intensity of electromagnetic energy in different wavelengths.

Band: This can be defined as a single layer of an image created using a specific range of wavelengths. For example, a colour digital image is composed of three bands that record red, green, and blue wavelengths of light. 'Channel' is typically used as a synonym of band.

Multispectral: Refers to multiple bands.

Optical sensor: This refers to a sensor that is sensitive to visible and infrared wavelengths of light.

Passive sensors: These detect natural radiation that is emitted or reflected by the object being observed. Reflected sunlight is the most common source of radiation measured by passive sensors: a famous example of a passive sensor is film photography. Passive sensors on board satellites measure natural radiation that is emitted or reflected by the Earth: these types of sensors are all sensitive to cloud cover.

Active sensors: These emit a pulse and later measure the energy bounced back to a detector.

RADAR (acronym for RAdio Detection And Ranging) is a famous example of an active remote sensing: in this case, the time delay between emission and return is measured, allowing determination of the location, height, speed, and direction of the object being observed.

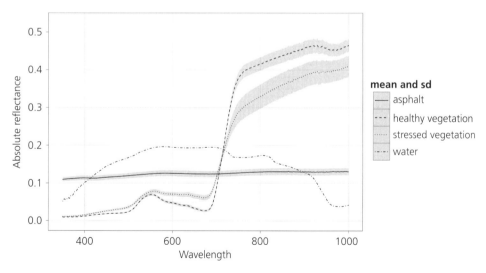

Figure 1.2 Average spectral reflectance signatures of healthy and stressed grass (lawn), asphalt, and water. Wavelength unit here is nanometres. Reflectance is expressed as a ratio of the reflected over the incoming radiation: therefore, reflectance only takes on values between 0 and 1. *In situ* measurements were obtained using an ASD FieldSpec Pro handheld spectrometer. Mean and standard deviation for absolute reflectance were calculated across twenty measurements for each target. The measurements and illustration are courtesy of B. Leutner and Dr M. Wegmann, Department of Remote Sensing, University of Wuerzburg and German Aerospace Data Center (DLR).

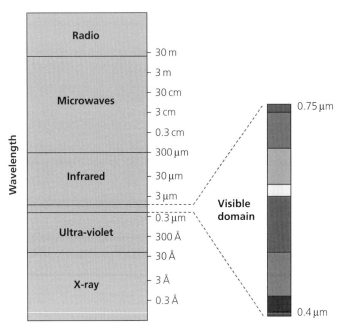

Figure 1.3 The electromagnetic spectrum. The human eye can perceive information only from a very small part of the electromagnetic spectrum, namely the visible spectrum (which ranges from ~400 nm to 750 nm). See also Plate 1.

the spectrum, and can acquire data in hundreds of different bands (Horning et al. 2010; Figure 1.3).

There are four primary variables that can be observed from space (Robinson 2003), namely colour (considered in its wider sense to include near to mid-infrared reflected radiation), temperature, roughness, and height. This information is gathered by sensors, which can be categorized as 'passive' or 'active'. 'Passive sensors' measure radiation from a given target without the sensor first transmitting a pulse of radiation. In these cases, data are derived from measuring the electromagnetic radiation reflected, emitted, or scattered towards it (Campbell 2007). 'Active' sensors on the other hand emit a pulse and later measure the energy backscattered towards them (Turner et al. 2003; see also Box 1.3). In ecological studies, passive sensors that record visible and infrared wavelengths (from reflected and scattered radiation) and thermal wavelengths (from emitted radiation), and active ones recording microwave wavelengths, are currently the most frequently used (Kerr and Ostrovsky 2003; Turner et al. 2003; Gillespie et al. 2008; Pettorelli et al. 2011).

Sensors collect information at different spatial, radiometric, spectral, and temporal resolutions. Spatial resolution here refers to the size of the ground element represented by an individual pixel. This is generally determined by the optical system of the sensor used to capture the data and the height of the sensor. Radiometric resolution refers to the number of distinct levels into which the intensity of the signal is divided, so the lowest possible radiometric resolution would be a sensor producing only two levels of intensity, i.e. black and white. Spectral resolution refers to the ability of a sensor to detect fine wavelength intervals, so the narrower the range of wavelengths a sensor responds to, the better its spectral resolution. Temporal resolution refers to the time between successive images of the same features, usually set by the orbital repeat of a satellite (Jones and Vaughan 2010).

1.2.3 Remote sensing: a world of possibilities

There are many satellites currently in orbit, equipped with sensors collecting information at different spatial, radiometric, spectral, and temporal resolutions (see Table 1.1). The aim of this section is to highlight the diversity of opportunities associated with satellite monitoring by presenting a selection of relatively well-known types of sensor.

Table 1.1 Non-exhaustive list of relevant satellite sensors for biodiversity monitoring and assessment.

Sensor	Type	Spatial resolution	First launched
QuickBird	Passive	0.6–2.5 m	2001
IKONOS-2	Passive	1–4 m	1999
OrbView	Passive	1–4 m	1997
GeoEye	Passive	0.41–1.65 m	2008
Landsat/MSS	Passive	56–82 m	1972, 1975, 1978, 1982, 1984
Landsat (TM, ETM +)	Passive	15–60 m	1982, 1984, 1999
RESOURCESAT 1 IRS/P6 (including LISS-3, LISS-4, AWiFS)	Passive	5.8–56 m	1996, 2003
Terra/ASTER	Passive	15–90 m	2000
SPOT VEGETATION	Passive	1150 m	1986
SPOT 4/HRVIR	Passive	10–20 m	1998
SPOT 5/HRG	Passive	2.5–20 m	2002
IR-MSS/CBERS2, 2b	Passive	20 m	2003, 2006
NOAA (AVHRR)	Passive	1100 m	1981
Terra/MODIS	Passive	250–1000 m	2000
SeaWIFS	Passive	1100 m	1997
ENVISAT/MERIS	Passive	260–1200 m	2002
ENVISAT/ASAR C-band	Active	150 m	2002
ALOS/PALSAR	Active	7–88 m	2006
SRTM	Active	30–90 m	2000
QSCAT	Active	2500 m	1999
SIR-C	Active	10–200 m	1994
TRMM Microwave Imager	Active	18 000 m	1997
ERS-2	Active	26 m	1995

Sources: Strand et al. (2007) and Gillespie et al. (2008).

1.2.3.1 Monitoring at very high spatial resolution

It can be argued that monitoring at very high spatial resolution began with the Satellite Pour l'Observation de la Terre (SPOT), a system initiated by the Centre National d'Etudes Spatiales (CNES) in the 1970s and developed in association with the Belgian Scientific, Technical and Cultural Services and the Swedish National Space Board. The first satellite, SPOT 1, was launched in 1986 and was designed to produce multispectral pictures at a 20 m resolution. The current SPOT satellite, SPOT 5, was launched in 2002 with 2.5 m to 5 m (panchromatic mode) and 10 m (multispectral mode) spatial resolution capabilities.

But the real explosion of 'Google Earth' type satellite imagery occurred in early 2000, with the launch of IKONOS and QuickBird. IKONOS imagery began being sold in 2000, and IKONOS was the first satellite to collect publicly-available high resolution imagery at 1 m (panchromatic mode) and 4 m (multispectral mode) spatial resolutions. At such resolution, details such as buildings or other infrastructure are visible. QuickBird, on the other hand, was launched in 2001 and collects panchromatic imagery at 60–70 cm resolution and multispectral imagery at 2.4 and 2.8 m resolutions <http://www.digitalglobe.com>. The latest satellite from GeoEye

Inc., GeoEye-1, is currently the world's highest resolution commercial Earth-imaging satellite, with images collected at 41 cm (panchromatic mode) and 1.65 m (multispectral imagery) resolutions. Satellites such as QuickBird, IKONOS or GeoEye-1 can be perfect for gathering high resolution, spatially precise information, and may help to map the distribution of human infrastructures such as road or rail networks, or city delimitation. In very rare cases, these satellites can also help ecologists to detect and count wildlife: Fretwell et al. (2012) demonstrated how images from the QuickBird, WorldView-2, and IKONOS satellites could be used to detect emperor penguin *Aptenodytes fosteri* colonies and estimate colony size, at the scale of the whole continental coastline of Antarctica. In Australia, very high resolution SPOT 5 and Quickbird imagery helped in mapping the distribution of an invasive plant species, *Lantana camara* (Taylor et al. 2011). Because of the costs of very high resolution images (all of these satellites are commercial), and because of the number of images required to cover a typical study area, such satellites are rarely used in ecology and conservation.

1.2.3.2 Monitoring at lower resolution: the example of Landsat

The Landsat Program is a series of Earth-observing satellite missions jointly managed by the National Aeronautics and Space Administration (NASA) and the US Geological Survey (USGS). The Landsat system consists of spacecraft-borne sensors that observe the Earth and transmit information by microwave signals to ground stations that receive and process the data for dissemination. The Earth Resources Technology Satellite, also known as the first Landsat satellite, was launched in 1972, and the most recent, Landsat 8, was launched in May 2013. Landsat sensors have a moderate spatial resolution (Landsat 8 spatial resolution ranges from 15 to 100 m). Landsat provides access to an important spatial resolution that is coarse enough for global coverage, yet detailed enough to characterize human-scale processes such as urban growth. Landsat satellite data are the only record of global land surface conditions at a spatial scale of tens of metres spanning the last thirty years (Tucker et al. 2004).

The particularity of Landsat satellites is that they collect several images at once (all of the images are obtained at the same time, and at exactly the same location), with each image showing a specific section of the electromagnetic spectrum. Landsat TM, for example, collects information in seven bands, with each band maximizing the ability to differentiate particular objects or structures. Band 1 helps coastal water mapping, soil/vegetation discrimination, forest classification, and man-made feature identification. Band 2 helps vegetation discrimination and health monitoring, as well as man-made feature identification. Band 3 helps plant species identification and man-made feature identification, while Band 4 is useful when monitoring soil moisture and vegetation, as well as for water body discrimination. Band 5 is generally used for vegetation moisture content monitoring and Band 6 for monitoring surface temperature, vegetation stress, and soil moisture. Band 6 can also help with cloud differentiation and volcanic monitoring. Band 7 helps with mineral and rock discrimination, as well as with vegetation moisture content monitoring (see Table 1.2). It is the combination of the information collected in each band that allows the mapping of specific habitats.

Landsat data are probably among the most widely used satellite imagery in ecology and conservation. Examples of applications include coral reef (Ahmad and Neil 1994; Palandro et al. 2008), mangrove (Giri et al. 2011), and seagrass meadow (Wabnitz et al. 2008) detection and mapping; emperor penguin colony detection and mapping (Fretwell and Trathan 2009); or the mapping of Giant panda *Ailuropoda melanoleuca* habitat for the whole of China (De Wulf et al. 1988), as well as the monitoring of vegetation changes in the species' habitat (Jian et al. 2011). The success of Landsat in ecological and environmental monitoring may be attributed to the fact that access to Landsat imagery has been free for several years: this was recently discussed by Wulder et al. (2012a) who reported that, while the Earth Resources Observation and Science Centre provided about 25,000 Landsat images at a price of US$600 per scene in 2001, this number increased to about 2.5 million images distributed free in 2010.

1.2.3.3 RAdio Detection And Ranging (RADAR) and Light Detection And Ranging (LiDAR)

RADAR and LiDAR are active sensors (see Box 1.3 for a definition of these). The principle of RADAR

Table 1.2 Landsat Thematic Mapper (TM) sensor characteristics. The principal applications associated with each Landsat TM band are also detailed.

Band	Resolution	Wavelength (µm)	Principal applications
TM1	30 m	0.45–0.52 (blue)	Penetration of clear water: bathymetry; mapping of coastal waters; chlorophyll absorption; distinction between deciduous and coniferous vegetation.
TM2	30 m	0.52–0.60 (green)	Records the green reflectance peak of vegetation; assesses plant vigour; reflectance from turbid water.
TM3	30 m	0.63–0.69 (red)	Operates in the chlorophyll absorption region and is best for detecting roads, bare soil, and vegetation types.
TM4	30 m	0.76–0.90 (near-infrared)	Used to estimate biomass. Although it separates water bodies from vegetation and discriminates soil moisture, it is not as effective as TM3 for road identification.
TM5	30 m	1.55–1.75 (mid-infrared)	Band 5 is considered to be the best single band overall. It discriminates roads, bare soil, and water. It also provides a good contrast between different types of vegetation and has excellent atmospheric and haze penetration. Discriminates snow from clouds.
TM6	120 m	10.5–12.5 (thermal–infrared)	Responds to thermal (heat) radiation emitted by the target. Thermal radiation is closely related to soil moisture and is best for measuring plant heat stress and thermal mapping.
TM7	30 m	2.08–2.35 (mid-infrared)	Useful for discriminating mineral and rock types and for interpreting vegetation cover and moisture.

systems is based on a transmitter emitting microwaves or radio waves, which bounce off any object in their path. The object then returns a tiny part of the wave's energy to a dish or antenna (which is usually located at the same site as the transmitter) and the time it takes for the reflected waves to return to the dish enables a computer to calculate the distance from the object (Kasischke et al. 1997). Interpreting the returned signal can help to build digital elevation models (DEMs), as well as providing information on changes in water level and land cover. In synthetic-aperture RADAR (SAR), multiple radar images are processed to yield higher resolution images. These multiple RADAR images are obtained either by using a single antenna mounted on a moving platform (such as a spacecraft) or by using many low-directivity small stationary antennas scattered over an area near the target area. In both cases, the many echo waveforms received at the different antenna positions are post-processed to resolve the target. A number of RADAR/SAR satellites are currently in operation, including RADARSAT, ALOS/PALSAR, ENVISAT/ASAR, TerraSAR-X and SkyMed (Goetz et al. 2009; see also Table 1.1). SAR sensors can operate day or night, and can penetrate through haze, smoke, and clouds. Environmental and ecological applications of RADAR technology are relatively recent, and are therefore scarce. Notable recent examples include the monitoring of above-ground biomass in woodland ecosystems (Mitchard et al. 2012) and the detection of tropical deforestation (Whittle et al. 2012).

Contrary to RADAR/SAR sensors that make use of microwaves, LiDAR is a remote sensing technology that determines distance to an object using laser pulses (Dubayah and Drake 2000). Like RADAR technology, the distance to an object is determined by measuring the time delay between transmission of a pulse and detection of the reflected signal. Interpreting the returned signal can also help to build DEMs, as well as providing information on height and structure of the vegetation layer and other features. One main difference between LiDAR and RADAR/SAR is that LiDAR is highly sensitive to aerosols and clouds (Goetz et al. 2009). LiDAR data have been shown to be associated with many opportunities in forestry and agriculture (Turner et al. 2003), while recent work has highlighted how the information captured by LiDAR technology can be related to species richness and community composition (see Muller and Brandl 2009 on forest beetle assemblages, and Vierling et al. 2011 on spider distribution in Germany).

1.2.4 Why is remote sensing useful?

Data collected on the ground are generally difficult to use for mapping and predicting regional or global changes in climatic conditions, land cover distribution or vegetation dynamics, because such data are traditionally collected at small spatial scales and vary in their type and reliability. Furthermore, such data often come from a single time period during the year, which is usually not synchronized spatially, making it difficult to gather information on temporal changes and phenology. Collecting data on the ground, especially over large areas or at multiple times over a year, can be extremely costly. Imagine the human and financial effort required to map the extent of the forest in the Congo basin on a regular basis!

Remote sensing, on the other hand, offers a relatively inexpensive and verifiable means of deriving complete spatial coverage of environmental information for large areas in a consistent manner that may be updated regularly (Muldavin et al. 2001; Duro et al. 2007). It has considerable potential as a source of information on biodiversity at landscape, regional, ecosystem, continental, and global spatial scales (Roughgarden et al. 1991; Gillespie et al. 2008). It can help to monitor the occurrence of extreme events such as droughts, fires or storms (Horning et al. 2010), changes in ecosystem functioning (Kerr and Ostrovsky 2003; Alcaraz-Segura et al. 2009), changes in the distribution of natural habitats (Aplin 2005; DeFries et al. 2007), and land use change or land degradation (Goetz et al. 2009; Prince et al. 2009) across the world. Thus remote sensing makes it possible to collect data from dangerous or inaccessible areas. War zones, the middle of the Amazon basin, or the Antarctic can all be monitored, providing information on deforestation rates, desertification, ocean depth, or changes in the extent of the ice sheet cover. Satellite imagery can be applied retrospectively across wide regions, with some data having been collected relatively frequently over a long time period. For example, the Landsat Program, which is still running, gives ecologists access to nearly forty years of information on ecosystem distribution. This is an excellent opportunity to reanalyse old data and make use of previously unavailable information. Remote sensing can be applied to ecological research directly related to individuals, species or communities of interest (Turner et al. 2003; Aplin 2005; Pettorelli et al. 2005b). As detailed in the following chapters, remote sensing can enable study on the nature of the environmental conditions that shape individuals' reproductive outputs, species' distribution, or communities' complexity. Remote sensing can also help to quickly identify areas of concern on a global scale, supporting managers in their effort to design and apply adaptive management strategies. It can provide a cost-effective way to target monitoring effort, by identifying areas with rapid changes in the functional attributes of ecosystems where more intense monitoring might be required. Two advantages of remote sensing particular to vegetation studies are that it is non-invasive and non-destructive (Jones and Vaughan 2010).

Altogether, remote sensing technologies can help to support a dynamic approach to environmental and wildlife management (see Chapters 5–7), where the relevance and efficiency of management actions can be regularly evaluated. Useful monitoring tools require long-term commitment: programs such as Landsat have demonstrated that satellite-based monitoring can be sustainable. Satellite-based data enable projects, organizations, and nations to report standardized and transparent information (discussed in more detail in Chapter 10); free access to some of these data enables the reported information to be verified using ground-based methodologies. Moreover, developments in satellite and sensor technology are continuous, meaning that our ability to gather relevant environmental information will further improve, and new opportunities will arise (Gillespie et al. 2008).

1.3 Getting started: the remote sensing user toolkit

What are the tools and information required to start using remote sensing technologies for ecological research purposes? In this section, I deal briefly with some of the main issues that need to be considered when using remote sensing data. The first step is to define clearly the objectives of the study, in order to best match data requirements with data availability. A second important consideration is the realization that acquired remote sensing data will need to be stored, manipulated, and analysed: to do so will require users to have access to relevant hardware and software. Finally, in most cases remote sensing information will need to be complemented by ancillary data. In particular, land cover analyses based on remote sensing data generally benefit from being informed by relevant geo-referenced field data, which can help to ensure the accuracy of the interpretation, classification, and prediction of biological and physical parameters.

1.3.1 Defining your objectives

Navigating through the many available datasets (Table 1.1) is not easy, as various factors (such as the type of sensor used (passive versus active), the quality and consistency of the information collected, the spatial and temporal resolutions, the temporal coverage as well as the costs) shape the current availability and usefulness of datasets. The first step is to define the objectives of the study, namely the location, type of information sought, the appropriate spatio-temporal resolutions, as well as the pertinent temporal coverage. As acknowledged by Defries et al. (2007), a major criterion for choosing which datasets to consider is based on defining the geographic and temporal scales required. For example, in the case of large areas spanning across countries or even continents, very high resolution images are not ideal; if rapid temporal changes are to be detected then regular coverage of the Earth's surface will be required to obtain frequent updates (Defries et al. 2007). As a quick rule of thumb, Defries et al. (2007) have classified remote sensors into three main types (see Table 1.1 for details on the sensors):

- medium spatial resolution ≥250 m (e.g. NOAA/AVHRR, Terra/MODIS)
- high spatial resolution 10 m–60 m (e.g. Landsat ETM/TM, ASTER)
- very high spatial resolution 1 m–5 m (e.g. Quickbird, IKONOS, Orbview).

When it comes to terrestrial ecology and conservation science, satellite-based data are generally used for (i) land-cover classification; (ii) change detection; and (iii) integrated ecosystem measurements (Kerr and Ostrovsky 2003).

Land cover classification represents one of the most popular applications of remote sensing techniques in ecology. The idea behind land cover classification is to match land cover classes identified in the study region to particular features within the remotely sensed imagery (Aplin 2004). Such a process is based on the assumption that different land cover types are associated with radiations across different regions of the electromagnetic spectrum (Jensen 1996; Figure 1.2). For example, the spectral signature for water is virtually 0% reflectance in the infrared band (making this band particularly useful to map water features; see also Table 1.2). Spatial data at tens of metres are generally required to accurately map many areas, because of the low spatial autocorrelation of land surface features (Townshend and Justice 1988). Multispectral imagery collected by passive sensors at high spatial resolution (≤30 m) is generally used for this purpose, with Landsat satellite data being the only record of global land surface conditions at a spatial scale of tens of metres spanning the last thirty years (Tucker et al. 2004; Table 1.2). The choice of the imagery to be analysed will depend on the desired spatial resolution and extent, the period considered, as well as the available budget (since not all satellite imagery is free). Active sensors such as RADAR can also be useful for land cover classification purposes, especially in cloudy environments (Gillespie et al. 2008).

Change detection consists in tracking changes in landscape features' state, distribution, and dynamics using remote sensing information. Satellite data can indeed provide valuable information regarding land cover degradation and changes in forest distribution, composition, primary productivity, and phenology (Turner et al. 2003). These data can also help monitor variation of ice sheet extent, or changes in the spatial distribution of human disturbances such as night-time light brightness and road density (Campbell 2007; Jensen 2007). Such information can be derived from time series analysis, where the extent of the feature of interest is compared between imagery from different time periods.

Ecosystem-based approaches for assessing ecological integrity integrate structural and functional biological measures to evaluate the overall health of ecosystems. Good examples of such a type of approach generally involve the use of vegetation indices (see Chapter 2), which can be built from monitoring the Earth's reflectance to quantitatively evaluate biomass and vegetative vigour. Several examples of such approaches are discussed in Chapter 5.

1.3.2 Storing and manipulating satellite data

1.3.2.1 Geographic information systems

Information systems will aim to facilitate the organization, storage, access, manipulation, and synthesis of multi-sourced data (Longley et al. 2005). Geographic information systems (GIS) are a particular type of information system allowing the association of geographic locations with the records compiled (Longley et al. 2005). The beginnings of GIS can be traced back to the Egyptians c.1292 BC, or even earlier to the Babylonians, with the development of the first maps, which were subsequently improved across the centuries to assist trade, exploration, tax collection, and military purposes (Bernhardsen 2002). By the nineteenth century the development of modern infrastructure required the incorporation of geological, topographic, socio-economic, and even statistical information, expanding the common use of maps and increasing their specialization (Bernhardsen 2002). The rise of computing systems during the 1970s and 1980s led to unprecedented processing, storage, and visualization capabilities, and thus to the incorporation of GIS into a wide variety of disciplines, including ecology, biogeography, and wildlife conservation (Bernhardsen 2002; Longley et al. 2005). GIS are associated with multiple functions (Table 1.3), for example allowing the simultaneous display of satellite imagery with other data layers such as roads, rivers or political boundaries (many of which can now be sourced freely on the World Wide Web). This can be a very effective means for obtaining a fairly realistic view of a particular terrain (Horning et al. 2010).

There are multiple constraints that need to be considered when combining data from multiple sources in a GIS (Lillesand et al. 2008). All the data layers must cover much of the same geographic area and share the same geographic coordinate system (a GIS can convert data from one coordinate system to another). Another issue is linked to the fact that

Table 1.3 Classification of geographic information system functions.

Feature	Function
Data input and encoding	Data capture
	Data validation
	Interactive editing
	Quality control
	Data storage and structuring
	Topology
Data manipulation	Conversion
	Geometric correction
	Generalization and classification
	Enhancement
	Abstraction
	Reprojection
Data retrieval	Data selection criteria
	Browse facility
	Spatial analysis
	Statistical analysis
	Measurements and calculations
	Error analysis
	Extension into three spatial dimensions and time
Data display	Graphic display
	Map generation
	Text
	Report writing
Database management	Database organization
	Multi-user access
	Maintenance and security

Source: Barrett and Curtis (1999).

no two-dimensional map projection can accurately preserve equal distances, equal areas, shapes, and absolute directions relative to the central point of projection: it is therefore important to fit the choice of the projection to be used to the intended aim. Once the fundamentals of GIS are understood, this type of system is a practical and efficient tool when it comes to the storage and manipulation of satellite data.

1.3.2.2 Orthorectification

Raw satellite images usually contain significant geometric distortions and therefore cannot be used directly in GIS. These distortions can be severe: for example, if the IKONOS satellite sensor acquires image data over an area with 1 km of vertical relief, with the sensor having an elevation angle of 60°, the image product may have nearly 600 m of terrain

Table 1.4 Description of error sources associated with the use of remote sensing information, differentiating distortions linked to the acquisition system, and those linked to the 'observed' features.

Category	Sub-category	Description of error sources
Acquisition system	Platform	Variation of the movement Variation in platform attitude (low to high frequencies)
	Sensor	Variation in sensor mechanics Viewing/look angles Panoramic effect with field of view
	Measuring instruments	Time-variation or drift Clock synchronicity
Observed	Atmosphere	Refraction and turbulence
	Earth	Curvature, rotation, topographic effect
	Map	Geoid to ellipsoid Ellipsoid to map

Orthorectification refers to the process by which satellite data are corrected for such distortions. This process combines knowledge of the elevation of each image point with the precise viewing geometry at that point, which requires associating digital elevation data with the respective satellite data by matching coordinates (Tucker et al. 2004).
Source: Toutin (2004).

displacement. There are two broad sources of distortion: the acquisition system and the 'observed'. These sources are detailed in Table 1.4.

1.3.3 Obtaining the adapted algorithms and software

Most projects require personalized manipulation, analysis, and synthesis of remote sensing information (Figure 1.4). Sometimes, such processes require access to specialized remote sensing software, which supply sophisticated tools for visualizing imagery as well as image-processing tools for automated land cover classification or radiometric correction (Horning et al. 2010). Several possible software packages may be considered: commercial software such as ArcGIS <http://www.esri.com> and ERDAS IMAGINE <http://www.erdas.com> are examples of integrated collections of GIS software products with a standards-based platform for spatial analysis, data management, and mapping. Similarly, Idrisi Selva is an integrated GIS and

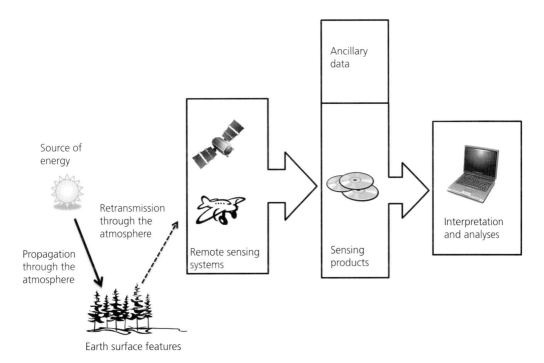

Figure 1.4 Schematic representation of the methodological steps allowing the remote sensing of the Earth, spanning from data collection to data interpretation.

image processing software solution with nearly 300 modules for the analysis and display of digital spatial information <http://www.clarklabs.org>. SPRING is a GIS and remote sensing image processing system with an object-oriented data model, enabling the integration of raster and vector data representations in a single environment <http://www.dpi.inpe.br/spring>. The Geographic Resources Analysis Support System, commonly referred to as GRASS GIS, is an open source GIS used for data management, image processing, graphics production, spatial modelling, and visualization of many types of data <http://grass.osgeo.org/>. Likewise, Quantum GIS (QGIS) is a user-friendly open source GIS licensed under the GNU General Public License. This free software allows users to visualize, manage, edit and analyse data, and to compose printable maps. It runs on Linux, Unix, Mac OSX, and Windows and supports numerous vector, raster, and database formats and functionalities <http://www.qgis.org/>.

1.3.4 Acquiring reference data – ground-truthing

Remote sensing and fieldwork are complementary (Horning et al. 2010), as geo-referenced field data acquisition can help to ensure the accuracy of the interpretation, classification, and prediction of biological and physical parameters from satellite-based data. Ground-truth data might be used: (i) to aid the analysis and interpretation of remotely sensed data; (ii) for sensor calibration; (iii) to help assess the accuracy of products created from remotely sensed data, and validate results based on satellite-derived information using independent data (Pinzon et al. 2004; Lillesand et al. 2008).

Land cover classification is a typical situation involving gathering geo-referenced data to aid the analysis and interpretation of remotely sensed data. Land cover distribution can be derived from unsupervised and supervised land cover classifications (Lu and Weng 2007; Figure 1.5). The choice of approach depends on *a priori* knowledge regarding the land-cover types across the study region: when no information is available, an unsupervised classification can be carried out to cluster the pixels within the imagery into classes with similar spectral characteristics, which can later be associated with existing land cover types, thus ensuring the collection of accurate data on this subject. If, on the other hand, the identity and geographic location of land cover types is known,

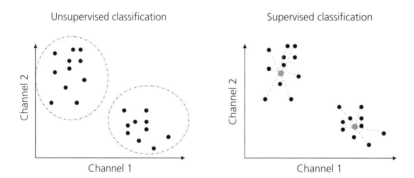

Figure 1.5 Unsupervised versus supervised classification: principles. Classification is a computational procedure that sorts pixels into groups (or classes) according to their similarities in terms of reflectance in the various bands captured (e.g. visible, infrared) by a given sensor. In this illustration, reflectance in channels 1 and 2 are supplied for all pixels (dots) encompassed in a given scene. The objective of an unsupervised classification is to group multiband spectral response patterns into clusters that are statistically separable. An unsupervised classification will thus classify pixels according to their intrinsic grouping or clustering, without any reference to prior knowledge about habitat characteristics associated with these pixels. Once the classification is made, it must be interpreted by the user as to what that actually represents in the real-world scene. This interpretation step requires some knowledge of the scene's feature content from general experience or personal familiarity with the area imaged. The strategy behind supervised classification is based on the user recognizing classes in a scene from prior knowledge. This familiarity allows the user to choose and set up discrete classes (thus supervising the selection), and then assign them category names. A supervised classification will thus classify pixels according to their similarity with certain predetermined references (appearing here as larger grey dots).

the spectral characteristics of these classes can be used to train the classification algorithm to illustrate their distribution across the imagery extent (Jensen 1996). In both cases, the incorporation of ancillary data (such as topography, climate or human footprint) has proven useful to imagery classification (Aplin 2004).

1.4 Conclusions

The main goals of this chapter have been to discuss and illustrate how remote sensing can help to tackle many of the local, national, regional, and global challenges currently faced by scientists all around the world; and to highlight the remote sensing principles needed to understand the possible limitations associated with the use of satellite-based data in ecology and conservation. An important issue arises: remote sensing cannot solve the entire information collection issue associated with biodiversity monitoring. To maximize monitoring effectiveness, remote sensing-based methods need to be integrated with geo-referenced field observations: whereas remote sensing-based techniques can address spatial and temporal domains inaccessible to traditional ground-based approaches, remote sensing cannot match the accuracy, precision, and thematic richness of *in situ* measurement and monitoring (Gross et al. 2009). In our attempts to tackle the many challenges posed by global environmental change, higher levels of collaboration between disciplines, including ecology and remote sensing, are fundamental.

1.5 Organization of the book

This book aims to provide a coherent review of the Normalized Difference Vegetation Index (NDVI), which is a vegetation index based on the information collected by passive sensors. The book now focuses on vegetation indices, in particular the principles and possible applications of the NDVI in ecology, environmental management, and conservation. The first section provides an overview of vegetation indices (Chapter 2), and then considers the NDVI (the principle behind it, the available NDVI datasets and the possible complications and errors associated with the use of this satellite-based index; Chapter 3) and its correlation with climate (Chapter 4). The second section discusses the possible applications of the NDVI in ecology, environmental, and wildlife management, as well as in conservation. In particular, it explores how the NDVI can be used to inform environmental monitoring (Chapter 5), plant ecology (Chapter 6), wildlife management (Chapter 7), and conservation in general (Chapter 8). As well as many successes, the NDVI has been associated with unsuccessful applications, and in Chapter 9 the situations where the NDVI has not worked will be assessed. Chapters 10 and 11 discuss the challenges ahead in terms of enhancing the usefulness of this vegetation index in ecology and management. Chapter 10 introduces the major international platforms and conventions dealing with the preservation of biodiversity and other ecosystem services, as well as discussing how the NDVI can help nations to report to these initiatives. Chapter 11 explores possible directions for future NDVI-related research.

CHAPTER 2

Vegetation indices

History is a gallery of pictures in which there are few originals and many copies.
Alexis de Tocqueville

Remote sensing-based approaches provide a wealth of opportunities for the monitoring of the Earth's surface. One such opportunity is the monitoring of vegetation, which represents the main component of the terrestrial biosphere and a crucial element in the climate system (Foley et al. 2000). Primary producers are indeed an essential vehicle for energy transfer from the Sun to consumers, influencing greenhouse gas concentrations in the atmosphere through photosynthesis; almost all life on Earth is directly or indirectly reliant on primary production. Primary production, in this case, can be broadly defined as the amount of energy that is converted to organic compounds by photosynthetic organisms during a given time period. Gross primary production (GPP) is generally distinguished from net primary production (NPP). GPP refers to the total amount of carbon dioxide fixed by terrestrial plants per unit time through photosynthesis. A substantial fraction of GPP goes into plant autotrophic respiration (R), with the remainder of the production (NPP) allocated to growth, survival, and reproduction of the plant: GPP = NPP + R.

There are various remote sensing-based methodologies relevant to the monitoring of vegetation and primary production dynamics: this chapter focuses on vegetation indices, which can be broadly defined as combinations of surface reflectance at two or more wavelengths designed to highlight a particular property of vegetation. Section 2.1 introduces the concept of vegetation indices, and highlights their potential role in ecological research and conservation. To date, more than 150 vegetation indices have been published in the scientific literature; Section 2.2 reviews some of the main indices that have been proposed and discusses their relative merits.

2.1 Vegetation indices: definition and value in ecological studies

2.1.1 Introduction

The amount of light reflected from a surface is determined by the amount and composition of irradiance that reaches it, and the reflectance properties of the surface itself. Green vegetation absorbs incoming solar radiation in the photosynthetically active spectral region, which provides the energy needed to power photosynthesis (Jensen 2007). More specifically, a typical reflectance spectrum of a vegetation canopy can be subdivided into three parts, namely the visible (400–750 nm), the near infrared (0.75–1.3 µm), and the mid-infrared (1.3–2.5 µm). The visible part of this reflectance spectrum is controlled by the chlorophyll pigments, which are the major absorber of radiation in the visible region. Chlorophyll pigments absorb strongly in the red and blue regions of the visible spectrum but not in the green region (Jensen 2007; Glenn et al. 2008). Other leaf pigments also absorb strongly in the visible part of the spectrum, for example carotene and xanthophyll in blue wavelengths. The optical properties in the near-infrared spectral domain are explained by leaf structure, as the spongy mesophyll cells located in the interior or underside of the leaves reflect light in the near-infrared spectrum (especially around 760–900 nm; Tucker 1979). Finally, the mid-infrared region contains information about the absorption of radiation by water, cellulose, and lignin (Curran 1989), allowing the identification of vegetation stress due to drought. The combination of absorption in the red and blue regions of the visible spectrum, and the very strong reflectance in the near-infrared, forms a very specific spectral signal of vegetation (Knipling 1970; Jensen 2007), enabling it to be easily distinguished in remote monitoring (Figure 2.1).

Vegetation indices exploit such properties of the vegetation and are formed from combinations of several spectral values that are added, divided, or multiplied in a manner designed to yield a single value that indicates the amount or vigour of vegetation within a pixel (Campbell 2007). They are intended to enhance the vegetation signal, while minimizing solar irradiance and soil background effects (Jackson and Huete 1991). Vegetation indices were first developed in the 1970s (Glenn et al. 2008), and there are now more than 150 of them (ENVI 2011), all of which have been claimed to show good (or improved) sensitivity for vegetation detection and monitoring (Barrett and Curtis 1999).

Two general classes can be distinguished: ratios and linear combinations. Ratio vegetation indices may be the simple ratio of any two spectral bands (for example the ratio of the red (R) and near-infrared (NIR) bands, NIR/R), or the ratio of sums, differences or products of any number of bands (e.g. NIR − R/NIR + R). Linear combinations are orthogonal sets of n linear equations calculated using data from n spectral bands (e.g. NIR-R; Jackson and Huete 1991). Indices may also be grouped into categories that calculate similar properties, for example, indices providing information on broadband greenness, narrowband greenness, light use efficiency, canopy nitrogen, dry or senescent carbon, leaf pigments, and canopy water content (ENVI 2011). Indices on broadband greenness originate from multispectral images, while indices on narrowband greenness, light-use efficiency, canopy nitrogen, dry or senescent carbon, leaf pigments, and canopy water content originate from hyperspectral images. Multispectral imagery is produced by sensors measuring reflected energy within several specific broad bands of the electromagnetic spectrum (Chapter 1). The satellites of the National Oceanic and Atmospheric Administration (NOAA), Landsat Program, Quickbird, and SPOT carry multispectral sensors (see also Table 1.1). Hyperspectral sensors measure energy in narrower and more numerous bands than multispectral sensors. With hyperspectral imagery, information is generally collected over a contiguous spectral range (Figure 2.2). As discussed in Chapter 1, sensors collect information at different spatial, radiometric, spectral, and temporal resolutions. Because there is a limit to the amount of information

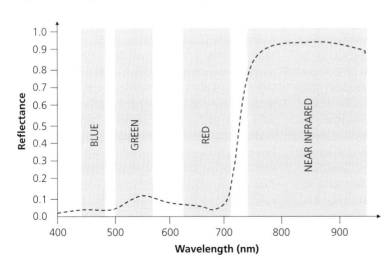

Figure 2.1 Theoretical vegetation reflectance across the visible and near-infrared portions of the electromagnetic spectrum. Most of the absorption occurs in the red and blue regions of the spectrum, while the near-infrared is not absorbed by any pigments within a plant. It therefore travels through most of the leaf and interacts with the spongy mesophyll cells. This interaction causes part of the energy to be reflected and the other part to be absorbed and transmitted by the leaf. In plants with healthy mesophyll cell walls and in dense canopies, more of the near-infrared energy is reflected and less is absorbed or transmitted by the leaf.

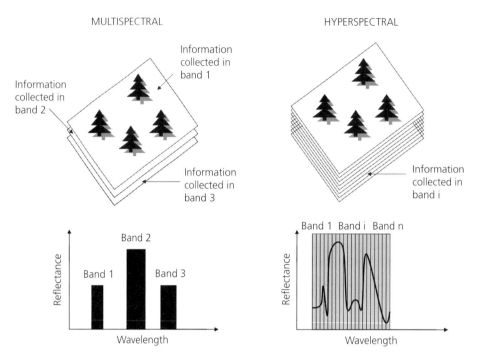

Figure 2.2 Differences between multispectral and hyperspectral imaging. Multispectral imagery is produced by sensors that measure reflectance within specific broad bands of the electromagnetic spectrum. Hyperspectral sensors measure reflectance in narrower and more numerous bands than multispectral sensors, over a contiguous spectral range and hence can detect subtle changes in the land surface.

that can be collected per time unit, currently no sensor collects information at high spatial, temporal, spectral, and radiometric resolutions, meaning that each sensor is associated with trade-offs in terms of available resolutions for these four parameters (Jensen 2007). As a general rule, multispectral imagery tends to acquire information at high temporal resolution, and hyperspectral imagery at low temporal resolution. Hyperspectral sensors can be found on board satellites such as Hyperion from NASA, but currently most of the sensors are airborne and only locally applicable. Due to the limited availability of hyperspectral data sets, vegetation indices based on multispectral data are more usual than those based on hyperspectral imagery, and are well established within landscape ecology and biodiversity studies.

One may wonder why so many vegetation indices have been developed. The great range and variety of vegetation indices is believed to be associated with the fact that a myriad of variables can affect satellite measurements of vegetation. The reflectance of light detected from a vegetated ground surface is determined by several factors, which include: atmosphere composition; sensor calibration; viewing conditions; solar illumination geometry; soil moisture and type; vegetation and soil colour; plant chemistry; vegetation type, condition, and behaviour; leaf geometry and morphology (Bannari et al. 1995; Barrett and Curtis 1999). Such factors need to be taken into account or corrected for (Bannari et al. 1995). A great deal of effort has been made to reduce the effects of such factors and develop vegetation indices that are more robust to these sources of interferences, partially explaining why there are currently so many different vegetation indices. But this great range of indices can also be attributed to technical developments in terms of sensor design and capabilities: vegetation indices that can be calculated using a specific dataset are determined by the available spectral bands. For example, an input dataset from a sensor that matches only the near-infrared and red spectral bands (such as the Advanced Very High Resolution Radiometer (AVHRR) sensors) is able only to calculate two of the most frequently used vegetation indices, namely the NDVI and the Simple Ratio (see Section 2.2.1). The increased number of sensors capturing information in more numerous and narrower bands have therefore likely favoured the emergence of new vegetation indices.

2.1.2 Importance of vegetation indices for ecological studies

Vegetation indices are dimensionless radiometric measures designed to allow reliable spatial and temporal inter-comparisons of terrestrial photosynthetic activity and canopy structural variation. They show better sensitivity than individual spectral bands for the detection of vegetation biomass (Asrar et al. 1984), and have been highly successful in assessing vegetation condition, foliage, cover, phenology, and processes such as evapotranspiration and primary productivity, related to the fraction of photosynthetically active radiation absorbed by a canopy (Kerr and Ostrovsky 2003; Pettorelli et al. 2005b; Glenn et al. 2008; see also Table 2.1).

Due to their simplicity and ease of application, and being consistently computed for all pixels in time and space, vegetation indices allow for a wide range of investigation. They can be used for globally monitoring vegetation conditions, or for producing imagery displaying land cover and land cover changes. Vegetation index data may be used as an input for modelling global biogeochemical and hydrological processes, and global and regional climate. These data may also be used for characterizing land surface biophysical properties, including land cover. Vegetation indices thus represent a useful tool when it comes to assessing how environmental changes might affect the distribution and dynamics of vegetation, particularly at large temporal and spatial scales, and in areas of limited availability of *in situ* data. These indices are also useful in the interpretation of remote sensing images: for example, some vegetation indices allow the detection of subtle land use change; the evaluation of vegetative cover density, crop discrimination and crop prediction; and, for the purpose of thematic mapping, classification improvement (Bannari et al. 1995).

2.2 A brief history of vegetation indices

This section focuses mainly on the history of indices relating to broadband greenness, which are combinations of reflectance measurements sensitive to the combined effects of foliage chlorophyll concentration, canopy leaf area, foliage clumping, and canopy architecture (ENVI 2011). They are designed to measure the overall amount and quality of photosynthetic material in vegetation, being well correlated with the fractional absorption of photosynthetically active radiation (Box 2.1).

Table 2.1 Studies on the quantitative link between the NDVI and ground-based measures of various vegetation parameters.

Parameter	Habitat/location	R^2	References
Photosynthetically active radiation absorption	Wheat field	0.965	Asrar et al. 1984
Fraction of photosynthetically active radiation intercepted by green foliage	Cordgrass plots	0.79	Bartlett et al. 1990
Gross ecosystem production	Wet sedge tundra	0.75	Boelman et al. 2003
Above ground biomass	Wet sedge tundra	0.84	Boelman et al. 2003
Ecosystem respiration	Wet sedge tundra	0.71	Boelman et al. 2003
Potential evapotranspiration	Canada	0.86	Cihlar et al. 1991
Actual evapotranspiration	Canada	0.77	Cihlar et al. 1991
Primary production	Savanna, Senegal	0.12–0.82	Diallo et al. 1991
Shrub cover	New Mexico	0.40	Duncan et al. 1993
Above-ground net primary production	Grassland, USA	0.89	Paruelo et al. 1997

R^2 refers to the proportion of the variance in the parameter considered explained by the NDVI.

> **Box 2.1 Some definitions**
>
> **Reflectance:** Ratio of the intensity of reflected radiation to that of incident radiation on a surface. Reflectance is expressed as a percentage and usually refers to a specific wavelength.
>
> **Irradiance:** The power of electromagnetic radiation per unit area at a surface. Total solar irradiance refers for example to the overall incoming solar radiation, which is the dominant driver of the global temperature of the Earth.
>
> **Leaf area index (LAI):** A dimensionless value obtained from the ratio between the total upper leaf surface of vegetation and the surface area of the land on which the vegetation grows. LAI can thus be defined as the total one-sided area of photosynthetic tissue per unit ground surface area. LAI ranges from 0 (e.g. bare ground) to >10 (e.g. dense conifer forests).
>
> **Photosynthetically active radiation (PAR):** The spectral range of solar radiation that photosynthetic organisms are able to use for photosynthesis. This spectral region approximates to the visible range (400–750 nm).
>
> **Primary production:** The production of organic compounds from atmospheric or aquatic carbon dioxide, principally through the process of photosynthesis.

2.2.1 From understanding reflectance to the emergence of the first vegetation indices

How did we go from measuring reflectance in various bands of the electromagnetic spectrum to producing vegetation indices using some of this information? NASA's 'Monitoring the Vernal Advancement and Retrogradation of Natural Vegetation' programme was one of the first aimed at establishing a close relationship between radiometric response and vegetation cover, either in the area of ground measurements or in the exploitation of the first generation Landsat satellite images (see Section 1.2). This and other research initiatives including the Large Area Crop Inventory Experiment (LACIE) first demonstrated that the use of the red and near-infrared channels on satellite sensors are well suited for the study of vegetation (Bannari et al. 1995).

Following these discoveries, Jordan (1969) and Pearson and Miller (1972) developed the first two indices in the form of ratios (the Ratio Vegetation Index (RVI) and the Vegetation Index Number (VIN)) using the NIR radiance divided by the red radiance (Jackson and Huete 1991; Bannari et al. 1995). These two indices were intended to enhance the contrast between the ground and vegetation, and reduced the effect of illumination conditions. The relationship between the reflectances of the two bands also eliminated disturbances from factors which manifest themselves equally in each band (Holben and Justice 1981). Yet these indices (particularly the RVI) were shown to be sensitive to atmospheric effects and ground optical properties (Jackson et al. 1983; Baret and Guyot 1991). Their discriminating power was reported to be weak when the vegetation cover was sparse (<50%; Jackson and Huete 1991; Bannari et al. 1995). The RVI was also shown to saturate in dense vegetation conditions, when the leaf area index (LAI; see Box 2.1 for a definition) became high (ENVI 2011).

Rouse et al. (1974) introduced the NDVI, computed as (NIR–R)/(NIR + R). As may be seen from their respective definitions, the RVI and NDVI are functionally equivalent, as there exists a bijective relationship between their respective elements (Perry and Lautenschlager 1984):

$$NDVI = (RVI - 1)/(RVI + 1)$$

There are differences between the NDVI and the RVI though, as normalization reduces the effect of sensor calibration degradation (Kaufman and Holben 1993). More generally, the combination of NDVI's normalized difference formulation and use of the highest absorption and reflectance regions of foliage makes this vegetation index robust over a wide range of conditions (ENVI 2011). As with the RVI, however, it has been suggested that the NDVI saturates in dense vegetation conditions (Gitelson 2004; Ünsalan and Boyer 2004).

At the time of the emergence of the NDVI, the Transformed Vegetation Index (TVI) was also proposed (Rouse et al. 1974; Deering et al. 1975): the TVI modifies the NDVI by adding a constant of 0.5 to all its values and taking the square root of the results (Table 2.2). The constant 0.5 was introduced to avoid operating with negative NDVI values, while the calculation of the square root was intended to correct NDVI values that approximate the Poisson distribution and introduce a normal distribution. The

Table 2.2 Vegetation indices frequently cited in the literature.

Name	Abbreviation	Formulae	References
Simple Ratio or Ratio Vegetation Index	RVI	NIR/R	Jordan 1969; Pearson and Miller 1972; Foran 1987
Vegetation Index Number	VIN	R/NIR	Pearson and Miller 1972
Normalized Difference Vegetation Index	NDVI	$(NIR - R)/(NIR + R)$	Rouse et al. 1974; Deering 1978; Tucker 1977, 1979
Transformed Vegetation Index	TVI	$\sqrt{(NDVI + 0.5)}$	Rouse et al. 1974; Deering et al. 1975
Corrected Transformed Vegetation Index	CTVI	$CTVI = ((NDVI + 0.5)/ABS(NDVI + 0.5)) * \sqrt{(ABS(NDVI + 0.5))}$	Perry and Lautenschlager 1984
Green/Red Ratio		G/R	Kanemasu 1974
Simple subtraction	SS	$NIR - R$	Pearson et al. 1976; Clevers 1986
Green vegetation Index	GVI	$-0.283MSS4 - 0.66MSS5 + 0.577MSS6 + 0.388MSS7$	Kauth and Thomas 1976
Soil Brightness Index	SBI	$0.332MSS4 + 0.603MSS5 + 0.675MSS6 + 0.262MSS7$	Kauth and Thomas 1976
Yellow Vegetation Index	YVI	$-0.899MSS4 + 0.428MSS5 + 0.076MSS6 - 0.041MSS7$	Kauth and Thomas 1976
Non-Such Index	NSI	$-0.016MSS4 + 0.131MSS5 - 0.425MSS6 + 0.882MSS7$	Kauth and Thomas 1976
Differenced Vegetation Index	DVI	$2.4MSS7 - MSS5$	Richardson and Wiegland 1977
Soil Background Line	SBL	$MSS7 - 2.4MSS5$	Richardson and Wiegland 1977
Perpendicular Vegetation Index (vertical)	PVI(V)	$\sqrt{[(Rsv - Rva)^2 + (NIRsv - NIRsa)^2]}$	Richardson and Wiegland 1977
Perpendicular Vegetation Index (angular)	PVI(A)	$\sqrt{[(Rva-Rsv)^2 + (NIRsa - NIRsv)^2]}$	Richardson and Wiegland 1977
Misra Green Vegetation Index	MGVI	$-0.386MSS4 - 0.53MSS5 + 0.535MSS6 + 0.532MSS7$	Misra et al. 1977
Misra Soil Brightness Index	MSBI	$0.406MSS4 + 0.6MSS5 + 0.645MSS6 + 0.243MSS7$	Misra et al. 1977
Misra Yellow Vegetation Index	MYVI	$0.723MSS4 - 0.597MSS5 + 0.206MSS6 - 0.278MSS7$	Misra et al. 1977
Misra Non-Such Index	MNSI	$0.404MSS4 - 0.039MSS5 - 0.505MSS6 + 0.762MSS7$	Misra et al. 1977
Complex Division	CD	$NIR/(R + \text{other wavelengths})$	Carter and Gardner 1977
Ashburn Vegetation Index	AVI	$2.0MSS7 - MSS5$	Ashburn 1978
Greenness Above Bare Soil	GRABS	$(GVI - 0.09178SBI + 5.58959)$	Hay et al. 1979
Multi-Temporal Vegetation Index	MTVI	$NDVI(\text{date 2}) - NDVI(\text{date 1})$	Yazdani et al. 1981
Greenness Vegetation and Soil Brightness	GVSB	GVI/SBI	Badhwar 1981
Adjusted Soil Brightness Index	ASBI	2YVI	Jackson et al. 1983
Adjusted Green Vegetation Index	AGVI	$GVI - (1 + 0.018GVI)YVI - NSI/2$	Jackson et al. 1983
Substraction Ratio	RAT	$(G - R)/(G - NIR_7)$	Foran and Pickup 1984
Soil Adjusted Vegetation Index	SAVI	$((NIR - R)/(NIR + R + L))*(1 + L)$	Huete 1988
Transformed SAVI	TSAVI	$[a(NIR - aR - b)]/(R + aNIR - ab)$	Baret et al. 1989

continued

Table 2.2 Continued

Name	Abbreviation	Formulae	References
Transformed SAVI	TSAVI	[a(NIR − aR − b)]/[R + aNIR − ab = X(1 + a²)]	Baret and Guyot 1991
Normalized Difference Greenness Index	NDGI	(G − R)/(G + R)	Chamard et al. 1991
Redness Index	RI	(R − G)/(R + G)	Escadafal and Huete 1991
Atmospherically Resistant Vegetation Index	ARVI	(NIR − RB)/(NIR + RB) where RB = R − γ(B − R)	Kaufman and Tanré 1992
Global Environment Monitoring Index	GEMI	$\eta(1 − 0.25\eta) − [(R − 0.125)/(1 − R)]$	Pinty and Verstraete 1992
Normalized Difference Index	NDI	(NIR − MIR)/(NIR + MIR)	McNairn and Protz 1993
Transformed Soil Atmospherically Resistant Vegetation Index	TSARVI	$[a_{rb}(NIR − a_{rb}RB − b_{rb})]/[RB + a_{rb}NIR − a_{rb}b_{rb} + X(1 + a_{rb}^2)]$	Bannari et al. 1994
Modified SAVI	MSAVI	$[2NIR + 1 − [\sqrt{(2NIR + 1)^2 − 8(NIR − R)}]]/2$	Qi et al. 1994
Angular Vegetation Index	AVI	$\tan^{-1}\{((\lambda_3 − \lambda_2)/\lambda_2)[NIR − R]^{-1}\} + \tan^{-1}\{((\lambda_2 − \lambda_1)/\lambda_2)[G − R]^{-1}\}$	Plummer et al. 1994
Enhanced vegetation index	EVI	$G_a * ((NIR − R)/(NIR + C_1R − C_2B + L))$	Huete et al. 2002

Rsv, vertical red soil reflectance; Rva, angular red vegetation reflectance; NIRsv, vertical near-infrared soil reflectance; NIRsa, angular near-infrared vegetation reflectance; R, red reflectance (0.6–0.7 μm); G, green reflectance (0.5–0.6 μm); NIR, near-infrared reflectance; NIR6, near-infrared reflectance as estimated by MSS6 (0.7–0.8 μm); NIR7, near-infrared reflectance as estimated by MSS7 (0.8–1.1 μm); C1, atmospheric resistance red correction coefficient; C2, atmospheric resistance blue correction coefficient; B, blue band reflectance; L, soil adjustment factor; MIR, middle infrared; γ: atmospheric self-correcting factor which depends on aerosol type; λ1, λ2, λ3, centre wavelengths of the green (1), red (2) and NIR (3) channels; Ga, gain factor.
Source: Bannari et al. (1995).

NDVI, however, can be as low as −1, which means that adding a constant of 0.5 to all NDVI values does not always eliminate the negative values in the square root. To tackle this issue, the Corrected Transformed Vegetation Index (CTVI) was introduced by Perry and Lautenschlager (1984). The CTVI was intended to resolve the situation by dividing the term [NDVI + 0.5] by its absolute value [ABS(NDVI + 0.5)] and multiplying the result by the square root of the absolute value [SQRT(ABS(NDVI + 0.5))] (Table 2.2). Resulting images of the CTVI were found to be very noisy due to an overestimation of the greenness (Thiam 1997). To resolve the issue, it was proposed that the first term of the CTVI equation should be ignored (Thiam 1997).

Amid these proposals for new indices, research efforts began to be targeted at identifying factors potentially affecting the information captured by sensors on board satellites. Changes in the direction of solar irradiance and its interaction with canopy biomass, cover, height, and row orientation were shown to be significant in determining the remotely sensed information (Holben 1986). Canopies were also found to be non-isotropic diffuse reflectors; that is, they do not uniformly reflect light in all directions. Therefore, canopy structure and the angle of the Sun and sensor were suspected to have an effect on the multispectral reflectance of canopies (Pinter et al. 1985). The position of the Sun in the sky controls the area and the darkness of the shadow, which is also a function of leaf transmittance: the position of the Sun has been shown not to affect all bands equally (Curran 1980; Holben 1986). Because of this phenomenon, it was also suspected that solar elevation angle, the relative solar azimuth angle between the plane of illumination and the plane of detection, and the look angle of the sensor may affect satellite-based measurements (Kimes et al. 1984; Figure 2.3).

Based on these research outcomes, several drawbacks for the NDVI were highlighted: the index was shown to be sensitive to viewing and illumination geometry, notably for low density covers (Holben and Fraser 1984). Tucker reported that the precision of reflectance measurements was inversely related to the cosine of the solar zenith angle, meaning that NDVI estimates were unreliable if solar zenith angle was >80° (Tucker 1980; Goward et al. 1991). The NDVI was also suggested to overestimate the percentage of vegetative cover at the beginning of

VEGETATION INDICES 25

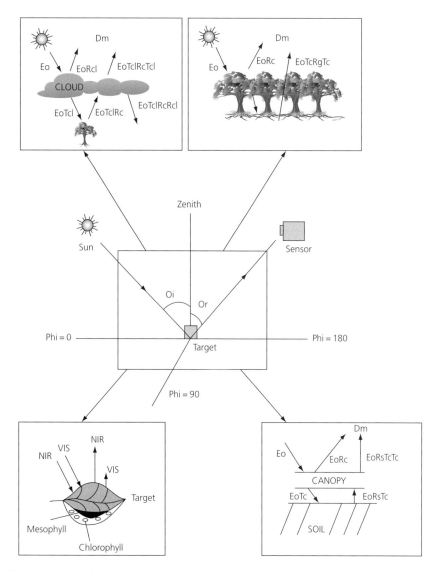

Figure 2.3 Satellite data collection. The sensor zenith angle is the angle between the sensor view vector and the local vertical to the target (Or). Very high values for this angle can lead to large errors in satellite measurements and extension of pixel width. The solar zenith angle is the angle between the solar vector and the local vertical to the target (Oi). All data with high solar zenith angles (>80°) can also lead to large errors in satellite measurements. The assessment of vegetation conditions from satellite data is based on the fact that incoming red visible wavelength is absorbed by chlorophyll whereas near-infrared wavelength is reflected by the spongy mesophyll leaf structure (lower left panel). However, many factors interact with that process. Cloud reflectance and transmittance diminish signal transmission to the sensor (upper left panel). The vegetation stratification also biases reflectance measurements, as dense canopies reduce ground vegetation response (upper right panel). Finally, soil reflectance may bias satellite measurements, especially in poorly vegetated areas (lower right panel). Eo, irradiance; T, transmittance from the canopy (Tc) and the cloud (Tcl); R, reflectance from the cloud (Rcl), the canopy (Rc), the ground vegetation (Rg) or the soil (Rs); Dm, measured response from the satellite; Oi, solar zenith angle; Or, sensor zenith angle; NIR, near-infrared light; VIS, visible red light.

the growth season, while underestimating it at the end of the season (Bannari et al. 1995). Alternative ratio-based vegetation indices therefore continued to appear. For example, Crippen (1990) suggested that only the sum of NIR and R was important in the use of the red band in the NDVI and so merely used the Percentage Vegetation Index (PVI = NIR/(NIR + R)) which constrains the values from 0 to 1, unlike the range of –1 to 1 for the NDVI.

2.2.2 Ratios versus linear combinations

Ratio vegetation indices such as the RVI or the NDVI appeared before linear combinations (Table 2.2). The first notable vegetation indices based on linear combinations of spectral bands can be attributed to Kauth and Thomas (1976), who proposed an orthogonal transformation of the original Landsat data space to a new four-dimensional space. The authors referred to this as the 'tassel cap transformation' and introduced four new axes, namely the Soil Brightness Index (SBI), Greenness Vegetation Index (GVI), Yellow Vegetation Index (YVI), and Non-Such Index (NSI). Similarly, Wheeler et al. (1976) and Misra et al. (1977) applied principal component analysis to Landsat MultiSpectral Scanner (MSS) data, with the structure of the resulting transformation and the interpretation of the principal components being comparable to those for the Kauth–Thomas transformation (Perry and Lautenschlager 1984). Around that time, the Difference Vegetation Index (DVI) and the Perpendicular Vegetation Index (PVI) were proposed by Richardson and Wiegand (1977), the idea being to use the perpendicular distance to the 'soil line' (a two-dimensional analogue of the Kauth–Thomas SBI) as an indicator of plant development. Other indices from the late 1970s include the Greenness Above Bare Soil (GRABS) proposed by Hay et al. (1979), which was an attempt to develop an indicator for which a threshold value could be specified for detecting green vegetation (Perry and Lautenschlager 1984).

2.2.3 Empirically-derived versus process-orientated generation of vegetation indices

As it can be seen from Table 2.2, two phases can be distinguished in the history of vegetation indices: the first phase is defined by a generation of vegetation indices based only on linear combinations or raw band ratios; the second phase is defined by the explicit consideration of the interactions between electromagnetic radiation, the atmosphere, the vegetative cover and the soil background (Bannari et al. 1995).

Such a switch is well illustrated by the history of suggested alternatives to the NDVI. By the end of the 1980s, the relationship between the NDVI and vegetation was reported to be biased in sparsely vegetated areas (e.g. arid to semi-arid zones in Australia) and in dense canopies (e.g. Amazonian forest; Huete 1988). To address such issues, the Soil-Adjusted Vegetation Index (SAVI) was proposed by Huete (1988). The SAVI was intended to minimize the effects of soil background on the vegetation signal by incorporating a constant soil adjustment factor L into the denominator of the NDVI equation (see Table 2.2). L was constructed to vary with the reflectance characteristics of the soil (including colour and brightness); it was expected to depend on the density of the vegetation to be analysed, and Huete (1988) provides a graph from which the values of L can be extracted. There, the use of an L factor of 1.0 is suggested for cases of very low density of vegetation, 0.5 is suggested for intermediate, and 0.25 is suggested for high density. For L = 0, the SAVI equals the NDVI. For L = 1, the SAVI approximates the PVI (Table 2.2). One drawback of the SAVI is that it requires a calibration factor to take into account local soil conditions. It may also be difficult to predict how soil effects will manifest across the range of spatial resolutions available. Several modifications were therefore subsequently proposed to the SAVI equation: the transformed SAVI (TSAVI; Baret et al. 1989; Baret and Guyot 1991), the Modified SAVI (MSAVI; Qi et al. 1994), the Optimized SAVI (OSAVI; Rondeaux and Baret 1996), and the Generalized SAVI (GESAVI; Gilabert et al. 2002).

The NDVI was then shown to be sensitive to attenuation and scattering by the atmosphere from aerosols (Carlson and Ripley 1997; Kaufman and Tanré 1992; Miura et al. 1998). Following this observation, the Atmospherically Resistant Vegetation Index (ARVI) was proposed by Kaufman and Tanré (1992): with the ARVI, aerosol effects are self-corrected by using the difference in blue and red reflectance to derive the surface red reflectance.

Building on the result that longer wavelengths are much less sensitive to smoke and aerosols, Miura et al. (1998) and Karnieli et al. (2001) later suggested minimizing atmospheric effects on the NDVI by using the middle-infrared wavelength region as a substitute for the red band.

As highlighted in Section 2.2.1, the NDVI was also shown to saturate in high biomass vegetated areas, and several methods have been suggested to overcome the saturation effects on this vegetation index. One proposed solution was to transform the NDVI using an inverse tangent function (Ünsalan and Boyer 2004). Another suggestion was to add weighting factors to the NIR reflectance term in the NDVI equation to adjust the relative contributions of the NIR and RED reflectances to the NDVI (Gitelson 2004; Vaiopoulos et al. 2004). More recently, the Enhanced Vegetation Index (EVI) was developed as a standard satellite vegetation product for the Moderate Resolution Imaging Spectroradiometer (MODIS), to take full advantage of the sensor capabilities (Huete et al. 1997, Huete et al. 2002). MODIS operates on board the Terra and Aqua satellites, which are both Sun-synchronous, near-polar orbiting satellites that provide repeated global coverage every two days at the equator (Tucker and Yager 2011). The Terra satellite was launched in December 1999, and image acquisition started in February 2000. Aqua was launched in May 2002, and image acquisition started in June 2002. The EVI, like the SAVI, was developed to improve the NDVI by optimizing the vegetation signal by using the blue reflectance to correct for soil background signals and reduce atmospheric influences, including aerosol scattering. The EVI was shown to be highly correlated with photosynthesis and plant transpiration in a number of studies (Glenn et al. 2008), being generally considered most useful in regions where the NDVI may saturate (Huete et al. 2002). Yet, whereas the NDVI responds mostly to red variation, the EVI is more sensitive to NIR variation than the NDVI: such sensitivity is associated with the EVI tending to present relatively low values in all biomes and also lower ranges over semi-arid sites. The EVI is also sensitive to variation in the blue band reflectance, with the consistency between the EVI from the different sensors being reduced when compared to the NDVI due to the different atmospheric correction schemes of the blue band (Fensholt et al. 2006b). These restrictions on inter-calibration opportunities for EVI are enhanced by the limited number of sensor systems designed with a blue band, in addition to the red and near-infrared bands (Jiang et al. 2008).

Factors affecting vegetation indices are summarized in Box 2.2.

> **Box 2.2 Factors affecting vegetation indices**
>
> Cloud contamination, atmospheric constituents such as aerosols, gases and water vapour, and solar zenith angle are major sources of errors that can heavily affect vegetation indices. Minor sources of errors include variation in viewing geometry, reflective properties of the soil background, inaccuracies in the location of data, and biases caused by pre-processing methods (such as Maximum Value Compositing; see Chapter 3 for discussion of this noise reduction method) (Sellers et al. 1994; Figure 2.3).

2.3 Characteristics of a successful vegetation index

2.3.1 Supplying global information

Most vegetation indices have been developed to satisfy quite specific applications in remote sensing, such as crop yield monitoring, forest exploitation monitoring, vegetation management, or vegetation detection in inundated regions (Bannari et al. 1995). During most phases of development, priority was given to identifying an index that would best correlate with the desired information in a given set of locations. In a few cases, this initial step led to analysis of the global performance of a given index and to a global product in which disturbances arising from sensor calibration, atmospheric contaminations, as well as viewing and illumination geometry were adequately taken into account. A first requirement for a vegetation index to be successful is thus to ensure that it is able to reliably inform vegetation dynamics across the globe.

2.3.2 Emanating from a sustainable programme

As we saw in Chapter 1, raw satellite data need to be processed before information on the spatio-temporal dynamic of any given vegetation index can be extracted with confidence. A given set of raw data can lead to multiple datasets for a given vegetation index, depending on the processing method applied. Gathering and processing raw data and maintaining datasets for end-users has a cost: if these costs are not met in the long term, the maintenance and production of a given dataset can be stopped. Therefore, a second requirement for a vegetation index to be successful is linked to the long-term sustainability of the programmes producing the associated datasets.

2.3.3 Possibility for inter-calibration

The ability to compare the information captured by a vegetation index across sensors boosts confidence in it considerably. This also allows for the production of consistent, continuous, and reliable datasets. In the case of the NDVI, for example, numerous studies have evaluated its continuity across sensors (see Jiang et al. 2008 for a review). Accordingly, a third requirement for a vegetation index to be successful is linked to the possibility of comparing datasets originating from different sensors.

2.3.4 Using frequently monitored spectral bands

There are many sensors on board satellites collecting information each day. The design and launch of these sensors are associated with major costs, and the funders behind these programmes naturally aim to maximize the return on their investments. Indices making use of the most frequently monitored spectral bands, such as the visible red or the near-infrared bands, are associated with opportunities for many of these programmes to produce vegetation index datasets derived from the raw information collected by their sensors. Therefore, successful vegetation indices can be expected to be based on reflectance information captured in the most frequently monitored spectral bands (Figure 2.4).

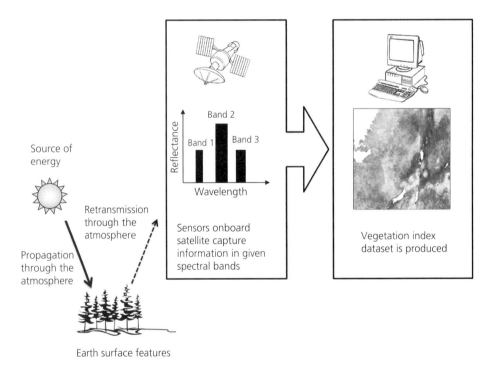

Figure 2.4 Processes involved in the creation of a vegetation index dataset. For a given satellite, information is monitored from specific spectral bands by the sensors. This raw information is then corrected and combined to create a dataset matching the specifications of the vegetation index targeted. If the vegetation index considered is based on information from spectral bands not acquired by the sensors on board that particular satellite, the dataset cannot be generated. This explains why not all vegetation index products are available from all multispectral sensors on board satellites (e.g. there is no Enhanced Vegetation Index product associated with the Advanced Very High Resolution Radiometer sensors).

2.4 Conclusions

Vegetation indices are satellite data products providing key information for monitoring terrestrial landscapes. The NDVI is only one of the many vegetation indices that have been proposed in recent decades. In trying to explain its success relative to the DVI or the SAVI (Table 2.2), for example, one should remember that the formulation of the NDVI is relatively simple, allowing the index to be calculated from different sensors on board various satellites. Such simplicity, associated with opportunities for inter-calibration and the ability to reliably inform vegetation dynamics across the globe, may well be responsible for the NDVI's success. However, its popularity does not always guarantee that it is the most relevant, and there are situations where some of the indices introduced in this chapter might be interesting alternatives to the NDVI. This point is discussed in more depth in Chapters 9 and 11.

CHAPTER 3

NDVI from A to Z

> Nothing tends so much to the advancement of knowledge as the application of a new instrument.
> **Sir Humphry Davy**

The NDVI has been associated with more applications, especially in ecology, than any other vegetation index. In all, 693 articles with the topic 'NDVI' and subject area 'ecology' have been published since 1990 and cited more than 12 500 times (ISI Web of Science search, 13 March 2013; Figure 3.1).

The NDVI's popularity can be attributed partially to the fact that it is easy to calculate, requiring only the information collected by the red and near-infrared bands (which are common to almost all passive space-borne sensors; Gillespie et al. 2008). The NDVI has been readily available from as early as the 1970s (through Landsat) at various spatial and temporal resolutions. However, there may be less easy access to pre-existing, processed data for other vegetation indices (Pettorelli et al. 2011). The popularity of the NDVI may also be linked to the wide array of disciplines in which it has been used successfully, ranging from environmental monitoring and agronomy, to macroecology, community ecology, animal behaviour or paleoecology. Examples of successful applications are detailed in Chapters 5 to 8. The breadth of possible applications has allowed the audience of potential users to increase steadily over the past decades, and to establish the NDVI as a reliable addition to the traditional set of environmental variables available to scientists.

This chapter gives an in-depth presentation of the NDVI, from detailing the rationale behind the NDVI's formulation, to reviewing the current set of available NDVI datasets and illustrating the diversity of measures that can be derived from these data. The main caveats and limitations associated with the NDVI in ecology are discussed, and the benefits of complementing NDVI datasets with ancillary information are highlighted.

3.1 How does it work?

3.1.1 Rationale behind the formulation of the NDVI

The absorption, reflection, and transmission of a given vegetation canopy are primarily controlled by the physiology and pigment chemistry of its leaves. Green leaves absorb incoming solar radiation in the photosynthetically active radiation spectral region, drawing from this the energy needed to power photosynthesis (see Chapter 2; Jensen 2007). Leaves from live green plants have also evolved to scatter solar radiation in the near-infrared spectral region, as the energy level per photon in that domain (wavelengths longer than about 700 nm) is not sufficient to synthesize organic molecules. A strong absorption at these wavelengths would only result in overheating the plant and denaturing its proteins (Jensen 2007). Because leaves have high visible light absorption and high near-infrared reflectance (Figure 3.2), green vegetation appears relatively dark in the photosynthetically active radiation spectral region and relatively bright in the near-infrared. By comparison, clouds and snow tend to be rather bright in the visible red band (as well as other visible wavelengths) and quite dark in the near-infrared. The rationale behind the NDVI is based on these characteristic patterns of vegetation absorption and reflectance in the red

The Normalized Difference Vegetation Index. First Edition. Nathalie Pettorelli.
© Nathalie Pettorelli 2013. Published 2013 by Oxford University Press.

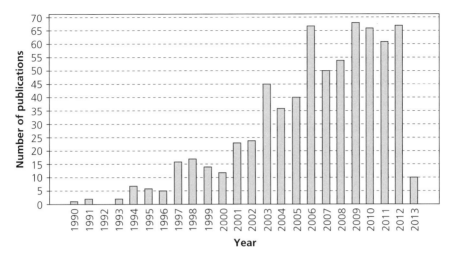

Figure 3.1 Number of ecological publications per year, from 1990 to 2013, with the topic 'NDVI'. The numbers presented have been sourced from an ISI Web of Science search, performed on 13 March 2013.

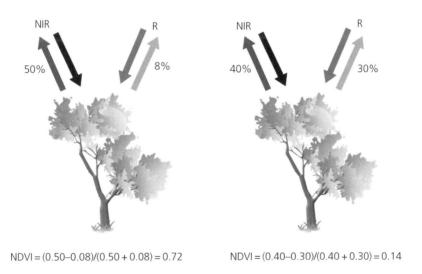

NDVI = (0.50–0.08)/(0.50 + 0.08) = 0.72   NDVI = (0.40–0.30)/(0.40 + 0.30) = 0.14

Figure 3.2 NDVI calculation for healthy (left) and senescent vegetation (right). Healthy vegetation absorbs incoming red light, while reflecting infrared radiation. Senescent vegetation reflects more visible light and less near-infrared light, leading to a reduced NDVI value (see also <http://earthobservatory.nasa.gov/Features/MeasuringVegetation/measuring_vegetation_2.php>). See also Plate 2.

and near-infrared, being computed as (NIR − R)/(NIR + R) (see Chapter 2 for more information on the principles guiding the construction of vegetation indices such as the NDVI). The NDVI can be seen as an index of 'greenness,' supplying information about the level of photosynthetically active material available in a given spatial unit.

The NIR and R spectral reflectance are both expressed as ratios of the reflected over the incoming radiation in each spectral band individually: therefore, NIR and R only take on values between 0 and 1. Thus, the NDVI itself can only vary between −1.0 and +1.0. Negative NDVI values correspond to an absence of vegetation (Justice et al.

Table 3.1 Examples of typical ranges of NDVI values for a selection of ecosystems.

Ecosystem	Typical NDVI values	Location	References
Boreal forest	0.6–0.8	Alaska	Parent and Verbyla 2010
Temperate forest	0.3–0.7	France	Pettorelli et al. 2006
Coastal rainforest	0.88–0.92	Solomon Islands	Garonna et al. 2009
Alpine pastures	0–0.35	Italy	Pettorelli et al. 2007
Annual grassland	0.15–0.45	California	Gamon et al. 1995
Desert	0.06–0.12	Sinai, Egypt	Dall'Olmo and Karnieli 2002

> **Box 3.1 Some definitions**
>
> **Evapotranspiration (ET):** The sum of evaporation and plant transpiration from the Earth's land surface to the atmosphere. Apart from precipitation, evapotranspiration is one of the most significant components of the water cycle. Actual evapotranspiration (AE or AET) is the quantity of water removed from a surface due to the processes of evaporation and transpiration. Potential evapotranspiration (PET) is a measure of the ability of the atmosphere to remove water from the surface through the processes of evaporation and transpiration assuming no limitation on water supply. PET is thus considered the maximum ET rate possible with a given set of meteorological and physical parameters.
>
> **Photosynthetic capacity:** A measure of the maximum rate at which leaves are able to fix carbon during photosynthesis.
>
> **Bidirectional Reflectance Distribution Function (BRDF):** This refers to a four-dimensional function that defines how light is reflected at an opaque surface. The function enables access to the reflectance of a target as a function of illumination geometry and viewing geometry. The BRDF is needed in remote sensing for the correction of view and illumination angle effects (for example, in image standardization and mosaicking), for deriving albedo, for land cover classification, for cloud detection, for atmospheric correction and other applications. The BRDF describes what we all observe every day: that objects look differently when viewed from different angles, and when illuminated from different directions.

1985). Very low values of NDVI (≤ 0.1) correspond to barren areas of rock, sand, or snow. Free-standing water (such as oceans, seas, lakes, and rivers) generally has a rather low reflectance in both R and NIR spectral bands, and thus NDVI values associated with free-standing water tend to be in the very low positive to negative values. Soils generally exhibit NIR reflectance somewhat larger than the R reflectance, and thus tend to generate rather small positive NDVI values (roughly 0.1–0.2). Sparse vegetation such as shrubs and grasslands or senescing crops may result in moderate NDVI values (~0.2–0.5). High NDVI values (~0.6–0.9) correspond to dense vegetation such as that found in temperate and tropical forests or crops at their peak growth stage (Jensen 2007; Neigh et al. 2008). Other examples of typical NDVI ranges for corresponding ecosystems can be found in Table 3.1.

3.1.2 NDVI as a proxy for greenness

The NDVI has been linked to various parameters of vegetation dynamics. Asrar et al. (1984) and Sellers (1985) both demonstrated a near-linear relationship between the NDVI and the intercepted fraction of photosynthetically active radiation, the driving energy for photosynthesis. Later the NDVI was shown to be highly correlated with photosynthetic capacity, net primary production, LAI (see Box 2.1 for definition), carbon assimilation, and evapotranspiration (Myneni et al. 1995; Buermann et al. 2002; Hicke et al. 2002; Wang, Q. et al. 2005; see Box 3.1 for definition of evapotranspiration (ET) and photosynthetic capacity).

In short, the NDVI thus provides a measure of 'greenness' (one value per temporal and spatial unit) for the whole world. NDVI data can be manipulated and aggregated in a number of ways: NDVI values for one time-step may be averaged across years to establish 'normal' growing conditions in a region. Figure 3.3 represents the average NDVI curves for two natural reserves in France, namely Trois Fontaines and Chizé (Pettorelli et al. 2006). For each year considered (1982–2003) and

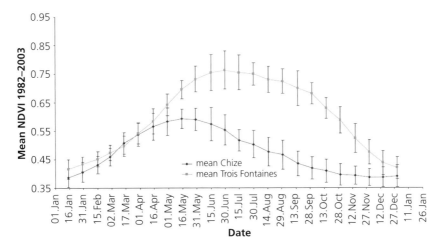

Figure 3.3 Average NDVI values over the period 1982–2003, with associated standard errors, in relation to date at Chizé and Trois Fontaines reserves, France. (Reproduced from Pettorelli et al. 2006 with permission from John Wiley & Sons.)

each pixel located in the two French reserves, the corresponding NDVI values were extracted. The NDVI data originated from the Global Inventory Modelling and Mapping Studies (GIMMS) dataset (see Table 3.2) with a spatial resolution of 8 km and a temporal resolution of 15 days (i.e. two NDVI values per month). Due to the size of the two study areas, only one pixel fell in each reserve. For each date, the NDVI values over the whole period were averaged. This led to the production of one average NDVI value per date for all years, with an associated standard error (yielding information on the level of inter-annual variability in NDVI values for each date).

Properties of NDVI time-series can also be summarized in a variety of related indices. These include measures of overall productivity and biomass, such as the Integrated NDVI (INDVI) or the annual maximum NDVI value; measures of variability in productivity (such as the relative annual range of the NDVI); and a variety of phenological measures (Reed et al. 1994; see also Figure 3.4). Examples of phenological measures include: the rate of increase and decrease of the NDVI; the dates of the beginning, end, and peak(s) of the growing season; the length of the growing season; and the timing of the annual maximum NDVI (Table 3.3). Changes in NDVI-based indices over time can be used for

Table 3.2 Frequently used NDVI datasets in ecology.

Name	Satellite	Instrument	Period covered	Spatial resolution	Temporal resolution
PAL/PAL II	NOAA	AVHRR	1981–2001	~8 km	10 day
GVI	NOAA	AVHRR	1982–present	~16 km	Weekly, monthly, seasonal
GIMMS	NOAA	AVHRR	1981–present	~8 km	Bi-monthly
LTDR	NOAA	AVHRR	1981–1999	~0.05°	Daily
FASIR	NOAA	AVHRR	1982–2000	0.25–1°	10 day/monthly
MOD13	TERRA	MODIS	2000–present	250 m to 1 km	Bi-monthly
	TM/ETM	Landsat	1984–present	15–60 m	Up to 16 days
	VGT	SPOT	1998–present	~1 km	10 day

34 THE NORMALIZED DIFFERENCE VEGETATION INDEX

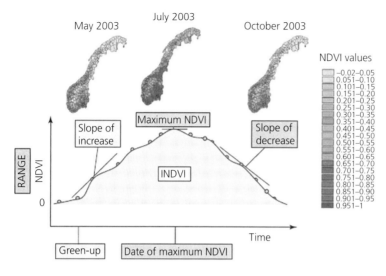

Figure 3.4 Presentation of the different indices (the slopes of increase (spring) and decrease (autumn), the maximum NDVI value, the Integrated NDVI (INDVI, i.e. the sum of NDVI values over a given period), the date when the maximum NDVI value occurs, the range of annual NDVI values, and the date of green-up (i.e. the beginning of the growing season) that may be derived from NDVI time-series over a year. Maps presenting NDVI values (ranging from 0 to 1) for Norway in May, July, and October 2003 are also shown. See also Plate 3. (Reproduced from Pettorelli et al. 2005b with permission from Elsevier.)

Table 3.3 Examples of phenological measures that can be derived from NDVI time-series.

Measure	Biological meaning	Reference
Annual maximum NDVI	Maximum photosynthetic capacity of the system under consideration	Alcaraz-Segura et al. 2009
Annual minimum NDVI	Minimum photosynthetic capacity of the system under consideration	Alcaraz-Segura et al. 2009
Integrated NDVI (INDVI)	Primary productivity proxy for the period when the NDVI values are integrated	Pettorelli et al. 2005c
Annual relative range (RREL)	Descriptor of the intra-annual variation of light interception in the system under consideration (this can be interpreted as a measure of seasonality)	Alcaraz-Segura et al. 2009
Maximum slope between any two successive bimonthly NDVI values during this period	This can supply information on the speed of the green-up phase	Pettorelli et al. 2007
Dates of the beginning (or end) of the growing season	This can yield information on the timing of the green-up	Mysterud et al. 2008
Length of growing season	Proportion of the year with significant green biomass production	Herfindal et al. 2006

many purposes, including the assessment of ecological and ecosystem responses to global warming (Pettorelli et al. 2005b; Alcaraz-Segura et al. 2009), phenological change (White et al. 2009), crop status (Tottrup and Rasmussen 2004), land cover change (Hüttich et al. 2007), or desertification (Symeonakis and Drake 2004). Several of these potential applications are discussed in Chapters 5 to 8.

How are these metrics useful for understanding the functioning of the area under consideration. Looking at the differences in NDVI dynamics between the two reserves in France (Figure 3.3), it becomes evident that (i) annual net primary production is higher in Trois Fontaines than in Chizé; (ii) seasonality in primary production is more marked in Trois Fontaines than in Chizé; (iii) the period of maximum photosynthetic capacity occurs earlier in Chizé than in Trois Fontaines; (iv) vegetation senescence also starts earlier in Chizé than in Trois Fontaines. Such conclusions are supported by independent data on climatic conditions and wood production: Trois Fontaines is known to experience a continental climate with relatively severe winters, whereas Chizé has an oceanic climate with mild winters and hot dry summers. Spring is therefore known to start earlier in Chizé, and hot dry summers known to limit primary productivity in summer in this area. Trois Fontaines is, on the other hand, recognized as being a highly productive forest lying on rich soils: wood production in Trois Fontaines reaches a long-term average of 5.92 m^3 of wood produced per hectare per year; in Chizé, wood production reaches only 3.77 m^3 per hectare per year (Pettorelli et al. 2006).

Table 3.3 and Figure 3.4 detail classical measures generally extracted from NDVI curves. However, recent studies have demonstrated that non-classical indices derived from NDVI time-series can be useful according to the situation and the issue considered. For example, determining annual variation in the INDVI over the vegetation onset period can yield the same type of information as estimating annual variation in the date of the beginning of the growing season: such a conclusion was reached by Pettorelli et al. (2005c) as they were trying to link vegetation onset in Norway to red deer *Cervus elaphus* body mass and climatic conditions in winter and spring. In Norway, vegetation dynamics are highly seasonal and highly predictable (Loe et al. 2005), which leads to a high correlation between the INDVI over the vegetation onset period (which was roughly occurring in May for the area considered) and the estimated date of the beginning of the growing season.

Another example where non-classical phenological measures derived from NDVI curves can supply ecologically relevant information comes from mountainous areas (Pettorelli et al. 2007), where the timing of snowmelt and the timing of vegetation onset are expected to affect the life histories of alpine wildlife (Rutberg 1987). Because plant phenology is the major factor affecting forage quality, it is frequently described as the driving force in habitat use by herbivorous vertebrates (Albon and Langvatn 1992). Higher forage quality is indeed associated with early phenological stages (Crawley 1983), while feeding patch choice and forage selection by ungulates are positively associated with plant quality (White 1983). A shorter period when high-quality forage is available should thus lower herbivore performance (Albon and Langvatn 1992). Because forage quality peaks during early phenological stages, slow vegetation growth should prolong access to high-quality forage. Moreover, spatial heterogeneity in snowmelt may lead to spatial heterogeneity in the timing of vegetation green-up onset, which may lengthen the period when high-quality forage is accessible to herbivores (Mysterud et al. 2001). Rapid temporal changes in plant productivity might thus correlate both with fast vegetation growth and reduced spatial heterogeneity in the timing of vegetation onset in alpine areas. Interestingly, the continuous nature of NDVI time-series actually allows partitioning of the effect of an early start of vegetation growth from that of a rapid rate of changes in vegetation phenology—a technique that was applied by Pettorelli et al. in 2007. The rate of change in plant productivity during green-up can be defined as the rate of increase between two fixed dates: the dates considered are generally the estimated date when vegetation starts growing and the estimated date when vegetation biomass reaches a plateau. For all sites considered, these correspond to early May and early July (Pettorelli et al. 2007). Considering the slope between early May and early July as an index of the rate of vegetation changes during green-up, however, is associated with a major constraint: such an index would not capture any deviation from a linear increase in the NDVI between those two dates and would average the rate of change during green-up. For example, a linear and a logarithmic increase between the two dates would yield the same slope. The authors therefore indexed the rate of vegetation change during green-up as the maximum slope between any two consecutive bimonthly NDVI values (maximum temporal resolution given the dataset considered) from early May to early July. Higher maximum increases indicated faster changes in vegetation growth and higher deviations from a linear increase in NDVI during green-up (Pettorelli et al. 2007).

Information on habitat structure can also be derived from NDVI images by compiling texture measures, which are defined as the variability of pixel values in a given area. NDVI texture analyses aim to capture heterogeneity in the amount of vegetation (Hepinstall and Sader 1997). High texture can be induced by high horizontal variability among plant growth forms: habitats that are heterogeneous either in terms of plant species composition, or in terms of the spatial distribution of plants, can be expected to display high texture; sometimes, these habitats can also be expected to be associated with an increased number of ecological niches that species can exploit. In other words, high texture can be expected to correlate with high species richness. Such an expectation was supported by Hepinstall and Sader (1997), who showed that image texture calculated from the variance in NDVI values can help to explain the occurrence of seven bird species in Maine, US. Three years later, Gould reported similar results, with the texture of NDVI accounting for up to 65% of the variability in plant species richness in the Canadian Arctic (Gould 2000). More recently, texture of the NDVI was reported to account for up to 82.3% of the variability in bird species richness in the northern Chihuahuan Desert of New Mexico (St-Louis et al. 2009).

3.2 Available datasets

Many sensors carried on board satellites measure red and near-infrared light waves reflected by land surfaces (Table 3.4). Yet reliable NDVI time-series are not readily available to ecologists for all optical sensors, and several NDVI datasets can originate from the same raw data. There are several explanations for this. First, raw data can be altered by several sources of bias and noise, and not all optical sensors capture the information required to make the required corrections. Second, correcting raw information to produce reliable NDVI datasets is costly and not necessarily part of the agenda for the agencies behind the collection of these data. Third, not everybody agrees as to how raw data should be corrected to produce reliable NDVI values, explaining the diversity of datasets than may be associated with the same raw data.

The NDVI datasets most often used in ecology are generally those that have been made freely available to the end-users. These can be distinguished according to the time-periods they cover as well as their spatial and temporal resolutions (Tables 3.2 and 3.4).

Without a doubt, the most frequently used NDVI datasets originate from the AVHRR sensor on board the NOAA satellites (Gutman 1999; Tucker et al. 2005; Pettorelli et al. 2005b, 2011). The first NOAA satellite was launched in 1979. Several others have been launched since, with lifetimes of up to seven years. The spatial resolution of the NDVI data originating from the NOAA satellites is 1.1 km nominal, the highest temporal resolution available is daily, and the spectrum spans from visible red to thermal

Table 3.4 Non-exhaustive list of relevant optical satellite sensors that can be used to derive NDVI datasets.

Sensor	Launch year	Spatial resolution	Repeat frequency	Coverage
MSS on board Landsat	1972, 1975, 1978, 1982, 1984	56–82 m	16–18 days	185*185 km
AVHRR on board NOAA	1979	1.1 km	1 day	3000 km sw
TM and ETM+ on board Landsat	1982 (TM 4) 1984 (TM 5) 1999 (ETM+)	15–60 m	16 days	185*170 km
SeaWIFS on board OrbView-2	1997	1.1 km	1 day	1500*2800 km
HRVIR on board SPOT 4	1998	10–20 m	2–3 days	60*60 km
IKONOS-2	1999	1–4 m	1–3 days	11.3 km sw
ASTER on board Terra	2000	15–90 m	16 days	60*60 km
MODIS on board Terra	2000	250 m to 1 km	~1 day	2330 km sw
QuickBird	2001	61 cm to 2.44 m	1–3 days	16.5 km sw
HRG on board SPOT 5	2002	2.5–20 m	2–3 days	60*60 km
MERIS on board ENVISAT-1	2002	~300 m	3 days	1150 sw
GeoEye-1	2008	41 cm to 1.65 m	2–8 days	15.2 km sw

sw, swath width.
Source: Horning et al. (2010).

infrared. AVHRR data represent an invaluable and irreplaceable archive of historical land surface information: those data have literally revolutionized vegetation studies. It is interesting to note that the original primary aim and design of NOAA satellites was not to collect data on vegetation; yet, to date, this is the only freely available dataset that supplies daily information for an extensive time period (1981–present). There are several NDVI datasets that originate from the raw data collected by the AVHRR sensors: examples include the Pathfinder AVHRR Land product (PAL; James and Kalluri 1994); the Global Vegetation Index (GVI; Gutman et al. 1995; Kogan and Zhu 2001); the Fourier-Adjusted, Sensor and Solar zenith angle corrected, Interpolated, Reconstructed (FASIR) adjusted NDVI dataset (Los et al. 2000); the Land Long Term Data Record (LTDR) dataset (Pedelty et al. 2007); and the GIMMS NDVI product (Tucker et al. 2005). The differences among these datasets are linked to differences in the spatial and temporal resolutions available, to differences in the type of corrections applied, and to differences in the temporal periods covered (Table 3.2).

Other NDVI products commonly found in the literature are based on data collected by the moderate-resolution satellite sensors with the proper instrumentation for studying vegetation greenness. These sensors are MODIS carried aboard NASA's Terra and Aqua satellites, SeaWIFS on board Geo-Eye's OrbView-2, and the high-resolution visible and infrared (HRVIR) instrument on board the SPOT satellites. Data produced by the GIMMS group have shown good correlation with data from these higher quality sensors (Tucker et al. 2005; see also Box 3.2). NDVI data originating from MODIS are frequently used in ecological research and applications due to their ease of access.

Readily available, free-of-charge data gathered by Landsat Thematic Mappers' sensors can also be used to generate NDVI data (see, e.g., <http://glovis.usgs.gov>). The ≥30 years' record of data acquired by the Landsat satellites constitutes the longest continuous record of the Earth's continental surfaces. With a resolution of <100 m, TM/ETM+ data can be transformed into NDVI images that have greater spatial detail than those derived from AVHRR (Table 3.2). Importantly, Landsat's orbit repeats every 16 days, compared with AVHRR's daily coverage: because of

---

Box 3.2 Are GIMMS data reliable?

At present, the only updated global coverage NDVI dataset, covering the full period from 1981 to present, is the GIMMS 8 km resolution 15-day composite dataset (Tucker et al. 2005). This NDVI product is also currently the most frequently used for evaluating vegetation patterns and trends around the world. Yet several authors have been discussing its reliability for some regions of the world (e.g. Baldi et al. 2008; Parent and Verbyla 2010; Alcaraz-Segura et al. 2010a and 2010b). These authors pointed out that, although for some regions, the NDVI trends were consistent across the different datasets and sensors (e.g. humid Sahel or the Chilean arid zones), in other regions the use of different datasets could lead to conflicting findings. This view has been opposed by others, who reported good consistency between the GIMMS products and NDVI products from the Sea-viewing Wide Field-of-view Sensor (SeaWiFS), MODIS, and SPOT (Tucker et al. 2005; Brown et al. 2006; Fensholt et al. 2006a, 2009; Song et al. 2010). As a general rule, the comparative use of different NDVI products in situations where such a choice is feasible and meaningful is probably worth considering. It might also be useful to track the future performance of the LTDR dataset: the production of the LTDR data is associated with the NASA-funded project REASoN, which aims to produce a consistent long-term dataset from AVHRR, MODIS, and the Visible/Infrared Imager/Radiometer Suite sensors. This project aims to reprocess the entire original AVHRR data from 1981 to present by applying the pre-processing improvements identified by the Pathfinder AVHRR Land II project, and the atmospheric and Bidirectional Reflectance Distribution Function (BRDF; see Box 3.1) corrections used in MODIS pre-processing steps.

---

this difference in temporal resolution, creating cloud-free NDVI products from Landsat data can be more difficult (van Leeuwen et al. 2006)

3.3 Known caveats and limitations

Like any tool, the NDVI is associated with caveats and limitations which can sometimes reduce its reliability and/or usefulness. This section reviews the factors that may influence NDVI measurements and discusses situations where the NDVI has been shown to yield less reliable information on photosynthetic capacity, net primary production, and leaf area index.

3.3.1 Caveats

Atmospheric conditions such as the presence of clouds, water vapour or atmospheric contaminants have a strong, negative influence on NDVI values (i.e. clouds, atmospheric contaminants, smoke, and water vapour lower NDVI values; Forster 1984; Holben 1986; Gutman 1991). Detecting such a source of bias can be difficult: optically thick clouds may be quite noticeable in satellite imagery, yet thin clouds (such as the ubiquitous cirrus), or small clouds with typical linear dimensions smaller than the diameter of the area actually sampled by the sensors, can be harder to notice, leading to NDVI values becoming inexact representations of the vegetation status on the ground (Tanré et al. 1992; Achard and Estreguil 1995). Similarly, cloud shadows can affect NDVI values and lead to misinterpretations. In tropical ecosystems, smoke and cloud cover can lead to paradoxes: Saleska et al. (2007), for example, reported that reduced rainfall resulted in higher satellite-based primary productivity estimates in a South American wet tropical region. This result was also observed by Garonna et al. (2009), who reported lower NDVI values in Makira, Solomon Islands, during the wet season. In both cases, the results can be linked to the effect of clouds on NDVI values: with clouds being more likely to bias NDVI estimates in wet months than in dry months, and with more light reaching the canopy during the dry season, NDVI values appeared higher during the dry season.

Orbital degradation and the deterioration of sensors (Kogan and Zhu 2001), sensor calibration error, sensor radiometric resolution, sensor drift (i.e. sensitivity of sensors that changes with time), mistakes associated with signal digitization (i.e. transformation of the signal into digital numbers; Curran and Hay 1986), and transmission errors, such as line drop-out causing localized NDVI increases (leading to the appearance of abnormally high NDVI values in the dataset; Viovy et al. 1992), are other examples of abiotic factors that can affect NDVI measurements (James and Kalluri 1994). As briefly discussed in Chapter 2, soil may also influence NDVI values: soils tend to darken when wet, with their reflectance being a direct function of water content. If the spectral response to moistening is not exactly the same in the two spectral bands, the NDVI of an area can appear to change as a result of soil moisture changes (precipitation or evaporation) and not because of vegetation changes. Additional issues may occur when using the NDVI in winter in areas of high latitudes (>60°), as reflectance resolution in such areas can become coarse (Goward et al. 1991) and as a greater incidence of spuriously high NDVI values has been reported (Justice et al. 1985). Topography and altitude also affect NDVI measurements (Thomas 1997), and caution should therefore be taken when comparing NDVI measurements in topographically variable areas.

To eliminate much of these sources of noise in the data, processing algorithms are generally applied to the raw data (Markon and Peterson 2002; Tucker et al. 2005). Bias can be minimized by forming composite images from daily or near-daily images. In some cases, post-processing noise reduction procedures may be required to further reduce noise (Reed et al. 1994); for example, when the period chosen for the temporal aggregation was mainly cloudy or when transmission errors occur (Sellers et al. 1994), causing false NDVI increases (Viovy et al. 1992). To account for those problems and to 'correct' vegetation profile, various smoothing techniques have been proposed (Table 3.5 and Figure 3.5).

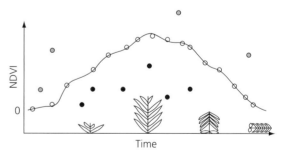

Figure 3.5 The three types of NDVI data: data collected during a cloudy day (dark circles), a clear day (open circles), and 'false high' NDVI values (grey circles) owing to transmission errors. Because of this diversity in the quality of information contained in NDVI values, time-series need to be smoothed. A typical smoother (black line) rebuilds the NDVI profile based mainly on clear-day estimates. (Reproduced from Pettorelli et al. 2005b with permission from Elsevier.)

Table 3.5 Non-exhaustive list of NDVI smoothing procedures.

Procedure	What it does	Advantages	Disadvantages	References
The Maximum Value Compositing (MVC)	NDVI values are temporally or spatially aggregated. The highest NDVI value for the considered period and area is retained.	Easy; often works well because most errors are negative	Temporal aggregates might still be contaminated by cloud cover. The procedure will be confused by a single false high.	Holben 1986; Box 3.3
Curve-fitting	Polynomial or Fourier functions are fitted to NDVI time-series	Easy; the trajectory can be predicted and the time-series can be summarized by several indices linked to the function	Medium-order polynomials can be too inflexible to recreate an entire seasonal NDVI pattern, and can smooth the data too much. Fourier analysis fails to characterize each annual NDVI trajectory separately; it can generate spurious oscillations in the NDVI time-series. Neither approach accommodates the skewed error structure, and is therefore heavily affected by false lows or highs.	Van Dijk et al. 1987; Verhoef et al. 1996; Olsson and Eklundh 1994
Stepwise logistic regression	A series of piecewise logistic functions are used to represent intra-annual vegetation dynamics. Four key transition dates are estimated: green-up, maturity (the date at which plant green leaf area is maximal), senescence and dormancy	Because the method treats each pixel individually without setting thresholds or empirical constants, it is globally applicable; it enables vegetation types to exhibit multiple modes of growth and senescence within a single annual cycle	The method does not accommodate the skewed error structure, and will therefore be heavily affected by false lows or highs.	Zhang et al. 2003
Best Index Slope Extraction method (BISE)	NDVI observations are judged as trustworthy or not depending on whether the rate-of-change in the NDVI is plausible	The algorithm is robust to false highs that cause implausibly rapid increases in the NDVI	The delicate purpose of this method is to estimate correctly to what extent a rate of change in the NDVI is plausible, according to the temporal resolution under consideration	Viovy et al. 1992
Weighted least-squares linear regression	A sliding-window combination of piecewise linear approximations to the NDVI time-series, placing more weight on 'local peaks' (NDVI values higher than the preceding and following observations). Tuning parameters are the weights affected to the local peaks and window widths	Works well when successive false lows are rare, so that local valleys occur separately, such as in the biweekly MVCs	When several false lows occur in sequence, they cause false local peaks, which bias the estimated value downwards. Thus, this approach might not be suitable for daily data, and its applicability will depend on the frequency of cloud contamination and the strength of seasonality	Swets et al. 1999; Chen et al. 2004

Reproduced from Pettorelli et al. (2005c) with permission from Elsevier.

Smoothing techniques do not always necessitate complex modelling approaches to remove contamination; sometimes – depending on the aims of the study and the level of contamination – targeted, intuitive approaches can work well. For instance, Pettorelli et al. (2012) smoothed NDVI time-series from Africa by identifying rapid changes in NDVI values (of ≥0.25 from one composite to the next) for each pixel, which were immediately followed by a return to the original values or higher. Once these contaminated values were identified, they were replaced by the average of the previous and following values, so as to 'smooth' the annual NDVI curve for that pixel. If two consecutive 'drop'

> **Box 3.3 A quick highlight on the Maximum Value Composite method (MVC)**
>
> Atmospheric contaminations, lowering NDVI values, have early been identified as a major source of error when using remote sensing data (Curran and Hay 1986). Yet atmospheric contaminations are extremely common: clouds, for example, cover about 60% of the land surface at any given time (Rossow and Schiffer 1999). One first attempt to correct for the systematic negative biases associated with atmospheric contaminations was proposed by Holben (1986) in the form of the MVC. The principle is as follows: if all contaminations depress NDVI values, then it makes sense to retain only the highest value for each pixel over a given period, as this means that one will be extracting the information from the most cloud-free day. This is the principle adopted by the MVC: the procedure creates a composite NDVI image where each pixel takes the highest NDVI value from the sequence of NDVI values available over the period considered. One problem with this method is the temporal resolution, which does not allow precise estimates of plant phenology. For example, when a green wave is advancing, NDVI values will tend to be chosen from the end of the time period, whereas at the end of a growing season, with the advance of a brown wave, NDVI values will tend to be chosen from the beginning of the time period. Moreover, sudden water stress, causing a reduction in plant biomass can occur during the growing season: depending on the original and achieved temporal resolutions, these will not be captured by a method such as the MVC (Townshend and Justice 1986; Taddei 1997).

values were present, the average of the closest higher NDVI values was calculated.

In some situations, the choice of method might not be critical. Loe et al. (2005) used the NDVI to characterize predictability of spring phenology, and used locally weighted regressions to smooth the NDVI time series, which placed low weight (in this case, 0.005) on local valley points (i.e. when NDVI values are lower than the previous measure) since they most likely occur due to cloud cover at the time of sensing rather than reversed plant phenology. The results obtained using non-weighted least-squares smoothing and cumulative maximum gave qualitatively the same results.

It is important to understand that eliminating the noise means inevitably removing information from the raw data – information which might be of particular interest in the detection of environmental change. Alcaraz-Segura et al. (2010b) recently compared different processing schemes of the same raw data (from the AVHRR sensor), showing that spatial and temporal inconsistencies exist between processing schemes. More research into the effect of different image processing on detection of environmental change is needed to optimize removal of noise from the data while retaining valuable variation stemming from actual environmental variability in the image.

3.3.2 Limitations

One major limitation to the use of the NDVI for environmental monitoring purposes is linked to the fact that the relationship between the NDVI and Aboveground Net Primary Production (ANPP) is not constant over the entire range of ANPP (Asrar et al. 1984; Sellers 1985; Paruelo et al. 1997). Studies on various vegetation types, such as agro-ecosystems (Cohen et al. 2003), grasslands (Friedl et al. 1994), shrublands (Law and Waring 1994), conifer forests (Chen and Cihlar 1996; Cohen et al. 2003), and broadleaf forests (Fassnacht et al. 1997) have indeed led to the general conclusion that vegetation indices such as the NDVI show considerable sensitivities to the LAI (Turner et al. 1999; Asner et al. 2003). In sparsely vegetated areas with LAI <3, the NDVI is strongly influenced by soil reflectance (Huete 1988), whereas for LAI >6 (in densely vegetated areas), the relationship between the NDVI and the near-infrared reflectance saturates (Asrar et al. 1984; Birky 2001). This change in the relationship between the LAI and the NDVI was also reported by Tucker et al. (1986), who demonstrated that the NDVI had an obvious tendency to reach a plateau at high LAI levels. More recent work further highlighted the existence of seasonal and annual variation in the relationship between the NDVI and the LAI (Wang, Q. et al. 2005). Results showed that the NDVI–LAI relationship varies in tune with variation in the phenological development of deciduous trees, as well as responding to temporal variation in environmental conditions. Strong linear relationships were obtained during the leaf production and leaf senescence periods for all years, but the relationship between the NDVI and the LAI in the French beech forest under study became poor during periods of maximum LAI. Altogether, these results suggest that the NDVI can underestimate the green biomass of stands with high production of green biomass and strong foliage density. Wang, Q. et al. (2005) also reported that the NDVI–LAI relationship was relatively weak when all data were pooled across the years, apparently due to different leaf area development patterns in the different years. This suggests that attention must be paid to the temporal scale when applying NDVI–LAI relationships.

A second limitation emanates from the fact that the NDVI integrates the composition of species within the plant community, vegetation form, vigour, and structure, the vegetation density in vertical and horizontal directions, reflection, absorption, and transmission within and on the surface of the vegetation or ground (Markon et al. 1995; Markon and Peterson 2002), which means that variation in NDVI values can stem from multiple sources. For example, Pinter et al. (1985) showed that reflectance of all wavebands is usually higher for planophile than for erectophile canopies of spring wheat, and that reflectance from erectophile canopies varies more with changing Sun zenith and azimuth. Heterogeneous habitats, such as those with interspersed woody and herbaceous vegetation or sparse vegetation and abundant bare ground, are therefore more likely to exhibit a weakened link between the NDVI and primary production (Elvidge and Lyon 1985; Huete et al. 1985; Huete and Tucker 1991). Likewise, dead material can also affect NDVI estimates (Tucker 1979). These factors (presence of dead material, canopy orientation, level of habitat heterogeneity) can therefore influence the ability of the NDVI to reliably index spatial variation in photosynthetic capacity, making it difficult to track subtle change in greenness across relatively small study areas. Because of such potential limitations, independent field measurements are generally recommended to validate the biological significance of NDVI measures (Hamel et al. 2009; Santin-Janin et al. 2009).

3.4 Complementing NDVI with other datasets

One way to reduce the likelihood of deriving incorrect information about the patterns in primary production from the NDVI time-series is to complement NDVI datasets with ancillary information. This section explores those ancillary data that can yield relevant information, which, if used together with NDVI data, may help to increase users' confidence in the interpretation of NDVI patterns.

3.4.1 With geographic information

Misregistration refers to situations when NDVI values are wrongly assigned to a point on the globe as a result of errors in back-calculating the position of the satellite at the time the images were taken. Because misregistration can occur, the accuracy of the downloaded data should always be checked by superimposing NDVI data on known maps using a geographic information system (Pettorelli et al. 2005b). The rise of GIS promoted the development of a variety of spatially explicit databases that have granted free access to information such as the distribution of biomes and ecoregions, climatic conditions, land-cover and vegetation types, or human population and footprint, among many others (see Table 3.6). Information on topography and elevation (derived from DEMs, such as the ones associated with the Shuttle Radar Topography Mission (SRTM)) and coastline can be particularly useful to help address misregistration issues (see Table 3.6).

3.4.2 With information on atmospheric conditions

Clouds may influence NDVI values (section 3.4.1) and methods have been developed to reduce the remnant noise in NDVI data that can be attributed to variation in cloud cover. When available, information on cloud cover can be used to better inform noise reduction approaches (Jönsson and Eklundh 2002). The Clouds from the AVHRR Extended

Table 3.6 Freely available geo-referenced datasets available on the World Wide Web.

Dataset	Description	url
VMAP0	VMAP is a vector-based collection of GIS data about Earth at various levels of detail	<http://geoengine.nima.mil/ftpdir/archive/vpf_data/v0noa.tar.gz>
GSHHG	Global Self-consistent, Hierarchical, High-resolution Geography Database	<http://www.ngdc.noaa.gov/mgg/shorelines/gshhs.html>
GADM	Spatial database of the location of the world's administrative areas	<http://www.gadm.org/>
Natural Earth Data	Supplies a variety of spatial, cultural and physical datasets (e.g. administrative boundaries, roads)	<http://www.naturalearthdata.com/>
SRTM 90 m Digital Elevation data	Elevation data at the global scale	<http://srtm.csi.cgiar.org/>
ASTER DEM	DEM based on ASTER, offering a spatial resolution of 30 m with a global coverage	<http://asterweb.jpl.nasa.gov/gdem.asp>
Global Land Cover Facility	Portal supplying access to land cover data for local to global systems	<http://landcover.org>
GlobCover	Global 300 m landcover classification approach based on MERIS onboard ENVISAT for 2004–2006 and 2009	<http://dup.esrin.esa.int/globcover/>
GLC2000	Global landcover dataset is available on a 1 km spatial resolution and for the year 2000 based on VEGETATION onboard SPOT 4	<http://bioval.jrc.ec.europa.eu/products/glc2000/products.php>
GLWD-2	The database for global lakes and wetlands supplies spatial datasets for inland waters which are >1 km²	<http://geonode.twap.iwlearn.org/data/geonode:glwd_2>
Geodata	Portal with links to a wide variety of environmental databases	<http://geodata.grid.unep.ch/>
WorldClim	Portal with access to global climate data	<http://www.worldclim.org/>
TRMM	Tropical Rainfall Measuring Mission by NASA supplies daily precipitation information about each location. However, data is only available for locations between 35°N and 35°S.	<http://trmm.gsfc.nasa.gov/>
Last of the Wild	Dataset aiming to capture human influence on terrestrial ecosystems	<http://sedac.ciesin.columbia.edu/>

(CLAVR-x; Stowe et al. 1991, 1999) processing system is NOAA's operational cloud processing system for the AVHRR on the NOAA-POES and EUMETSAT-METOP series of polar orbiting satellites. CLAVR is derived from an algorithm that uses reflected and thermal AVHRR wavelength bands to classify pixels into clear, mixed, and cloudy categories (Stowe et al. 1991; Gutman and Ignatov 1996). Another means of gathering information about cloud cover comes from the geostationary Meteosat satellite and its sister satellites. In tropical regions it can be assumed that areas with temperatures lower than a certain threshold are covered with rain clouds. Based on this assumption, the cumulated number of hours in a dekad (a period of 10 days) with this low temperature can be calculated and stored in a dataset called 'Cold Cloud Duration' (CCD; Dugdale et al. 1991).

3.4.3 With field data

As Pinzon et al. (2004) wrote, 'users . . . are strongly encouraged to validate their results using independent data' (p. 18). Remote sensing and fieldwork are not irreconcilable alternatives, they are complementary. Field data can help to ensure the accuracy of NDVI interpretation, or to validate results based on satellite-derived information. For example, geo-referenced vegetation data collected on the ground using a Global Positioning System (GPS) device can enable assessment of the biological significance of NDVI-derived phenological measures; these data can also help to determine whether the biological signal of large-scale NDVI time-series is representative of the variation observed at smaller scales. Whenever possible, NDVI data should thus be complemented with relevant geo-referenced field data.

3.5 Conclusions

The NDVI is among the most intensely studied and frequently used vegetation indices in ecology. Not only is the NDVI the vegetation index associated with the highest number of applications in ecological research, but various other data products also use the NDVI as primary input, e.g. global land-cover maps (DeFries et al. 1995, 1999), net primary production datasets (Prince and Goward 1995), burned area product (Barbosa et al. 1999), fraction of absorbed photosynthetically active radiation, leaf area index (Myneni et al. 1997), land surface temperature (Otterman and Tucker 1982; Jin 2004), and air temperature (Prihodko and Goward 1997). Despite the many limitations of the NDVI in capturing the spatio-temporal variability in primary productivity, remote sensing-based indices remain the only means of obtaining direct, quantified measures of this parameter at such spatial and temporal extents, as well as at such spatial and temporal scales. The NDVI is best applied through understanding what the index can and cannot do.

CHAPTER 4

Climate and the NDVI: a complex story

> Whenever people talk to me about the weather, I always feel quite certain that they mean something else. And that makes me quite nervous.
> **Oscar Wilde**

'Climate' refers to the set of meteorological conditions (including temperature, humidity, atmospheric pressure, wind, and precipitation) in a given region over long periods of time, usually more than thirty years. 'Weather' refers to the present condition of these elements and their variations over shorter periods of time. There are many reasons to study and understand climate patterns and their variability in time and space: basic knowledge of climate can be used for weather forecasting. A good understanding of the factors structuring the distribution and variation in climatic conditions can help to predict future climate and future changes in the distribution of biological diversity, provided that quantified links can be established between changes in climatic conditions and changes in species distribution.

This chapter focuses on reviewing the literature exploring the links between the NDVI and climatic conditions at various spatial scales. In particular, it summarizes available information on how average precipitation and temperature and inter-annual variability in these parameters influence NDVI dynamics around the world. As we saw in Chapter 3, the NDVI is an index of 'greenness,' which has been shown to be highly correlated with photosynthetic capacity, primary production, leaf area index, carbon assimilation, and evapotranspiration. The NDVI therefore provides a mean of monitoring vegetation dynamics around the world, and investigating the relationships between vegetation dynamics and wildlife, raising for example significant opportunities in behavioural ecology, habitat selection studies, movement ecology, population dynamics, or macroecology. Understanding how climatic conditions influence the NDVI enables prediction of primary productivity changes under different climate change scenarios. Because the NDVI can be linked to wildlife ecology (as detailed later on in Chapter 7), such an understanding is key for assessing past and future biodiversity consequences of changes in climate. In addition to this, vegetation plays a crucial role in regulating climate through the exchange of energy and water vapour (Pielke et al. 1998). Vegetation influences atmospheric carbon dioxide concentration, being thought to be currently absorbing about one-third of anthropogenic fossil fuel emissions into the atmosphere (Wang, X. et al. 2011). A small change in global photosynthetic capacity could therefore have a large effect on the capacity for vegetation to act as a carbon sink. Since patterns in NDVI variability yield information on changes in vegetation dynamics and photosynthetic capacity, understanding the links between climate and the NDVI can help to refine our predictions regarding the impact of increased greenhouse gas concentrations in the atmosphere on climatic conditions on Earth.

4.1 Climatic variability and the NDVI: global trends

4.1.1 Precipitation and temperature

Temperature and precipitation can be expected to influence plant growth, greenness levels, and therefore the NDVI, through various direct and indirect pathways. Temperature is known to influence photosynthesis, transpiration, and respiration, while being key in breaking dormancy. Precipitation and temperature directly influence water balance, causing changes in soil moisture regime which, in turn, influences plant growth (Wu et al. 2011). Global studies on the correlations between NDVI dynamics, precipitation, and temperature showed that temperature generally constrains vegetation growth where the amplitude of the annual temperature cycle is high, such as in the northern hemisphere mid to high latitudes (Nemani et al. 2003). There, vegetation growth is particularly responsive to spring temperatures (Menzel et al. 2006; Schwartz et al. 2006). Significant positive correlations between inter-annual NDVI and temperature variation have been reported for northern mid to high latitude areas (e.g. Europe, Asia), whereas negative correlations have been detected in the southern hemisphere (e.g. South America, southern Africa, and northern Australia), as well as in Texas and in Mexico (Los et al. 2001; Ichii et al. 2002; Figure 4.1). Anomalies in vegetation dynamics and temperature have been positively linked in coastal temperate climates such as in Europe and eastern Asia (Los et al. 2001).

In relatively warm regions, on the other hand, temperature generally plays little role in driving vegetation dynamics since it exceeds the minimum necessary for vegetative growth (Schultz and Halpert 1993). Precipitation apparently dictates vegetation dynamics where the amplitude of the annual temperature cycle is reduced (Schultz and Halpert 1993), such as in many tropical and subtropical biomes,

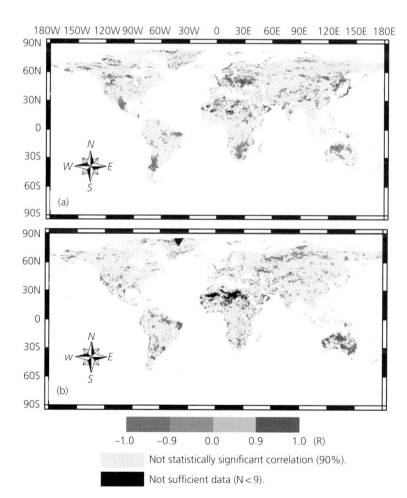

Figure 4.1 Global maps of correlation coefficients between annual averages of (a) NDVI and temperature and (b) NDVI and precipitation, for 1982–1990. Coloured areas indicate statistically significant associations (90% significance level). Grey areas are not statistically significant and black areas represent insufficient data ($n < 9$). See also Plate 4. (Reproduced from Ichii et al. 2002 with permission from Taylor & Francis.)

including grasslands, savannas, and forests. Positive NDVI–precipitation correlations are commonplace in Central Asia, southern Sahara Desert, southern Africa, Australia, South America, and Mexico (Ichii et al. 2002). Anomalies in vegetation dynamics have moreover been reported to be positively related to anomalies in precipitation throughout the tropics and subtropics and in mid-latitudes in the central parts of continents (Los et al. 2001). Interestingly, in several northern high-latitude regions, as well as mountainous regions, the NDVI has been observed to decrease with increased precipitation: this negative correlation has been interpreted as being indicative of a seasonal link between precipitation and snow cover extent (Los et al. 2001). In such places, increased precipitation during the cold season can indeed lead to increased snow cover, which decreases NDVI values. Temperature and precipitation can sometimes play an equal role in shaping NDVI dynamics: in northern and southern semi-arid regions, for example, significant correlations were identified between the NDVI, temperature, and precipitation (Ichii et al. 2002; Figure 4.1).

4.1.2 Sea surface temperature

Sea surface temperature (SST) refers to the temperature of the water close to the ocean's surface – 'close' being interpreted, depending on the measurement method used, as anything between 1 mm (which is the depth of the layer monitored by MODIS on the Aqua satellite) and 20 m below the sea surface. SST has a large influence on climatic conditions: locally, fluctuations in ocean temperature may influence the development of tropical cyclones; regionally, SST in the equatorial Pacific can affect precipitation (and therefore plant growth) over much of the North American continent, with cold water temperatures in the Pacific leading to decreased rain over western North America; globally, fluctuations in SST can generate El Niño events, which change rainfall patterns around the globe. Because SST plays such a key role in determining climatic conditions, several studies have explored the link between this parameter and NDVI dynamics.

Two famous global studies on the links between SST and the NDVI are those by Myneni et al. (1996) and Los et al. (2001). Myneni et al. (1996) explored how tropical Pacific SST anomalies influenced rainfall and NDVI anomalies in Africa, Australia, and South America using data collected between 1982 and 1990. The outcomes of the analyses carried out by Myneni et al. suggested that tropical Pacific SST warming leads to drought conditions and reduced NDVI in northern Brazil, while being associated with greater than average rainfall in southern South America. The study also highlighted how both tropical Pacific SST warming and cooling episodes can lead to negative NDVI anomalies in eastern Australia, while detailing areas of Africa most strongly associated with linked tropical Pacific SST and coincident NDVI anomalies. The study by Los et al. (2001) considered the same time period but took a more global approach, searching for linked, low-frequency inter-annual variation in global vegetation greenness, SST, land surface air temperature, and precipitation anomalies. The authors reported statistically significant modes of inter-annual variation in anomalies of SST, land surface NDVI, precipitation, and air temperature, with signals centred at 2.6 years per cycle and explaining about 28% of the variance in these anomalies.

The link between the NDVI and SST is not, however, merely about the importance of SST in generating El Niño events, a point well illustrated by the example of tropical forests. Tropical plant ecosystems are very sensitive to fluctuations in available soil moisture and absorbed radiation energy (Myneni et al. 1996; Huete et al. 2006), and soil water and radiation are closely dependent on atmospheric conditions such as cloudiness. The cloud cover in tropical areas located near the ocean is significantly affected by SST, as this parameter can modify the pattern and intensity of atmospheric moisture transport to the land regions (Knight et al. 2006; Good et al. 2008). Therefore, spatio-temporal variability in SST and its effects on cloudiness can also drive, to some extent, NDVI variability and the dynamics of tropical ecosystems (see e.g. Cox et al. 2004; Li et al. 2007).

4.1.3 Large-scale climatic indices and the NDVI

Large-scale climatic indices are associated with large-scale climatic patterns. They can be best understood as proxies for the overall climate condition, representing a 'package of weather' for a given period of time (Stenseth et al. 2003). The most notorious large-scale climatic patterns in ecology are probably the El Niño Southern Oscillation (ENSO; Trenberth 1997) and the North Atlantic Oscillation (NAO; Hurrell 1995; see also Table 4.1). The term ENSO is used to describe the atmosphere–ocean interactions throughout the tropical Pacific, characterized by variations in the temperature of the surface of the tropical eastern Pacific Ocean—warming or cooling known as El Niño and La Niña respectively—and air surface pressure in the tropical western Pacific—the Southern Oscillation (Stenseth et al. 2003). The NAO refers to a north–south alternation in atmospheric mass between the subtropical Atlantic and the Arctic. During positive phases of the NAO index, the prevailing westerly winds are strengthened and moved northwards causing increased precipitation and temperature over northern Europe and south-eastern USA, while causing dry anomalies in the Mediterranean region. Positive phases of the NAO index in winter result in warm and wet winters in Europe, mild and wet winter conditions in the eastern part of the USA, and cold and dry winters in northern Canada and Greenland. During negative phases, the patterns are reversed, and roughly opposite conditions occur. Negative phases of the NAO index thus bring moist air into the Mediterranean and cold air to northern Europe, while bringing snowy weather conditions to the east coast of the USA.

There are good reasons to explore the links between NDVI dynamics and the variability in these large-scale climatic indices. Although local weather conditions are the main driver of plant growth and phenology, indicators of large-scale climate processes can sometimes better account for ecological processes than local weather variables: indeed they provide a more holistic account of the climate systems, while reducing complex space and time variability into simple measures (Stenseth et al. 2003; Hallett et al. 2004). Another advantage of large-scale climatic indices is that their states can sometimes be predicted several months ahead (Rodwell 2003; Chen et al. 2004). Of course, there are disadvantages, mostly related to the fact that the link between global climate indices and local climate is spatio-temporally variable (Stenseth et al. 2003).

Now we turn to the links between NDVI dynamics and the variability in these large-scale climatic indices. Los et al. (2001) explored the link between the NDVI, the NAO, and the ENSO at selected sites in Eurasia, North America, South America, Africa, and Southeast Australia. The authors reported: (i) a close link between NDVI anomalies and variation in precipitation on the east coast of North America, both at ENSO and NAO timescales, with NAO signals dominating; (ii) strong ENSO signals in anomalies of NDVI, precipitation, and temperature for the Brazilian site; (iii) the existence of an 'ENSO effect' on the link between precipitation and NDVI anomalies in Africa; (iv) the possible interaction of the ENSO and the NAO in determining rainfall patterns in Southeast Australia; and (v) inverse relationships between NDVI and precipitation anomalies at both ENSO and NAO frequencies in the central northern parts of Russia (Los et al. 2001). That same year, Behrenfeld et al. explored the response

Table 4.1 Examples of large-scale climate patterns and their corresponding acronyms.

Acronym	Large-scale climatic pattern
AAO	Antarctic Oscillation
AO	Arctic Oscillation
EA	East Atlantic pattern
EAWR	East Atlantic/Western Russia pattern
ENSO	El Niño Southern Oscillation
EP	East Pacific pattern
IOD	Indian Ocean Dipole
MEI	Multivariate ENSO Index
NAM	Northern Annular mode (identical to AO)
NAO	North Atlantic Oscillation
NP	North Pacific Oscillation
PDO	Pacific Decadal Oscillation
PNA	Pacific–North American
PSA	Pacific–South American
SAM	Southern Annular mode (as AAO, but opposite sign)
SCAN	Scandinavian pattern
SO	Southern Oscillation
TNH	Tropical/Northern Hemisphere pattern
WP	West Pacific pattern

Source: Stenseth et al. (2003).

of biospheric net primary production to a major El Niño–La Niña transition, which occurred between September 1997 and August 2000 (Behrenfeld et al. 2001). The authors showed that, although biospheric net primary production varied from 111 to 117 Pg of carbon per year between 1997 and 2000, global land net primary production did not exhibit a clear ENSO response. They acknowledged, however, that regional changes could be detected and that these changes were substantial. Later that decade, Li (2006) carried out a global analysis of influence of the NAO on the NDVI, concluding that it is likely to be concentrated in Europe and North America.

Links between climatic conditions (as indexed using large-scale climatic indices) and the phenology of primary production have also been explored at the global scale. For example, Gong and Shi (2003) investigated the relationship of the inter-annual variations in the spring NDVI to nine large-scale climate indices. Of these nine, four were based on sea-level pressure fields, namely, the Southern Oscillation Index, the NAO, the Arctic Oscillation (AO), and the North Pacific index. The Southern Oscillation Index here refers to the standardized sea-level pressure difference between Tahiti and Darwin, while the AO is the first expansion coefficient of empirical orthogonal function analysis of sea-level pressure over the northern hemisphere. The NAO, as detailed above, is the difference of normalized sea-level pressures between the Azores and Iceland, while the North Pacific index is the area-weighted sea-level pressure over the region 30–65° N, 160° E to 140° W (Trenberth and Hurrell 1994). The other five indices are based on the middle troposphere geopotential height. These were the Eurasian pattern, the western Pacific pattern, the western Atlantic pattern, the Pacific/North American pattern, and the eastern Atlantic pattern (see Wallace and Gutzler (1981) for details on these patterns). All nine indices were shown to contribute to more than half of the inter-annual variations in spring NDVI, and were reported to account for a large portion of the trends in the NDVI as observed in five regions, namely, north-west North America (climate-related trend was 18.2% per ten years), south-eastern North America (5.8% per ten years), Europe (6.9% per ten years), high-latitude Asia (12.4% per ten years), and East Asia (8.0% per ten years).

4.2 Climatic variability and the NDVI: from global to regional and national trends

The studies mentioned in Section 4.1 highlight some of the major processes shaping NDVI dynamics around the world, relying on low-resolution averages and coarse correlations. Yet overall mean averages may actually be associated with large spatial variation in climatic conditions, for instance due to localized climate–topography interactions that influence NDVI dynamics locally. These interactions, which are generally dismissed by global studies, may also explain part of the global NDVI variability. In this section, I will therefore focus on studies that have explored the links between the NDVI and climatic conditions at regional scales.

4.2.1 Africa

Africa appears to be a favoured part of the world for people working with the NDVI, and many studies have used this index to monitor the response of vegetation to climatic conditions on this continent (e.g. Tucker et al. 1983; Justice et al. 1986; Townshend and Justice 1986). Work in the Sahel zone (Tucker et al. 1985b; Hielkema et al. 1986; Malo and Nicholson 1990), Botswana (Prince and Tucker 1986), East Africa (Boutton and Tieszen 1983; Davenport and Nicholson 1993) and Tunisia (Kennedy 1989) all demonstrated, positive, direct relationships between the NDVI derived from the AVHRR satellites, rainfall, vegetation cover, and biomass. Some of these studies have highlighted that a strong linear (or log-linear) relationship between the NDVI and precipitation can be identified in areas where monthly or annual precipitation is within a certain range (Malo and Nicholson 1990; Davenport and Nicholson 1993). For example, where precipitation is greater than a given threshold (such as 1000 mm per year in East Africa or in the Sahel of West Africa), the NDVI has been shown to increase only very slowly with increased precipitation (Davenport and Nicholson 1993; Nicholson and Farrar 1994). Several studies have also demonstrated that a lag between peak precipitation and peak NDVI response exists (Justice et al. 1986), with the best correlation between the NDVI and precipitation being in the concurrent month plus two previous months (Malo and Nicholson 1990; Davenport and Nicholson 1993).

As discussed in Section 4.1, global studies suggested early on that both the NAO and the ENSO influence NDVI dynamics in Africa (Los et al. 2001), but the relative importance of each large-scale climatic index in determining greenness variability has been a topic of disagreement. Continent-wide explorations of the links between vegetation and the variability in these large-scale climatic indices have revealed the relative importance of a particular index in a given region, highlighting for example how NDVI variability has been shaped by the ENSO in eastern parts of Africa, and by the NAO in the western parts of the continent (Anyamba and Eastman 1996; Anyamba et al. 2001; Stige et al. 2006). More recent work by Brown et al. (2010) started to explore the spatio-temporal relationship between NDVI-based phenology metrics and the NAO, the Indian Ocean Dipole (IOD), the Pacific Decadal Oscillation (PDO), and the Multivariate ENSO Index for Africa (MEI). As may be expected from previous continent-wide studies focusing exclusively on the ENSO and the NAO, the particular climate index and the timing showing highest correlation depend heavily on the region examined. In eastern Africa, for example, the start of the June–October season was reported to correlate strongly with the PDO in March–May, whereas the PDO in December–February was shown to correlate with the start of the February–June season. The cumulative NDVI over this last season, however, was best related to the MEI of March–May (Brown et al. 2010). Although more research is probably needed to identify which combination of large-scale climatic indices is best to predict NDVI variability across Africa, all studies point towards climate indices being key tools to anticipate future changes in primary production resulting from the impacts of climate change in these sensitive regions.

One geographical focus of many studies exploring the link between NDVI dynamics and climate in Africa is the Sahel, a region that has undergone the largest variation in rainfall and NDVI through the last three to four decades (Nicholson 2005). Such variability is an asset for exploring the impact of changing climatic conditions on net primary production dynamics and desertification rates. This requires reliably assessing current and past vegetation trends in this area, and the factors driving these trends; both have been hotly debated (e.g. Tucker et al. 1991; Anyamba and Tucker 2005; Herrmann et al. 2005).

The debate started when assertions of widespread irreversible desertification in the region were challenged by results reporting an increase in greenness level over large areas of the Sahel since the mid-1980s, which, at a coarse scale, was shown to occur concomitantly with an overall increase in rainfall (Herrmann et al. 2005; Figure 4.2). The great Sahelian droughts in the 1970s and 1980s were believed to be caused primarily by the impact of human activity on the vegetation (Clark et al. 2001; Haarsma et al. 2005), a hypothesis not supported by these new reported patterns, which seemed to indicate that desertification and vegetation recovery were solely driven by climate fluctuations. This in turn was subsequently challenged by Herrmann et al. (2005), who suggested that rainfall was possibly not the only factor driving vegetation recovery. The authors reported a good match between (i) the spatial distribution of the areas in which the vegetation had been greening up more than explained by rainfall alone; and (ii) the spatial distribution of 'hotspots' where the desertification crisis was believed to have hit hardest in the 1970s and 1980s. This led them to hypothesize that increased investment and improvements in soil and water conservation techniques, in response to the drought crisis experienced by farmers, was also an important factor driving vegetation recovery in the region.

Another debate then began over the causes of the large inter-annual and decadal fluctuations in rainfall in this region and the possible importance of regional and global SST conditions in determining rainfall patterns in the Sahel, leading to discussions about the impact of fluctuations in large-scale climatic indices on climatic conditions and NDVI fluctuations. Several studies have reported little or no influence of ENSO events on the Sahelian climate (e.g. Anyamba et al. 2001; Anyamba and Tucker 2005; Philippon et al. 2007; Propastin et al. 2010), but others have highlighted the importance of this climate index in determining climatic conditions in the region (e.g. Rowell et al. 1995; Janicot et al. 2001;

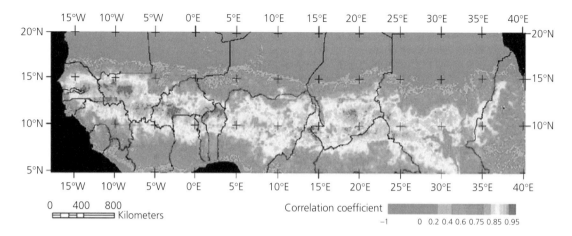

Figure 4.2 Linear correlations of monthly NDVI with three-monthly cumulative rainfall based on the Global Precipitation Climatology Project (GPCP) estimates for the period 1982–2003. The NDVI dataset used here is the GIMMS time-series. Both variables are highly correlated in the Sahel region. See also Plate 5. (Reproduced from Herrmann et al. 2005 with permission from Elsevier.)

Camberlin et al. 2001; Oba et al. 2001). Because the NAO was shown to have a significant impact on the ecosystem during the 1980s, and because the variability in this index has been reported to influence rainfall patterns, it has been suggested that both the NAO and the ENSO may be predictors of the response of the Sahelian ecosystem to global climate variability (Oba et al. 2001). This suggestion has been challenged: statistical analyses carried out by Wang, G. (2003) indicated that the influence of the NAO on vegetation productivity occurs exclusively through its impact on precipitation; yet the relationship between the NAO index and precipitation in the Sahel was shown to vary substantially with time, preventing useful predictions about the impact of changes in NAO on NDVI dynamics. This level of contradiction in the Sahelian vegetation response to changes in climate indices has been hypothesized to be caused partly by spatial mismatches associated with the definitions of climate indices and the size and location of the region studied, and a closer exploration of the links between the NDVI and climate indices across various sub-regions of the Sahel showed that vegetation dynamics in the western and central Sahel are better correlated to predefined climate indices (namely, IOD, MEI, NAO, and PDO) than in the eastern Sahel (Huber and Fensholt 2011).

4.2.2 Europe

Given the wide array of climatic conditions and habitat types across Europe, plus the amount of information collected on the ground for this continent and available for analysis, Europe has been a major focus for studies on the impact of climate change on vegetation phenology (e.g. Ahas et al. 2002; Chmielewski and Rötzer 2002; Stöckli and Vidale 2004; Menzel et al. 2006). Studies exploring the links between NDVI dynamics and climatic conditions have highlighted: (i) the role of winter and spring temperature in determining the timing of plant growth in spring; and (ii) the link between summer drought conditions and reduced greenness during this period in the Mediterranean region (Stöckli and Vidale 2004; Mao et al. 2012). Studies linking recent climate variability with long-term trends in NDVI data have shown that (i) southern Europe is slowly becoming more arid; (ii) the rest of Europe is seeing an increase in greenness levels; and (iii) seasonal amplitude in northern Europe is decreasing, while a general shift to earlier and prolonged growing periods is being observed across the continent, especially in central Europe (Zhou et al. 2001; Stöckli and Vidale 2004; Julien et al. 2006).

Many studies have explored the links between the NAO and NDVI dynamics in Europe. In general, strong positive phases of the NAO tend to be associated with above-normal temperatures across northern Europe and below-normal temperatures in Greenland (and often across southern Europe and the Middle East). Strong positive phases of the NAO are also associated with above-normal precipitation over northern Europe and Scandinavia and below-normal precipitation over southern and central Europe. Opposite patterns of temperature and precipitation anomalies are typically observed during strong negative phases of the NAO. NAO indices can be calculated for different times of the year, and it has been shown that the wintertime NAO exhibits significant inter-annual and decadal variability (Hurrel 1995; Hurrell et al. 2003): this level of variability has been expected to shape greenness dynamics in Europe (Trigo et al. 2002). Gouveia et al. (2008) carried out a detailed analysis of the link between the NAO and NDVI dynamics: the authors reported that, over the Iberian Peninsula, positive values of winter NAO induce low vegetation activity in the following spring and summer seasons. For north-eastern Europe, the authors reported a different pattern than that observed in the Iberian peninsula, with positive values of winter NAO inducing high values of NDVI in spring, but low values of NDVI in summer (Gouveia et al. 2008; Figure 4.3). High values of the NAO tended to be associated with warmer and wetter winters, especially in north and central Europe; moreover, significant correlations between the number of days with snow cover and the NAO index were reported for central Europe (Bednorz 2004). Unsurprisingly, spring phenology, as assessed using the NDVI, was shown to correlate with anomalies in the winter NAO index (Stöckli and Vidale 2004). Although the authors found strong correlations between winter weather (temperature/precipitation) anomalies and NAO, they reported weaker correlations ($r^2 = 0.46$) between the NAO and their estimates of the start of the season.

The results from these continental-wide analyses held only partially at the national scale: Vicente-Serrano and Heredia-Laclaustra (2004) for example reported a negative correlation between vegetation activity and the NAO in winter as dominant in the south-west part of Spain, and that years with a negative winter NAO index were associated with increases in the Integrated NDVI (INDVI; see Chapter 3 for definition). In the eastern part of Spain, on the other hand, this correlation was reported to be positive (Vicente-Serrano and Heredia-Laclaustra 2004). Likewise, the influence of the NAO in winter on the vegetation onset was reported to be a function of the topography in Norway, with winter-NAO being reported to interact with local topography in determining patterns of spatial snow accumulation (Pettorelli et al. 2005a; Figure 4.4). A negative correlation between the winter NAO and snow depth

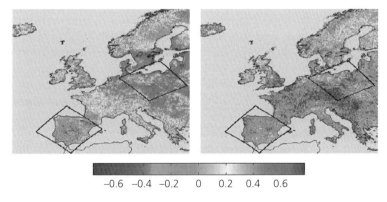

Figure 4.3 Point correlation fields of NAO versus NDVI in spring (left panel) and NAO versus NDVI in summer (right panel) for the period 1982–2002. The NDVI dataset used here is the GIMMS time-series. The NAO index used in this study is defined, on a monthly basis, as the difference between the normalized surface pressure at Gibraltar, in the southern tip of the Iberian Peninsula, and Stykkisholmur, in Iceland. The black frames identify north-eastern Europe and the Iberian Peninsula; the colour bar denotes the strength of the correlation: values above 0.42 and below −0.42 are associated with significant correlations between the NAO and the NDVI. See also Plate 6. (Reproduced from Gouveia et al. 2008 with permission from John Wiley & Sons.)

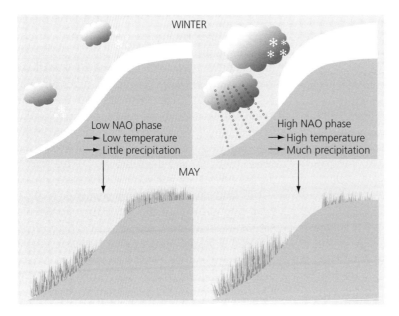

Figure 4.4 Linking the North Atlantic Oscillation (NAO) in winter and NDVI dynamics in Norway. At low altitudes, the vegetation was found to start earlier during high-winter NAO phases than low-winter NAO phases. At high altitudes, high-winter NAO phases mean more snow, which tends to delay the vegetation onset. See also Plate 7. (Adapted from the original figure in Pettorelli et al. 2005a with permission from The Royal Society.)

in March was found at low altitude, but a positive correlation between the winter NAO and snow depth in March was reported for high altitudes. Consequently, vegetation onset (as assessed from NDVI dynamics) was observed to be more delayed with increasing altitude following high NAO winters than following low NAO winters. Due to this contrasting effect of the winter NAO on snow accumulation at high and low altitude, no overall effect of the winter NAO on spring phenology across all sites could be detected (Pettorelli et al. 2005a).

The NAO is not the only climate index suspected to influence climatic conditions and vegetation dynamics is Europe: Gong and Shi (2003) reported that nine of the most important climate indices (including the NAO) are needed to explain more than half of the inter-annual variability in NDVI magnitude for the northern hemisphere. El Niño events, for example, tend to be accompanied in late winter by a negative NAO index, low temperatures in north-eastern Europe and a change in precipitation patterns (Brönnimann et al. 2007), whereas statistical associations between the ENSO and precipitation in the Mediterranean basin during the September–December wet season have been reported (Rodó et al. 1997; Shaman and Tziperman 2011). The AO is also suspected to shape NDVI dynamics across many parts of Europe: although highly positively correlated, the NAO index and the AO index are distinct indices, especially outside of the winter months (Rogers and McHugh 2002). The importance of considering such a distinction is well illustrated by Buermann et al. (2003): in this study, the authors showed that, under high AO, warmer and greener spring conditions prevail in Europe and Asian Russia.

4.2.3 Asia

4.2.3.1 Central Asia

Central Asia is dominated by arid lands, temperate deserts, and semi-deserts: NDVI dynamics in these areas are strongly influenced by precipitation (Liu et al. 2011). Linkages between the onset of vegetation green-up (as estimated using the NDVI) and the beginning of spring precipitation in the dry region of the Mongolian Plateau have been clearly established (Yu et al. 2004). Decreases in precipitation, such as those evoked by the IPCC (see Chapter 1, Box 1.1 for more information on this Panel) for this region (IPCC 2007; but see also Lioubimtseva and Henebry 2009), are thus expected to drive reduced primary productivity and desertification. Following this line of thought, changes in desert boundaries in Asia (focusing on the Gobi, Karakum, Lut, Taklimakan, and Thar deserts) during the growing season (April—October) over the 1982–2008 period have recently been investigated by Jeong et al. (2011). The authors reported that, although bare soil areas (or non-vegetated areas) inside the desert boundaries contracted by 9.8% per decade in the 1990s, they expanded by 8.7% per decade in the 2000s. In desert boundaries located along 40°N (Gobi, Taklimakan, and Karakum), these decadal changes were shown to be related mainly to: (i) precipitation increasing in the 1990s, and then decreasing in the 2000s; and (ii) intensified warming in the 2000s.

4.2.3.2 East Asia

Not only does climatic variability influence NDVI dynamics, but vegetation is also known to influence climate at local, regional, and global scales through exchanges of energy, moisture, and momentum between the land surface and the overlying atmosphere, causing changes in regional and global atmospheric circulations (Pielke et al. 1998). Thus the link between the NDVI and climate should be a two-way affair, with NDVI variability likely sometimes to be an important factor shaping subsequent climatic conditions (Wang, W. et al. 2006; Hua et al. 2008). Zhang et al. (2011) investigated the role of spring vegetation conditions over East Asia for the summer monsoon variation and prediction, by calculating the correlation coefficients between the East Asian summer monsoon index and March, April, May, and spring mean NDVI for the period 1982–2006 respectively. May vegetation greenness on the south-eastern Tibetan Plateau was the variable most closely linked to the East Asian summer monsoon, accounting for about half of the total variance in this climatic process. The authors suggest that increased vegetation greenness on the Tibetan Plateau enhances surface thermal effects, which subsequently warm the atmospheric temperature, as well as strengthen ascending motion, convergence at the lower layers and divergence at the higher layers, and summer monsoon circulation.

4.2.3.3 South East Asia

South East Asia is a particularly interesting region in which to explore the links between climatic conditions and vegetation dynamics, with some areas being strongly impacted by quasi-periodic climate fluctuations such as the ENSO. In Indonesia, for example, the effect of ENSO events is well pronounced during the dry period, leading to considerable reductions in precipitation (Gunawan et al. 2003; Hendon 2003). These droughts, occurring during the warm events of ENSO, are considered to be the result of a weakening of the Walker circulation, cooling of SST in the western tropical Pacific, and shifting of convection eastward. ENSO droughts can have major impacts on the environment, leading to uncontrolled forest fires and the associated losses of and modifications to carbon resources, biodiversity, and habitats (Fuller and Murphy 2006). These atmospheric and oceanic extremes associated with ENSO warm events can also have detrimental consequences for the economic and societal sectors (Glantz 1996; McPhaden et al. 2006). Erasmi et al. (2009) studied spatial patterns of NDVI variation over Indonesia, and their relationship to ENSO warm events during the period 1982–2006: in general, anomalies in vegetation productivity over Indonesia appear to be associated with an anomalous increase in SST in the eastern equatorial Pacific and to decreases in the Southern Oscillation Index respectively. The net effect of these variations is a significant decrease in NDVI values throughout the affected areas during the ENSO warm phases (Erasmi et al. 2009).

4.2.4 The Americas

4.2.4.1 North America

Localized, detailed studies carried out in North America have provided valuable insights into the links between climate and NDVI dynamics. Wang, J. et al. (2003) explored the link between the NDVI, precipitation, and temperature for the whole of Kansas, making use of biweekly and monthly precipitation data derived from 410 weather stations and biweekly temperature data derived from seventeen weather stations inside and around the borders of Kansas. This study represents one of the most comprehensive analyses on the topic, in terms of the number of weather stations used, the length of time analysed, and in terms of the systematic and complete design used to understand temporal patterns. The results illustrate mainly the existence of a temporal sequence in terms of the relative importance of precipitation and temperature in determining NDVI dynamics. The study also highlighted how temporal resolution can influence the reported results. The authors indeed showed that average growing season NDVI values were highly correlated with precipitation received during the current growing season and seven preceding months (15-month duration), while biweekly NDVI values were correlated with precipitation received during two to four preceding biweekly periods. Interestingly, response time of NDVI to a major precipitation event was shown to be as short as two to four weeks. Temperature was then shown to be an important factor for plant growth, but its variation appeared to be a contributing factor only at specific times during the growing season. Temperature was indeed reported to be positively correlated with the NDVI both early and late in the growing season, but only a weak negative correlation between temperature and the NDVI was detected in the mid growing season. By comparison, Yang et al. (1997) studied relations between the NDVI and temperature in Nebraska, observing strong correspondence between the NDVI and accumulated growing degree days. The authors concluded that total energy accumulation most strongly influenced plant growth, whereas Wang, J. et al. (2003) (i) found that the NDVI was more strongly correlated with minimum temperature within the growing season than with accumulated growing degree days; and (ii) concluded that precipitation was the major driver of NDVI variation and therefore of plant growth.

4.2.4.2 South America

Bordered on the west by the Pacific Ocean, on the north and east by the Atlantic Ocean and on the north-west by the Caribbean Sea, South America represents a truly amazing collection of ecosystems, ranging from the Amazonian rainforest, one of the most humid places on Earth to the Atacama desert, the driest place on Earth. The tropical Amazon region has been a particular centre of attention for many interested in identifying and comparing the relative importance of the drivers of NDVI dynamics. Such an interest is motivated not only by genuine curiosity: the response of vegetation processes in tropical forests to climate change is not yet fully understood, and the Amazon provides ideal settings for investigation. Studies showed that a variety of major environmental factors shape the functioning of this amazing ecosystem, and that their importance is spatio-temporally variable. During its rainy season, for example, the Amazon is not under soil moisture stress; rather solar radiation is the limiting factor for plant growth. But during the dry season, soil moisture decreases whereas radiation increases, meaning that the rainforest experiences different environmental stress in different seasons. Such a seasonal functioning was well illustrated by Saleska et al. (2003): carbon uptake by the forest was higher in the dry season than in the rainy season. Similarly, Huete et al. (2006) showed how a 'greening up' phase could be detected by satellites during the dry season. Large-scale climatic indices such as the El Niño–Southern Oscillation have been shown to correlate strongly with climatic conditions and forest dynamics in the area: the control of the ENSO on the Amazon rainfall operates via the warming of Pacific Ocean SST driven by the changes in wind and ocean–atmosphere heat exchange (Ronchail et al. 2002). Yet the influence of Atlantic SST on Amazon rainfall has also been found to be critical when ENSO is typically weak (Zeng et al. 2008), and climate models have been pivotal in helping demonstrate how the relative warming in the north tropical Atlantic Ocean could cause drying over part of the Amazon region (Knight et al. 2006). More recently, Cho et al. (2010) explored the effect of Atlantic SST on the spatial and temporal changes in the NDVI using satellite remote sensing data for the period of 1981–2001. Their work helped to demonstrate the existence of a strong correlation between the NDVI and SST for certain regions during the 1980s and 1990s. They also reported strong correlations with NDVI lagging behind SST for two months and one year, respectively, identified from the inter-annual December–February (rainy season) variation during 1981–2001. As acknowledged by the authors, however, these significant results are largely misunderstood and much work remains before we can unveil the mechanisms in operation here.

Other parts of South America have also been considered when studying the links between climate and NDVI dynamics. Positive NDVI anomalies over south-eastern America have been linked to tropical Pacific SST warming (Myneni et al. 1996), and significant quasi-biennial signals (2.2–2.4-year periods) between NDVI variability and rainfall variability have been identified for the continent (Tourre et al. 2008). In particular, at the national level, dominant NDVI patterns in Uruguay were reported to correlate with global ENSO (eight-month lag) whereas the dominant NDVI patterns in southern Entre-Rios/northern Buenos Aires provinces were shown to correlate with central equatorial Pacific SST (three-month lag). The dominant NDVI pattern in Santa Cruz Province in Patagonia is thought to be most correlated with the Pacific South America index and SST patterns along the Antarctica circumpolar current (three-month lag). Dominant NDVI patterns in southern Brazil, Uruguay, northern-central Argentina, the southern Paraguay–northern Argentina border, and the Santa Cruz Province were all reported to be positively correlated with global SST (Tourre et al. 2008).

4.3 Conclusions

Global climate has been experiencing prominent changes during the past decades: average climatic conditions, seasonal patterns and frequency/severity of climatic extremes are all being altered, resulting in remarkable biological consequences. Because primary productivity level is one of the key determinants of the distribution of biodiversity on Earth, understanding how climate relates to primary productivity dynamics is one of the key tasks required to enable prediction of how biodiversity and ecosystem services are likely to be further impacted by climate change. The studies reviewed in this chapter revealed that, although general correlations can be derived from global studies, the coupling between climatic variability and primary production dynamics is best explored at smaller spatial scales. These studies also highlighted that large-scale climatic indices can sometimes be particularly helpful in predicting the influences of climatic variations on the NDVI. Satellite-derived information on primary production can be used to predict subsequent climatic conditions, such that the link between the NDVI and climate may be expected to operate in a two-way fashion.

Much remains to be understood regarding how the impacts of changing climatic conditions on NDVI dynamics can be reliably predicted across the globe. However, predictive studies are starting to appear: for example, Los (in press) has reconstructed NDVI time-series using precipitation and temperature information to explore the relative role of CO_2 fertilization and climate in explaining the observed increase in vegetation greenness in mid-to-high northern latitudes. Similar methods are needed to enable prediction, in a reliable and quantifiable manner, of short- to medium-term impacts of expected changes in climatic conditions on biodiversity and ecosystem services.

CHAPTER 5

NDVI and environmental monitoring

> Usually, the main problem with life conundrums is that we don't bring to them enough imagination.
>
> **Thomas Moore**

Chapters 1–4 gave a general introduction to remote sensing, satellite-based approaches to vegetation monitoring, and the NDVI, and discussed the factors shaping NDVI variability. Chapters 5–11 focus on opportunities for the NDVI to support ecological research, wildlife, and environmental management as well as conservation; the potential limitations associated with these opportunities are also discussed.

There are many opportunities associated with using the NDVI for environmental monitoring. When focusing on terrestrial biomes, land surface monitoring over long time-scales is necessary to discern ecosystem responses to environmental change. Effectively monitoring ecosystems worldwide, however, involves several technical and financial challenges: monitoring methods need to be inexpensive, systematic, repeatable, and verifiable. The relevant metrics of condition should be frequently recordable, available over a long timeframe, and offer a rapid response allowing the early detection of changes (Alcaraz-Segura et al. 2009). Satellite-based approaches potentially meet all these criteria, providing in some cases a cost-effective, repeatable, and verifiable way to monitor ecosystem distribution and functioning.

NDVI is related to biophysical variables such as leaf area, canopy coverage, productivity, and chlorophyll density as well as providing information on vegetation phenology (see Chapter 3; Goward et al. 1985; Justice et al. 1985; Tucker et al. 1985a; Townshend and Justice 1986; Spanner et al. 1990; Yoder and Waring 1994; Peters and Eve 1995; Prince et al. 1995). There are many opportunities for the NDVI to inform environmental management, ranging from mapping ecosystem distribution to improving our predictions and impact assessments of disturbances such as drought, fire, flood, and frost. Section 5.1 introduces the NDVI as a tool to map ecosystem distribution. Section 5.2 discusses how it can help to predict the occurrence of environmental disturbances, such as droughts and wild fires, and help to assess their impacts on vegetation. Section 5.3 presents NDVI-based methods to monitor changes in functional attributes of ecosystems, such as changes in radiation intercepted by the vegetation or changes in the level of predictability in vegetation dynamics. In Section 5.4, examples are discussed in which the NDVI has been used to monitor habitat degradation.

5.1 Mapping ecosystem distribution

Reliable, geographically referenced information on global land cover is essential for many aspects of global change research, as predictions of changes in ecosystem physiology and structure in response to global change require knowledge of global distributions of land cover types. One of the first large-scale applications of the NDVI was to generate land cover maps, such as those for vegetation distribution and productivity in Africa by Tucker et al. (1985a). Because specific patterns in primary productivity dynamics tend to be associated with specific ecosystems, the NDVI can be used to differentiate ecosystem functional types (Soriano and Paruelo 1992; Paruelo et al. 2001), making it possible to map

The Normalized Difference Vegetation Index. First Edition. Nathalie Pettorelli.
© Nathalie Pettorelli 2013. Published 2013 by Oxford University Press.

land cover and land cover change at the continental (Townshend et al. 1987; Running 1990) and global (Defries and Townshend 1994; Nemani and Running 1997; Hansen, M.C. et al. 2000) scales.

The NDVI can help to differentiate savannahs from dense forests, non-forests, or agricultural vegetation (Achard and Blasco 1990; Achard and Estreguil 1995; Figure 5.1). Phenological characteristics can be used to determine evergreen forest versus seasonal forest types (Achard and Estreguil 1995; Van Wagtendonk and Root 2003; Clerici et al. 2012) or trees versus shrubs (Senay and Elliott 2002). The NDVI in coniferous or broad-leaved evergreen trees is indeed high and changes little throughout the year; the pattern of seasonal NDVI in grass is convex—the NDVI first increases and then decreases slowly; the pattern in coniferous or broad-leaved deciduous trees is trapezoidal—first the NDVI increases rapidly, then stabilizes, and then it slowly decreases (Peterson 1992; Kharuk et al. 1992; Gamon et al. 1995; Blackburn and Milton 1995; Shibayama et al. 1999). Other vegetation types might be distinguished using this vegetation index: Li et al. (2006) used nineteen maximum NDVI composite images in combination with an unsupervised classification method to map the distribution of evergreen broad-leaved forest, coniferous forest, bamboo forest, shrub-grass, aquatic vegetation and agricultural vegetation in East China. In most cases, spatial accuracy ranged between 68% and 95%. Changes in NDVI dynamics can also be used to track land cover change: in Mongolia, for example, multi-temporal NDVI data were recently combined with a ground-based ecological survey to address the issue of desertification over the period 1998–2006 (Sternberg et al. 2011). The authors were able to establish a significant correlation between same-year field observation (line transects) and NDVI data, enabling an historical land cover perspective to be developed for the considered period.

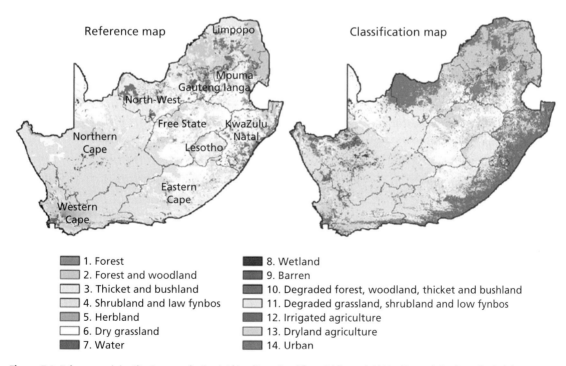

Figure 5.1 Reference and classification maps for South Africa. (Reproduced from Colditz et al. 2011 with permission from Elsevier.) The automated land cover classification methodology employed in this study makes use of MODIS time-series datasets. Input data are annual time-series of the spectral and emissive bands and the NDVI. Digitized national cartographies derived from Landsat 30 m data were employed as reference sets. Overall classification accuracy reached 80%. See also Plate 8.

There are yet limitations to what can be achieved: NDVI-based classifications of land cover types require, among other things, that each cover type has a distinguishable spectral signature (Defries and Townshend 1994). Differentiating between forests with, for example, different dominant species is not possible using the NDVI, as several assemblages of plant species can produce similar values or similar temporal trends. Even with data of sufficiently high spectral and spatial resolutions, few plant species, if any, can be identified accurately (Nagendra 2001). Limitations to NDVI-based assessments of vegetation cover have been well illustrated by Kasischke and French (1997) in the boreal forests of Alaska. The authors explored the usefulness of the NDVI AVHRR data (see Chapters 1 and 3) to map land cover types in boreal regions. They showed that (i) clouds and haze are factors shaping the intra-seasonal NDVI signature; (ii) there are significant inter-seasonal variations in NDVI signatures caused by variation in the length of the growing season, as well as variation in precipitation and moisture during the growing season; (iii) disturbances can affect large areas in interior Alaska, and forest succession after fire can result in significant variations in the inter-seasonal NDVI signatures; (iv) much of the landscape in interior Alaska consists of heterogeneous patches of forest that are much smaller than the resolution cell size of the AVHRR sensor, resulting in significant sub-pixel mixing (Kasischke and French 1997). Altogether, these results suggest that limitations to NDVI-based land cover assessments can be particularly severe when trying to map land cover types at relatively small spatial scales.

5.2 Predicting disturbances and assessing their impacts on ecosystems

Environmental disturbances resulting from events such as droughts, wildfires, hurricanes, or frost form an important component of the environmental variability experienced by living organisms. These natural disturbances (Table 5.1) can be associated with major levels of habitat loss and degradation. They can also severely impact the condition of numerous organisms and generate increased mortality rates. Such consequences are expected to directly and indirectly affect the dynamics and overall viability of wildlife populations, potentially leading to biodiversity loss and a reduction in the delivery of ecosystem services (Ameca y Juárez et al. 2012). Predicting the occurrence of environmental disturbances and monitoring the extent of the damages associated with these disturbances is therefore of high interest to environmental and wildlife managers. In this section, I will review how the NDVI can help to map the occurrence, and risk of occurrence, of several environmental disturbances (namely droughts, wild fires, floods, frost, and insect infestations); assess the extent of the associated damages; and monitor the recovery of the areas impacted by these environmental disturbances.

Table 5.1 Natural hazards that can be predicted and monitored using the NDVI.

Natural hazards	Extreme events	References
Hydrometeorological. Process or phenomenon of atmospheric, hydrological or oceanographic nature that may cause loss of life, injury or other health impacts, property damage, loss of livelihoods and services, social and economic disruption, or environmental damage.	Drought Flood Wildfire	Kogan 1997, 1998, 2000; Maselli et al. 2003; Wang, Q. et al. 2003; Diaz-Delgado et al. 2003; Segah et al. 2010
Biological. Process or phenomenon of organic origin or conveyed by biological vectors, including exposure to pathogenic micro-organisms, toxins and bioactive substances that may cause loss of life, injury, illness or other health impacts, property damage, loss of livelihoods and services, social and economic disruption, or environmental damage.	Insect infestation	Ceccato 2004; Ma et al. 2005; Eklundh et al. 2009; Jepsen et al. 2009

Categories and definitions follow the terminology of the United Nations International Strategy for Disaster Risk Reduction (UNISDR 2009).

5.2.1 Droughts

A drought is an extended period of time (generally months or years) during which a region experiences a deficiency in its water supply. Droughts originate from the combined effect of lack of precipitation over a certain period with other climatic anomalies, such as high temperature, high wind, and low relative humidity over a particular area. When drought conditions end, the recovery of vegetation generally follows (Nicholson et al. 1998), although the recovery process may take time (Diouf and Lambin 2001). Periods of drought can have significant environmental, agricultural, health, economic, and social consequences. Prolonged period of water deficit can indeed lead to reduced food production, increased food insecurity, and sometimes political instability. Ecological impacts thus generally include reduced availability of water, reduced green vegetation cover, and all the associated short- and long-term consequences for wildlife and the functioning of the disturbed ecosystem. With regard to the impact of droughts on wildlife, direct lethal effects (through reduced food and water availability) may be distinguished from more indirect effects on individuals (e.g. droughts leading to increased predation risk; increased risks of disease outbreaks; increased poaching rates or increased human-wildlife conflicts). The severity and broadness of these consequences explain why there has been so much interest in being able to anticipate drought occurrence early enough for all relevant stakeholders to be able to mitigate its impacts. As climatic extremes such as droughts are expected to become more frequent and more severe in most arid and semi-arid ecosystems (Easterling et al. 2000; IPCC 2007, 2012), this interest has become even stronger.

Because droughts directly affect vegetation dynamics, the NDVI can help to predict and monitor the impact of droughts, as well as helping document vegetation recovery. The NDVI itself does not reflect drought or non-drought conditions. NDVI anomalies, however, can inform early warning systems for drought (Kogan 1997; Kogan 2000; see also Box 5.1). Reduced water availability indeed generates vegetation stress, which can be identified by unusually large deviation from the 'normal' growing conditions in a given region for a given time of

> **Box 5.1 The Famine Early Warning System Network—FEWS NET**
>
> The Famine Early Warning System Network (FEWS NET; <http://www.fews.net>) illustrates how the NDVI can inform environmental monitoring and improve emergency response capabilities in Africa. FEWS NET was created in 1985 by the United States Agency for International Development (USAID, the agency that coordinates American foreign aid projects), in response to the 1984–1985 famines in Sudan and Ethiopia. Its aim is to provide monitoring and early warning support to decision-makers responding to famine and food insecurity (Brown 2008). Analyses underpinning the outputs generated from this system are carried out in partnership with: (i) implementing team partners (such as the National Oceanic and Atmospheric Administration's Climate Prediction Centre, the USGS, the United States Department of Agriculture's Foreign Agriculture Service) and a private-sector company managing field operations; and (ii) a large number of operational partners in host countries, as well as regional and international organizations. The early warning system relies upon vegetation, temperature, and rainfall data derived from remote sensing, atmospheric models, and available local measurements to identify abnormally wet and/or dry periods. Rainfall data are used by a variety of models that allow investigation of the direct effect of rainfall amount on crop production. Vegetation index data derived from satellite imagery (i.e. NDVI data), on the other hand, may enable insights into vegetative cover response to rainfall. Because vegetation and rainfall images measure different parameters, both types of satellite observations are needed to identify areas where famine events may occur (Ross et al. 2009). The system categorizes the severity of food insecurity levels according to a global Integrated Phase Classification-like scale, which allows comparable estimations of severity across countries and continents (Brown 2008).

the year. In other words, areas likely to experience droughts or experiencing droughts may be characterized by NDVI values that display unusually large deviation from their long-term NDVI mean.

Monitoring the response of vegetation to drought using the NDVI does not, however, work everywhere (Atkinson et al. 2011). Focusing on the Amazon, the authors showed that the response of the vegetation to the two last major drought events, in 2005 and 2010, was not detectable through satellite-observed changes in vegetation greenness: the standardized anomalies and vegetation index values for drought years were indeed of similar magnitude to those for non-drought years.

5.2.2 Wild fire

Wild fires generally refer to uncontrolled, large or destructive conflagrations that occur in 'wild' areas of grassland, woodlands, bushland, scrubland, peatland, and other wooded areas. Forest loss and degradation by wild fire can have profound impacts on land cover, land use, and biodiversity, while contributing to increased greenhouse gas emissions (the burning of biomass indeed generates the release of several gases such as carbon dioxide, carbon monoxide, methane, nitric oxide, tropospheric ozone, or methyl chloride). Wild fires can also have significant implications for human health, while sometimes severely impacting the socio-economic system of affected countries. To reduce the risks posed by wild fires, early fire detection is essential, especially in remote areas without water where fire suppression is extremely difficult. Mapping the extent of the damages caused by fire can help: (i) to assess the economic costs associated with these events; (ii) to evaluate greenhouse gas emissions related to fires; (iii) to inform ecosystem degradation assessments and plant regeneration dynamics.

Satellite-based information has proven crucial to the evaluation of fire risk, to early fire detection, to fire monitoring as well as to the mapping of ecosystem degradation associated with fire events. Remote sensing has made it possible to track fires in various locations across the globe: for example, the Office of Satellite Data Processing and Distribution, which manages and directs the operation of the central ground facilities that ingest, process, and distribute environmental satellite data and derived products to domestic and foreign users, produces freely available, real-time maps of fire locations for North and Central America <http://www.firedetect.noaa.gov/viewer.htm>. Although fires occur in most ecoregions, fire activity is not randomly distributed: fire activity is greatest in tropical grasslands and savannas, decreasing significantly towards the extreme of the productivity gradient (Pausas and Ribeiro 2013). These authors also showed that both the sensitivity of fire to extreme temperature and above-ground biomass is a function of productivity (as indexed by the NDVI), with fire being more sensitive to high temperatures in highly productive ecosystems.

With regard to mapping fire risk at local to regional scales, the use of the NDVI relies on the sensitivity of the index to vegetation dryness, which is a major predisposing factor for fire occurrence. Maselli et al. (2003) reported consistently negative correlations between fire probabilities and standardized NDVI levels of previous or contemporaneous decades using sixteen years of data on fire occurrence in Tuscany. The assessment of the extent of damage associated with a fire is generally carried out using ground surveys or Landsat images which are then classified using ground-based information (see Chapter 1). In some situations this assessment can be made using NDVI data; Segah et al. (2010) reported a strong association between assessments of land cover change derived from satellite-based NDVI data and ground-based data over the tropical peat swamp forest area in Central Kalimantan, Indonesia (an area affected by two forest fires in 1997 and 2002; Figure 5.2). Several remote sensing studies have found a good correlation between the NDVI and fire severity estimates based on biomass loss (e.g. Salvador et al. 2000; Chafer et al. 2004; Hammill and Bradstock 2006); some studies have suggested that vegetation type may influence markedly the detection of fire severity (Hammill and Bradstock 2006). Other alternatives for fire severity assessments include the normalized burn ratio (NBR) technique, developed at the end of the twentieth century and now widely used (Key and Benson 2004).

NDVI data may inform plant regeneration monitoring processes after a fire event as well as post-event environmental management. Diaz-Delgado et al. (2003) used NDVI measurements to analyse the interactions between fire severity and plant regeneration after the large fire that occurred in July 1994 in Barcelona province (Spain). Patterns of recovery for each dominant species (*Quercus ilex*, *Pinus halepensis*, *Pinus nigra*, *Pinus sylvestris*, *Pinus pinea*, shrubland, and crops) differed according to the level of fire severity experienced by the vegetation.

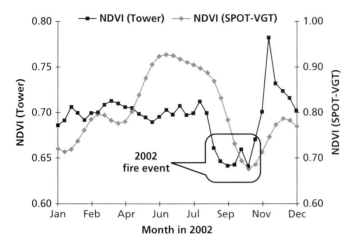

Figure 5.2 Ten-day average NDVI values of SPOT-VEGETATION and 10-day mean of ground-based NDVI (here referred as NDVI Tower) in 2002. (Reproduced from Segah et al. 2010 with permission from Taylor & Francis.) Peat and vegetation fires broke out in July 2002 in Central Kalimantan, Indonesia, and lasted for several months, as indicated from the decreasing NDVI values from both measurements. There is strong relationship between SPOT-VEGETATION data and ground-based NDVI measurements. This research demonstrated how the NDVI can be used to identify land-cover changes and vegetation recovery in tropical peat swamp forests affected by forest fires.

This study helped to establish a remote sensing-based protocol to analyse the influence of fire severity on regeneration at large spatial scales, and to support restoration efforts.

5.2.3 Flood

A flood generally refers to a covering by water of land not normally covered by water. Floods can have serious economic, societal, and environmental impacts, and are expected to become more frequent in various parts of the world as greenhouse gas concentrations in the atmosphere increase (IPCC 2007, 2012). The value of assets at risk of flooding can be tremendous: in Europe, for example, more than 10 million people live in the areas at risk of extreme floods along the Rhine, and the potential damage from floods in this area has been estimated at approaching €165 billion. Coastal areas are also at risk of flooding: focusing only on the economic aspect, the total value of assets located within 500 m of the European coastline, including beaches, agricultural land, and industrial facilities, is currently estimated at €500 to €1,000 billion <http://ec.europa.eu/>.

Images from Earth-observing satellites can provide near-instantaneous information necessary for the accurate mapping of flood extent, and are therefore widely used in monitoring flood extent and evaluating flood damage (Kogan 1998). Because water has a much lower NDVI value than other surface features, inundated areas can be distinguished by changes in the NDVI value before and after the flood, after eliminating the effects of other factors on the NDVI. This method was used in China to assess flood damage in 1998, and the results showed high correlation with flood damage, which was also estimated using other methods (Wang, Q. et al. 2003). The NDVI can also be used to assess the impact of floods on the dynamics of vegetation. In Queensland, Australia, flood frequency was plotted against 'the vigour of vegetation growth', as indexed using the NDVI (Sims and Thoms 2002). The largest proportion of 'vigorous' vegetation occurred where flood frequency was between 1.25 and 1.75 years; the smallest proportion of vigorous vegetation occurred where flood frequency was low. This type of approach could help to assess the impact of the predicted increase in flood occurrence on primary productivity dynamics.

5.2.4 Frost

Weather constitutes an important factor in agriculture, influencing food production, marketing, transportation, and therefore local, national, and regional economies. Freezing temperatures and early frost can be detrimental to plants development, tree recruitment, and crop production. In autumn, an earlier-than-normal frost can damage actively growing shoots and trigger a higher susceptibility of certain trees to diseases. In spring, a later-than-normal frost can cause differing kinds of damage depending upon the developmental stage reached, most leading to reduced yields (Snyder et al. 2005).

Because vegetation dynamics and local climate are intrinsically linked, the NDVI can be expected to provide information on conditions associated with frost occurrence, as well as helping to quantify frost damage on plants. In New Zealand, the NDVI was able to explain a significant amount of variation (from 10% to 20%) in the date of the first and last frosts, as well as the length of the frost-free period (Tait and Zheng 2003). In Japan, the NDVI was shown to be an effective tool in identifying frost-damaged tea fields (Ishiguro et al. 1994). Altogether, these studies show how the NDVI has the potential to support frost damage mapping exercises in various climatic regions and for various economically important plant species.

5.2.5 Insect infestations and associated disease outbreaks

Insect infestations (or outbreaks) generally refer to sudden increases in population sizes of certain insect species, such as defoliating insects or haematophagous arthropods. These outbreaks are comparable to pulsed disturbances, which affect ecosystem structure and function at a variety of temporal and spatial scales. Defoliating insects, for example, once present in a food-endowed environment, can quickly multiply to produce a devastating outbreak consuming their primary resource: insect damage can then disturb the growth of forests, affecting the production and quality of timber over large areas. These can sometimes lead to important economic losses. Defoliated trees can become more prone to attacks by insects and pathogens, generating a negative feedback loop on the functioning of the whole ecosystem. There are many economic, ecological, societal but also human health consequences associated with insect infestations: outbreaks of haematophagous arthropod vectors such as mosquitoes, ticks, and flies, which are responsible for transmitting bacteria, viruses, and protozoa between vertebrate hosts, can cause outbreaks of deadly diseases such as malaria, dengue fever or trypanosomiasis (Kalluri et al. 2007). The impact of insect infestations and disease outbreaks on the terrestrial carbon balance should not be overlooked (Cook et al. 2008): in Canada, moderate-to-severe insect defoliation affected an

Table 5.2 Use of the NDVI to predict or to monitor insect infestations and outbreaks.

Insect	Area	References
Geometrid moth	Fennoscandia	Jepsen et al. 2009
Pine sawfly (*Neodiprion sertifer*)	Norway	Eklundh et al. 2009
Culicoides imicola	Morocco	Baylis and Rawlings 1998
Locust (*Locusta migratoria manilensis*)	East China	Ma et al. 2005; Zha et al. 2005
Desert locust (*Schistocerca gregaria*)	Africa, Middle East and Mediterranean basin	Ceccato 2004
Brown planthopper (*Nilaparvata lugens*)	Taiwan	Yang and Cheng 2001
European gypsy moth (*Lymantria dispar*)	Mid-Appalachian highland region of the USA	Spruce et al. 2011

average of nearly 16 million hectares (4%) of forest annually during 1990–2000 (Canadian Council of Forest Ministers 2002).

The NDVI is a key variable in the management of insect infestations and disease outbreaks. First, the NDVI may be used to identify areas likely to experience insect infestations in the near future; second, it can be used to monitor and map the extent of the infestations and their associated impacts (Table 5.2). Monitor insect outbreaks mostly depends on the NDVI's ability to capture the impact of defoliating insects on their environment. Defoliating insect outbreaks (such as locust or moth outbreaks) generally result in a rapid reduction in standing biomass, enabling vegetation indices such as the NDVI to help map the affected locations. Thus, predicting insect infestation is linked to our ability to identify environmental conditions favouring insect outbreaks, such as abnormal climatic conditions leading to abnormal vegetation greenness (Indeje et al. 2006). In Kenya, for example, malaria incidence in any given month was shown to be best predicted ($R^2 = 0.88$) by the average NDVI for the thirty days including the final two dekads of the previous month and first dekad of the current month (Fastring and Griffith 2009). Another example demonstrating the usefulness of the NDVI to help predict insect outbreaks and mitigate their impacts is provided by the study carried out by Anyamba et al. (2009). The authors of this study were interested in predicting outbreaks of Rift Valley fever, a viral disease of animals and humans that occurs throughout sub-Saharan Africa, Egypt, and the Arabian Peninsula. Transmission occurs by vector mosquitoes (*Aedes mcintoshi* and *Culex* spp.), and the successful development and survival of mosquitoes that maintain, transmit, and amplify this virus is closely linked with rainfall events, with very large populations of mosquitoes emerging from flooded habitats after above-normal and persistent rainfall. Previous work by the authors showed that NDVI data, in combination with other climate variables, can be used to map areas where Rift Valley Fever outbreaks have been reported to occur (Anyamba et al. 2002, 2006b), and with this new study the authors wanted to take a step further and start mapping areas at potential risk of outbreak for the Great Horn of Africa. Areas with ideal ecological conditions for mosquito vector emergence and survival were shown to correspond with areas where greener-than-normal conditions persist over a three-month period. These findings supported the development of an early warning system, which subsequently mitigated and prevented disease spread in Kenya (Anyamba et al. 2009).

5.2.6 Extreme winter warming events

Sudden winter warming events are extreme weather events typical of the polar regions, and can have catastrophic consequences on the population dynamics of living organisms inhabiting these areas (Aanes et al. 2002; Bartsch et al. 2010). Like several other extreme events, extreme winter warming events are expected to become more frequent in the Arctic with climate change: the potential for recurring vegetation damage could be high, pointing up the need to understand how these events affect vegetation in the short and medium time-scales. NDVI data can help to inform vegetation recovery processes after an extreme winter warming event, as demonstrated by Bokhorst et al. (2012) for the sub-Arctic region. In December 2007, a twelve-day period with temperatures 2–10°C caused snow to melt across at least 1400 km^2 in northern Scandinavia. This event, followed by much colder temperatures to which vegetation was exposed due to the lack of snow cover, led to a 26% decline in NDVI between the summer of 2007 and 2008 (Bokhorst et al. 2009). This NDVI decline was associated with major shoot mortality of dwarf shrubs, and this decline was suggested as indicating a significant impact of this extreme event on ecosystem structure and function (Street et al. 2007). Bokhorst et al. (2012) produced a quantitative assessment of the time needed for the ecosystem to recover from the 2007 event, as this information could help to forecast and to mitigate the potential impact of an increase in the frequency of extreme winter warming events. Their results showed that vegetation had completely recovered after two years, suggesting that extreme winter warming events, if occurring infrequently, were unlikely to cause major vegetation shifts (Bokhorst et al. 2012).

5.3 Monitoring changes in the functional attributes of ecosystems

Functional attributes of ecosystems can be defined as variables informing different aspects of the exchange of matter and energy between the biota and the atmosphere. They can be differentiated from structural attributes, which refer to variables related to the structure of ecosystems (e.g. proportion of forested areas in an ecosystem). There are several reasons as to why monitoring functional attributes can be useful to environmental managers. First, variables describing ecosystem functioning have a faster response to disturbances because structural inertia might delay the perception of disturbances and stress (Milchunas and Lauenroth 1995). Second, functional attributes allow the quantitative and qualitative characterization of ecosystem services (e.g. carbon sequestration, nutrient, and water cycling; Costanza et al. 1997). Additionally, they are thought to be easier to monitor than structural attributes using remote sensing (Foley et al. 2007).

5.3.1 Monitoring changes in the dynamics of radiation intercepted by the vegetation

The radiation intercepted by the vegetation is an important functional attribute of ecosystems (Virginia and Wall 2001), being the main control of carbon gains (Monteith 1981): such an attribute can be indexed using the NDVI (Chapter 3; Wang, Q. et al. 2004). The benefits of monitoring changes in the radiation intercepted by the vegetation for environmental management at large spatial scales have been shown by Alcaraz-Segura et al. (2009), who describe the reference conditions and temporal trends of total amount, seasonality, and phenology of the radiation intercepted by the vegetation in areas included in the Spanish National Park Network. The aims were to gather information useful for protected area management in the country, which might serve as a reference for other areas with similar environmental conditions. The first step was to characterize the reference conditions of the protected areas by studying the seasonal dynamics of the NDVI. From the mean NDVI seasonal curve, six attributes were derived, which, they argued, might capture important features of ecosystem functioning: (i) the annual INDVI; (ii) the annual maximum (MAX); and (iii) minimum (MIN) NDVI values; (iv) the annual relative range (RREL = [MAX − MIN]/INDVI); (v) the date of maximum; and (vi) minimum NDVI values (see Chapter 3 for more information on these indices). Second, they assessed the temporal trends of these six descriptors, to identify directional changes in radiation interception, seasonality, and land-surface phenology.

Alcaraz-Segura et al. identified important functional changes occurring in the protected area network, significantly in terms of vegetation phenology and strength of seasonality. Across the parks, radiation interception increased, the contrast between the growing and non-growing seasons tended to diminish, and the date of maximum and minimum interception tended to occur earlier in the year. Their methodology illustrates ways for environmental managers to identify protected areas where rapid changes in the amount of radiation intercepted by the vegetation are occurring, and where on-the-ground monitoring might be required to elucidate what drives such changes.

Another similar example of the benefits of monitoring changes in the radiation intercepted by the vegetation for environmental management comes from Pettorelli et al. (2012): three hypotheses were tested regarding the impact of climate change on the dynamics of vegetation in category I and II protected areas in Africa (see also Section 8.1). Category I and II areas are designated to protect ecological integrity, and human impact in these areas is strictly controlled (IUCN 1994): focusing on these protected areas was thus expected to help maximize chances of solely detecting climatic signals. The focus on Africa was then supported by multiple arguments. First, Africa offers an adequate scenario in relation to: (a) the number of protected areas; (b) the availability of regional evaluations of the impact of human pressure; (c) the high diversity of ecosystems for testing the proposed hypotheses. Second, Africa is a biodiversity-rich continent expected to be hit hard by climate change. A non-negligible proportion of this biodiversity is encompassed by the protected area network, which is at risk of experiencing massive changes in ecosystem representation and functioning over the coming decades. These changes could seriously impact the ability of protected areas to conserve biodiversity, deliver ecosystem services and store carbon. Parts of these negative consequences could be mitigated by adapted management actions, if environmental and wildlife managers, governmental institutions and the international community are supplied with timely, actionable information.

The research team expected a lower annual net primary productivity and higher seasonality in net primary productivity in eastern and southern Africa; changes in net primary productivity dynamics to coincide with changes in precipitation; no correlation between changes in net primary productivity dynamics and human development. In general, these expectations were met, as (i) there was a trend towards lower annual net primary productivity and higher seasonality in eastern and southern Africa; (ii) a link was demonstrated between changes in rainfall and changes in net primary productivity dynamics across the continent; (iii) in most cases there was no significant correlation between the spatial distribution of significant changes in net primary productivity dynamics and human development or poverty levels. All of the protected areas with significantly reduced greenness and amplified seasonal patterns were located in eastern and southern Africa, and more than 50% of the protected areas from these regions showed similar trends in vegetation

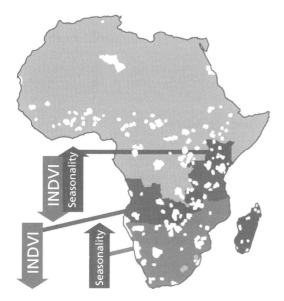

Figure 5.3 Map of Africa showing the 168 protected areas (white polygons). (Reproduced from Pettorelli et al., 2012 with permission from Elsevier.) Countries with protected areas displaying significant increase in NDVI seasonality are shown in orange. Countries with protected areas displaying significant decrease in the integrated NDVI (INDVI) are shown in dark green. Countries with protected areas displaying both a significant increase in NDVI seasonality and a significant decrease in INDVI during 1982–2008 are shown in purple. See also Plate 9.

dynamics (Figure 5.3). One out of ten of these protected areas showed significant decrease in primary production over the past three decades, and 15% of them were associated with significant increase in the strength of seasonality in vegetation dynamics. The reported patterns thus matched the IPCC (2007) projections for Africa (namely, increased seasonal patterns in temperature and precipitation in eastern and southern Africa). The main contribution of this study, in addition to that by Alcaraz-Segura et al., is the suggestion that the NDVI and protected areas could be used in concert to track the effect of climate change on ecosystems worldwide, widening opportunities associated with the NDVI to support environmental management.

5.3.2 Monitoring changes in vegetation dynamics predictability

As discussed in Section 5.3.1, functional attributes refer to variables informing different aspects of the exchange of matter and energy between the biota and the atmosphere. Focusing on detecting trends in annual changes in NDVI-based indices (such as INDVI, MAX, or RREL) may yield information on changes in an important functional attribute of ecosystems, namely the radiation intercepted by the vegetation. Approaches such as these, however, reveal nothing about the temporal nature of the dynamics of the radiation intercepted by the vegetation: the dynamics are indeed reduced to a set of NDVI-based indices, and much of the information encompassed in the NDVI curves is lost. Therefore the focus now turns to the dynamics of the radiation intercepted by the vegetation and how to assess the level of predictability in this functional attribute. Areas with a high level of predictability in vegetation dynamics are generally inhabited by organisms whose life histories are finely tuned to best match these expected variations in resource availability (Boyce 1979; Rutberg 1987): as such, the level of predictability in the radiation intercepted by the vegetation may be an important characteristic of an ecosystem. Being able to assess this level of predictability makes it possible to explore how global environmental change may or may not impact the level of consistency in vegetation dynamics; map areas likely to experience drastic changes in this parameter; and pinpoint places likely to experience disruptions in trophic chains and associated biodiversity loss.

Patterns of temporal fluctuation in NDVI can be captured by non-parametric indices such as those proposed by Colwell (1974): predictability (P) can be separated into two components termed constancy (C) and contingency (M). NDVI constancy will assess the importance of year-to-year stochastic variation in the NDVI. The lower the value of C, the more unpredictable the NDVI dynamics is among years. Contingency measures how much of the overall predictability P is due to a seasonal pattern repeated within each year, meaning that contingency reflects the strength of the seasonality (see Box 5.2 for calculation of P, M, and C). Few studies have so far made use of the metrics introduced by Colwell to explore spatio-temporal patterns in NDVI predictability, which is disappointing given the expected links between vegetation dynamics and life-history traits of consumers. To my knowledge, the only two examples come from Loe et al. (2005), who studied

> **Box 5.2 Measurement of predictability, constancy, and contingency using the method proposed by Colwell (1974)**
>
> Measures of predictability, constancy, and contingency are derived from the mathematics of information theory, more precisely from the Shannon information statistics. Imagine a frequency matrix, where there are t columns representing times within a cycle (in our case 24 as we considered two NDVI values per month) and s rows representing the states of the phenomenon (in our case ten different NDVI classes; 0–0.1, 0.1–0.2, ... 0.9–1). Let N_{ij} be the number of cycles for which the phenomenon (in this case the NDVI value) was in state i at time j. Let us also define the column totals (X_j), row totals (Y_i) and the grand total (Z) as
>
> $$X_j = \sum_{i=1}^{s} N_{ij}$$
>
> $$Y_i = \sum_{j=1}^{t} N_{ij}$$
>
> $$Z = \sum_j X_j = \sum_i Y_i = \sum_j \sum_i N_{ij}$$
>
> The uncertainty with respect to time is
>
> $$H(X) = -\sum_{j=1}^{t} \frac{X_j}{Z} \log\left(\frac{X_j}{Z}\right)$$
>
> The uncertainty with respect to state is
>
> $$H(Y) = -\sum_{i=1}^{s} \frac{Y_i}{Z} \log\left(\frac{Y_i}{Z}\right)$$
>
> And the uncertainty with respect to the interaction of time and state is
>
> $$H(XY) = -\sum_i \sum_j \frac{N_{ij}}{Z} \log\left(\frac{N_{ij}}{Z}\right)$$
>
> Predictability (P) can then be defined as
>
> $$P = 1 - \frac{H(XY) - H(X)}{\log(s)}$$
>
> Constancy is maximized when all rows but one are zero, while being minimized when all row totals are equal. A measure of constancy (C) with range (0–1) is given by
>
> $$C = 1 - \frac{H(Y)}{\log(s)}$$
>
> Contingency represents the degree to which time determines state, or the degree to which they are dependent on each other. An adjusted measure of contingency (M) with range (0–1) is given by
>
> $$M = \frac{H(X) + H(Y) - H(XY)}{\log(s)}$$
>
> In this scenario, predictability (P) is simply the sum of constancy (C) and contingency (M), with $P = C + M$.

the links between environmental predictability (as assessed by NDVI predictability) and breeding synchrony in red deer *Cervus elaphus* across Europe; and English et al. (2012), who examined the links between environmental predictability and breeding synchrony in ungulates.

5.3.3 Monitoring water availability

Freshwater availability is a strong determinant of terrestrial ecosystem functioning, and, in some very particular situations, NDVI variability can provide relevant information on this functional attribute (Aguilar et al. 2012). Focusing on barrier island shrub thickets on Hog Island, USA, these authors showed how patterns in the NDVI indicated cumulative rainfall and mean water table depth for this ecosystem. Positive linear adjustments were obtained between maximum ($R^2 > 0.9$) and mean NDVI ($R^2 > 0.87$) and the accumulated rainfall in the hydrological year and the mean water table depth from the last rainfall event till the date of the image acquisition. However, the link between water table depth and the NDVI was a function of the type of year considered (wet versus dry years): in dry years, productivity was shown to be closely related to water available from the recent past, as opposed to throughout the year for wet years. Good fits (i.e. $R^2 > 0.88$) were only obtained between monthly decrease in water table depth and NDVI variables in dry years. This example demonstrates the important feedback existing between climate, woody vegetation responses, and changes in the freshwater lens, and points to the need for more research on the potential for the NDVI to reliably help monitor water availability in other systems.

5.4 Monitoring habitat loss and degradation

Habitat loss and degradation are among the main threats posed to the preservation of biological diversity (Secretariat of the Convention on Biological Diversity 2010). These processes directly and indirectly influence the delivery of ecosystem services, through their effects on biodiversity and ecosystem functioning (Andren 1994; Kerr and Currie 1995; Fahrig 2003; Lepczyk et al. 2008). Land degradation is also thought to be a serious threat to the sustainability of human development, as it may lead to higher food insecurity, the disruption of the surface water balance, reduced carbon sequestration, and the release of carbon through soil erosion. Land degradation can moreover impact regional climatic conditions (Reynolds and Stafford Smith 2002). Being able to map the extent and severity of degradation at various spatial scales is thus of considerable interest to environmental managers as well as governmental agencies and policy-makers. Section 5.2 cited studies in which the NDVI was used to map degradation originating from an identified environmental disturbance (such as drought or wild fire). We now examine how the NDVI can help to assess land degradation, regardless of the factor causing it. To this end, degraded land refers to land that has suffered a change relative to its previous condition set by its climate, soil properties, topography, and expectations of land managers.

The first example comes from the Solomon Islands—highly forested, remote habitats that are increasingly impacted by anthropogenic forces. The Islands are located in the tropics, north of Australia and east of Papua New Guinea in the South West Pacific, between latitudes 5°S and 12°S and longitudes 152°E and 170°E. The Islands number more than 300, and are among the few places on Earth where large tracts of coastal rainforest cover remain at the start of the twenty-first century (Bayliss-Smith et al. 2003). The island of Makira is located in one of the wettest regions of the globe: the average annual rainfall reaches 3600–4000 mm, with no dry month (Allen et al. 2006). The region has some of the highest species endemism and diversity in the world (Green et al. 2006; Lamoreux et al. 2006; Danielsen et al. 2010); Makira is 'home to more endemic species than any other area in the country' and possesses 'the most significant remaining block of unlogged rainforest in the Solomon Islands, being of international importance because of its many endemic birds and mammals' (Conservation International 2007). Historically, communities in Makira have been affluent in subsistence resources due to both relatively low population densities and a large terrestrial and marine natural resource base. These communities, however, are now experiencing changes that are affecting the resilience of socio–ecological systems (Fazey et al. 2011). Two main factors are thought to drive these changes. The first is demographic: in the last decade, the population growth rate has been as high as 3.4% per annum, which is among the ten highest rates in Asia and in the Pacific (United Nations Economic and Social Commission for Asia and the Pacific 2005). In order to meet the requirements of this rapidly growing population, many households have started prioritizing short-term income-generating schemes, such as cash crop production (Fazey et al. 2007). These land-use conversions have resulted in great biomass loss, and have started to affect hydrological and biogeochemical processes (Feddema et al. 2005). The second factor is global anthropogenic climate change, which is beginning to have an impact on the Solomon Islands (Ebi et al. 2006). Altogether, these recent changes have been particularly intensive and rapid, yet few spatially explicit, quantified data are available on the extent and distribution of land degradation locally (Garonna et al. 2009).

To fill this gap, Garonna et al. (2009) decided to evaluate and map changes in primary productivity using NDVI data collected by MODIS in a region called Kahua, located at the eastern end of Makira Island and characterized by forested slopes that rise steeply from the coral and rocky-fringed coast to around 1000 m. Forty-two communities inhabit the Kahua region, the majority along the coast (Fazey et al. 2011 for more information on the area and the communities). Garonna et al.'s results indicated a rapid decrease in primary productivity, suggesting that significant ecological change in the form of greater clearance of vegetation may be occurring, despite the lack of commercial logging in the region. Such a spatially explicit analysis of primary productivity variation in Kahua yielded information on the state of vegetation in the region, allowing hotspots of change to be identified. This example is particularly interesting because NDVI-based approaches would not be the recommended way to assess degradation in this situation: the NDVI is indeed very sensitive to cloud cover; generally it cannot discriminate between different vegetation types; and in tropical regions where canopy cover is dense the linear relationship between the NDVI and primary productivity is weaker (see Chapter 3 for more on NDVI caveats and limitations). Other satellite-based methods were even less adequate. Landsat imagery might have helped mapping potential deforestation and land degradation at a higher spatial resolution and with higher reliability; however, the authors quickly discovered that all available Landsat images were highly contaminated with clouds, making it impossible to map land degradation. Multiple uncontaminated very high resolution images and high resolution data collected by active sensors such as LiDAR and RADAR (see Chapter 1) could have been another option to monitor above-ground biomass changes in this area, yet these data have only been collected for very recent years; are generally not available in the public domain; and are, in the case of LiDAR, still likely to be severely limited by the frequent cloud cover in Kahua (Garonna et al. 2009). So, even though NDVI-based approaches may not always be the preferred option, they can still provide relevant data for habitat degradation monitoring.

The second example comes from Zimbabwe, where land degradation has been an issue for several decades. Prince et al.'s (2009) NDVI-based study aimed to inform land management by supplying reliable information at the national scale on the spatial distribution and extent of land degradation. The authors sought to develop a methodology allowing the quantification and mapping of degradation in areas from a few square kilometres to a national scale, as well as to inform land management. To assess and map degradation, the authors estimated potential production in homogeneous

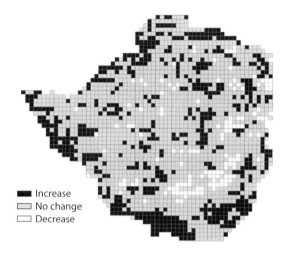

Figure 5.4 Change in degradation between 1980 Zimbabwe land degradation survey and 2000 local net production scaling (LNS) degradation map. (Reproduced from Prince et al. 2009 with permission from Elsevier.) The LNS method estimates potential production in homogeneous land capability classes and models the actual productivity using remotely sensed observations.

land capability classes and modelled the actual productivity, using, among other things, the NDVI as a surrogate of net primary productivity (Prince et al. 2009; see also Figure 5.4). This allowed the production of a map of the location and severity of land degradation for the whole country at a 250 m spatial resolution. These maps agreed with known areas of degradation and with an independent degradation map. Based on such a methodology, it was found that about 16% of the country was at its potential production. The total loss in productivity due to degradation, however, was estimated to be 13% of the entire national potential. The authors were also able to shed light on the potential drivers of land degradation at the national scale, as the locations of degraded land were unrelated to natural environmental factors such as rainfall and soils. In Zimbabwe, degradation was reported to be caused by human land use, concentrated in the heavily utilized, communal areas. Altogether, this example clearly shows the potential for the NDVI to help countries monitor their land and inform future land management decisions.

Our final example comes from south-eastern Australia, where a method referred to as Breaks

For Additive Seasonal and Trend (BFAST) was used to detect and characterize spatial and temporal changes in a forested landscape (Verbesselt et al. 2010). The method integrates the decomposition of NDVI time series into trend, seasonal, and remainder components with methods for detecting change within time-series. This yields data on the number, time, magnitude, and direction of changes in the trend and seasonal components of a time-series. To validate their approach, Verbesselt et al. decided to apply their new method to a pine plantation in Australia. To do so, they applied BFAST to the 16-day MODIS NDVI composites with a 250 m spatial resolution (MOD13Q1 collection 5) from February 2000 to the end of 2008. They then compared the detected changes with a spatial validation dataset. In 2002 and 2006 the study area experienced a severe drought, which caused the pine plantations to be stressed (and therefore the NDVI to decrease significantly). The method was able to detect both of the tree mortality events, and to highlight the severity of the 2006 event compared with that in 2002 (with changes detected in 2006 being larger (magnitude of the abrupt change) and the recovery (slope of the gradual change) slower than the changes detected in 2002). Again, this research demonstrates the potential for the NDVI to inform a national and regional landscape health-monitoring framework, where NDVI data such as those produced by the GIMMS group or derived from MODIS can be used as a 'first pass' filter to identify the areas and timing of major change activity. These areas could then be targeted for more detailed investigation using ground and aerial surveys, and finer spatial and spectral resolution imagery.

5.5 Conclusions

Based on the examples in this chapter, it is evident that the potential for the NDVI to support environmental monitoring and management is huge. From mapping ecosystem distribution and habitat degradation to monitoring ecosystem functioning and predicting the occurrence of environmental disturbances, using the NDVI may help to improve management of resources. There are, of course, limitations to what can be achieved, and the NDVI is not always the optimal remote sensing product; depending on the situation and expertise at hand, alternatives should be considered. Many of the correlations reported in this chapter need further research, enabling appropriate selection of NDVI data and metrics depending on the circumstances. Chapters 6 and 7 examine the potential for NDVI data use by plant and animal ecologists, and in supporting wildlife monitoring and management.

CHAPTER 6

NDVI and plant ecology

> If you think in terms of a year, plant a seed; if in terms of ten years, plant trees; if in terms of hundred years, teach the people.
> **Confucius**

This chapter illustrates the potential for the NDVI to enhance our understanding of plant ecology: the distribution and abundance of plants, the effects of environmental factors upon the abundance of plants, and the interactions among and between plants and other organisms (Keddy 2007). Relevant information from the NDVI includes distribution of plant species richness, plant distribution, plant growth patterns, and plant physiological status. Some of these relationships are particularly relevant to animal ecologists; for example, variation in animal species richness can be expected to vary as a function of the richness of plants with which the animals are associated (MacArthur and MacArthur 1961).

First, the literature on the links between the NDVI and plant richness is reviewed. Second, we look at studies exploring how this vegetation index can be used to inform plant distribution, for example, in mapping species or communities' distribution. Third, we focus on the relationship existing between the NDVI and plant demographic attributes, such as growth rate or mortality rate. This leads on to a review of the links between the NDVI and plant physiological attributes, such as water content and nitrogen content. The final section focuses on NDVI-based applications in agriculture.

6.1 NDVI and plant richness

Species richness is a basic indicator of biodiversity, underlying many ecological models and conservation policies. Identifying the biophysical controls of species richness patterns has long been of interest to biogeographers and community ecologists (e.g. MacArthur 1965; Pianka 1966; Brown 1988; Begon et al. 1990). Proposed explanations for the distribution of species richness are numerous and include climate and energy availability, habitat heterogeneity, interactions among species, evolutionary processes, and disturbance regimes (Diamond 1988; Currie 1991; see also Table 6.1). Most of these factors have been shown to influence (with various strength and at certain spatial scales) species richness (Begon et al. 1990).

Of particular relevance to the NDVI–plant richness link is the hypotheses relating species richness to patterns in energy availability. Energy availability and the level of spatial heterogeneity in its distribution are indeed assumed to be major determinants of species richness (Wright 1983; Abramsky and Rosenzweig 1984; Owen 1988; Adams and Woodward 1989). If these assumptions are true and if we consider the NDVI to be a good proxy for energy availability, then one can expect (i) a positive correlation between the NDVI and plant richness; and (ii) a positive correlation between spatial heterogeneity in the NDVI and plant richness.

Support for the expected correlations between NDVI patterns and patterns in the distribution of plant species richness can be found in several studies. Bawa et al. (2002) reported a high positive correlation between tree species richness and the NDVI in the Biligiri Rangaswamy hills in the Western Ghats,

Table 6.1 Some factors hypothesized to influence species richness, and some associated variables that have been linked to species richness.

Factor	Rationale	Variables
Climate and climatic variability	Benign conditions favour high species richness. Climatic stability favours specialization.	Rainfall, precipitation, air temperature, sea surface temperature, sunshine hours, maximum rainfall, seasonality strength, water deficit
Environmental heterogeneity	More niches available in heterogeneous habitats	Number of microhabitats, landscape diversity index, biotope diversity, aspect diversity, patchiness, substrate heterogeneity, topographic relief, altitudinal range
Area	Larger areas are associated with higher levels of species richness	Region area, plot area, forest area, habitat area, reserve area, pool volume, shelf area, number of caves
Energy	Species richness is limited by the partitioning of energy among species	Actual evapotranspiration, NDVI, accumulated respiration sum, potential evapotranspiration
Biotic interactions	Competition favours reduced niche breadth, but competitive exclusion can reduce species richness. Predation delays competitive exclusion.	Number of prey taxa, grazing intensity, invertebrate richness
Disturbance	Moderate levels of disturbances can delay competitive exclusion	Abiotic disturbance, time since abiotic disturbance, Pfankuch stability index

Sources: Begon et al. (1990), Currie (1991), and Field et al. (2009).

India. Gould (2000) found that in the Hood River region of the Central Canadian Arctic, ground-based measures of vascular plant species richness were positively correlated with variation in the NDVI extracted from Landsat Thematic Mapper satellite imagery. For this region, the author found that variation in the NDVI and weighted abundance of mapped vegetation types explained up to 79% of the variance in ground-based measures of species richness. Levin et al. (2007) highlighted the existence of a significantly positive linear relationship between plant species richness and NDVI values from Landsat 7 ETM+, Aster, and Quickbird in 34 quadrats along an elevation gradient in Mount Hermon, Israel (Figure 6.1). Plant richness in this study site was significantly correlated with the standard deviation of NDVI values (but not with their coefficient of variation) within quadrats and between images, suggesting that average conditions in energy availability, as well as the level of spatial and temporal heterogeneity in energy availability, may help to explain species richness distribution. Average NDVI was then shown to correlate significantly and positively with plant species richness in southeastern Spain (Alados et al. 2011), while Parviainen et al. (2009) demonstrated how NDVI conditions at the landscape scale integrated with local NDVI information could improve predictive assessments of

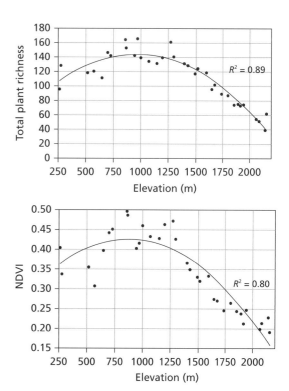

Figure 6.1 Plant richness, sampled in the 1000 m² quadrats, a function of elevation along with a second-order polynomial fit (upper), and mean NDVI within the 1000 m² quadrats as a function of elevation (lower). (Reproduced from Levin et al. 2007 with permission from John Wiley & Sons.)

plant species richness over extensive and inaccessible areas in high-latitude landscapes. Plant species richness was also reported to be linked to the NDVI in urban areas: Gavier-Pizarro et al. (2010) examined the factors shaping exotic plant species richness in areas of New England, USA, and reported a positive link between richness of invasive exotic plants and plant productivity (as indexed using the NDVI). This link was also shaped by other factors, such as the size of the area of wild land–urban interface, the extent of low-density residential areas, the change in number of housing units between 1940 and 2000, the mean income, as well as the altitudinal range and rainfall.

Not all results, however, point towards the existence of a positive link between plant richness and the NDVI. Oindo and Skidmore (2002), for example, examined the relationship between inter-annual NDVI parameters and species richness of vascular plants in Kenya. Their analyses revealed that higher average NDVI is associated with lower species richness of plants, whereas standard deviation of maximum NDVI and coefficient of variation correlated positively with species richness. The authors reasoned that as nutrients increase, light becomes an increasing problem for competing plants; thus higher productivity may be associated with increasingly intense competition for light, potentially explaining the negative relationship they reported between plant richness and the NDVI. Likewise Fairbanks and McGwire (2004) showed that, although plant richness of chaparral in California was positively related to NDVI heterogeneity and spring greenness, plant richness of coastal sage scrub was positively yet non-linearly related to annual NDVI and heterogeneity, while yellow pine forest richness was negatively related to spring greenness and positively related to heterogeneity. Schödelbauerová et al. (2009) reported a positive correlation between species diversity in orchids and ln(NDVI) in Africa, but there was a negative correlation between these variables for tropical America and no correlation at all between these variables for any other continent. In South Florida, USA, Gillespie (2005) investigated the usefulness of Landsat Enhanced Thematic Mapper Plus (ETM+) NDVI satellite images to predict stand- and patch-level richness of woody plant species in tropical dry forest. He demonstrated a significantly positive relationship between mean NDVI and stand species richness, compared with a significantly negative relationship between species richness and standard deviation of NDVI. Outcomes from this study thus supported the species–energy theory at the level of a forest stand and patch. Remarkably, though, the author also reported that combining forest patch area with NDVI significantly improved the prediction of patch species richness, giving weight to the species–area rule (see Box 6.1 for more information on the species–area relationship).

> **Box 6.1 The species–area relationship**
>
> The species–area relationship is often referred to as the closest thing to a rule in ecology. In its basic form, the rule states that the number of species in an area is a function of the size of the area. This can be described as: $S = cA^z$, where S represents the number of species, A the size of the area, and c and z are constants. Along a gradient of ecosystems of increasing size, the numbers of species inhabiting those ecosystems increases—rapidly at first, but then more slowly for the larger ecosystems (Lomolino 2000).

6.2 NDVI and plant distribution

Knowing where species occur is key to our understanding of ecological and evolutionary processes, and strongly influences our ability to adequately inform management and conservation planning. This section discusses how the NDVI can: (a) be used to map plant species distribution; (b) be used as an environmental variable to map areas likely to be occupied by a given plant species; and (c) help to map plant community composition reliably.

6.2.1 NDVI to inform species distribution

Few plant species can be identified and mapped accurately using the NDVI (Nagendra 2001; see also Chapters 3 and 5). However, exceptions do exist, and several studies have highlighted the potential for the NDVI in mapping the distribution of invasive plant species at multiple spatial scales (Morisette et al. 2006; Wilfong et al. 2009; Hoyos et al. 2010). A whole section on the NDVI as a tool to map invasive species distribution can be found in Section 8.5.

For species that cannot be identified and mapped accurately using the NDVI, there are other solutions. When information on species occurrence is patchy and rare, species distribution models (SDMs) can be used to estimate areas likely to be occupied by a given species (Elith and Leathwick 2009; see also Box 7.2). Broadly speaking, SDMs can be defined as numerical tools that combine observations of species occurrence with environmental variables to predict areas most likely to be suitable for the species considered. The NDVI, as an environmental variable, has been shown to improve several models of plant species distribution. In Brazil, SDMs of five *Coccocypselum* species were built with and without the NDVI as a predictor of species occurrence (Amaral et al. 2007). Whereas all models indicated a good fit with the data, SDMs including the NDVI were slightly better than models without the NDVI. Likewise, maximum NDVI was among the most important factors accounting for the geographical distribution of *Carapa guianensis* (Meliaceae), a high-value timber tree species inhabiting the Amazon Basin (Da Conceicao Prates-Clark et al. 2008). However, not all studies report a consistent positive association between plant species distribution and NDVI values, as illustrated by de Souza Gomes Guarino et al. (2012), who attempted to model the occurrence of nine threatened plant species in Brazil. The distributions of *Araucaria angustifolia* (Araucariaceae) and *Clethra scabra* (Clethraceae) were significantly negatively correlated with the NDVI, whereas the distributions of *Podocarpus lambertii* (Podocarpaceae) and *Trithrinax brasiliensis* (Arecaceae) were significantly positively correlated with NDVI fluctuations. No link between the NDVI and the occurrence or the abundance of the remaining five plant species could be established.

6.2.2 NDVI for plant community ecologists

Community ecology can be broadly defined as the study of patterns and processes involving natural assemblages of two or more species (Morin 2011). A key requirement for ecologists interested in exploring the factors shaping the spatial distribution of plant communities is the ability to map these communities—yet such a task may become challenging at the landscape scale. In such situations, remote sensing imagery and the NDVI might be useful. Dogan et al. (2009), for example, aimed to map the plant community composition of the entire Tersakan Valley in Turkey. They used georeferenced data from 1077 quadrats collected between 2000 and 2005 on the sociological characteristics of plant communities (as indexed using the Braun–Blanquet classification method; see Dogan et al. 2009) and assessed the possible relationship between these data and the NDVI. Socio-ecological groups, in this context, are defined as groups of plants that have similar ecological requirements. A strong link between NDVI classes and cover-abundance data from the field was reported, and the derived relationship used to determine the spatial distribution of 43 different plant community compositions for the study area as a whole.

Tuomisto et al. (2003) tested the hypothesis that species composition corresponds to environmental heterogeneity by studying the floristic variation of two phylogenetically distant plant groups along a continuous 43 km transect crossing *tierra firme* rain forest in northern Peru. They then compared the observed patterns to the patterns in the spectral reflectance characteristics of the forest, as recorded by Landsat. Floristic similarity patterns of the two plant groups were highly correlated with each other and with similarities in NDVI patterns. Thus significant spatial variation in species composition does exist, and this level of variability matches patterns in the spatial distribution of energy availability. The results therefore provided no support to the uniformity hypothesis (which states that species composition is not more or less uniform). More recently, He et al. (2009) highlighted the use of the NDVI to monitor species compositional changes at regional scales: using plant distribution data from North and South Carolina, USA, they investigated the correlations between species composition and the NDVI within defined ecoregions. Their results showed a significantly positive correlation between the NDVI distance matrices and the species compositional dissimilarity matrices, with correlation coefficient values decreasing from 0.40 at the species taxon rank to 0.30 at the family taxon rank. Remarkably, the species compositional dissimilarity matrices correlated slightly higher with yearly NDVI distance matrix than the summer matrix; the lowest correlation was associated with the winter NDVI distance matrix, as predicted. The authors speculated that the NDVI might be more related to species composition rather than the number of species found in the study sites, an idea that remains to be fully explored.

6.3 NDVI as a predictor of plant attributes

As detailed in Chapter 3, the NDVI is a vegetation index correlated strongly with absorption of photosynthetically active radiation, being widely adopted and applied as a proxy for leaf area index and net primary production. It is therefore not surprising to expect the NDVI to supply information on parameters linked to plant development and ecology, such as plant growth or plant mortality. This section reviews some of the key studies that have allowed us to establish and calibrate these relationships.

6.3.1 NDVI to monitor plant growth

The potential link between the NDVI and plant growth has been particularly well studied for trees and shrubs: in eastern Kansas, USA, for example, standardized tree ring width, diameter increase and seed production were all shown to be strongly correlated with the INDVI of the same growing season (Wang, J. et al. 2004) (Figure 6.2). Likewise, Malstrom et al. (1997) reported a strong correlation between tree ring data and NDVI-based net primary productivity estimates, while D'Arrigo et al. (2000) found that in boreal conifers, maximum latewood density of annual rings was significantly correlated with the NDVI. A spatial correlation between the trend for tree ring width (from 1980 to 1999) and the trend for forest NDVI during the growth season was also reported in Siberia (Shishov et al. 2002); in the mid- to high-latitude areas of North America and Eurasia, Kaufmann et al. (2004) reported a significant correlation between tree ring and forest NDVI in June and July. In the Komi republic (northwestern Russia), a significantly positive correlation was reported between NDVI data and tree ring width (adjusted $R^2 = 0.44$–0.59), while the increased integrated NDVI values from June to August over the period 1982 to 2001 was matched by on-the-ground observations of increased site productivity (Lopatin et al. 2006). More recently, Boelman et al.

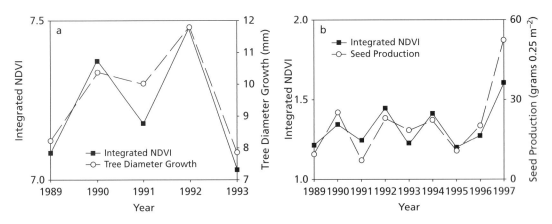

Figure 6.2 (a) Annual single-stem tree diameter growth and average NDVI (integrated from March to October) at Kansas Ecological Reserves; (b) annual seed production and average NDVI (integrated from mid March to late May) at Konza Prairie Research Natural Area, Kansas, USA. (Reproduced from Wang, J. et al. 2004 with permission from Taylor & Francis.) This research demonstrated strong links between NDVI and ground-based measurements of productivity for forest trees in the central Great Plains.

(2011) investigated the relationships between the NDVI and structural characteristics associated with deciduous shrub dominance in arctic tundra such as stature, branch abundance, aerial percentage woody stem cover (deciduous and evergreen species), and percentage deciduous shrub canopy cover. Using handheld spectrometers, the authors showed that (i) plot-level NDVI measurements made during the snow-free period prior to deciduous shrub leaf-out are best suited to capturing variation in the percentage woody stem cover, maximum shrub height, and branch abundance, particularly between 10 and 50 cm height in the canopy; (ii) plot-level NDVI measurements made during the period of maximum leaf-out are best suited to capturing variation in deciduous canopy cover; and (iii) plot-level NDVI measurements made at the point in the growing season when canopy NDVI has reached half of its maximum growing season value do not capture variability in any of our measures of shrub dominance.

Studies on the link between patterns in plant growth and fluctuations in the NDVI outside forests are rare. Using five well-replicated Qilian juniper tree-ring width index series from China's western arid region, He and Shao (2006) reported significant associations between tree-ring width index series and monthly NDVI of grassland from June to September, with the most significant association being in August. Several studies went on to assess the relationship between plant height and the NDVI for various species and in various locations. Payero et al. (2004), for example, compared the performance of eleven vegetation indices (including the NDVI) for estimating grass *Festuca arundinacea* and alfalfa *Medicago sativa* L. plant height. The authors found that, for alfalfa, good logistic growth relationships could be established between plant height and the NDVI. However, they also found that the NDVI became insensitive to additional plant growth when alfalfa reached a certain height. For grass, the NDVI was one of only a few indices for which a fairly good linear relationship could be established with plant height.

6.3.2 NDVI to monitor plant mortality

Changes in ecosystem structure and function resulting from plant mortality have the potential to significantly alter biogeochemical cycles, energy fluxes, and landscape patterns of vegetation composition (Adams et al. 2009; Amiro et al. 2010; van der Molen et al. 2011). When faced with disturbances such as fires or insect pest outbreaks, information on the extent and severity of plant mortality can be key for deciding whether and/or which management activities might be required (see also Chapter 5). Our ability to quantify the impacts of vegetation mortality on ecological, biogeochemical, and biosphere–atmosphere dynamics is currently limited due to a lack of information on where and when mortality events occur (Allen et al. 2010). In this respect, the analysis of satellite images has the potential to fill such gaps (Frolking et al. 2009).

Remote sensing information has been used previously to detect tree mortality at the level of individual forest stands and trees (Wulder et al. 2006, 2008; Goodwin et al. 2008): for example, Quickbird data have been successfully used to identify tree crowns that have red attack damage due to mountain pine beetle infestation (Coops et al. 2006). Coarse resolution NDVI data can be used to monitor forest health and detect mortality patterns, as demonstrated by Verbesselt et al. (2009), who aimed to map tree mortality in a *Pinus radiata* plantation in southern New South Wales, Australia. The authors reported a strong correlation between NDVI variability and tree mortality in this region, with up to 37% of the variability in tree mortality patterns explained by NDVI fluctuations (see also Chapter 5). As recently demonstrated by Garrity et al. (2013), tree mortality can also be detected using unsupervised clustering of bi-temporal NDVI and red:green ratio. With their technique, >95% accuracy could be reached in detecting tree mortality in a mixed species woodland in south-western USA.

6.4 NDVI and plant physiological status

Like all living organisms, plants require a number of elements, such as water, carbon, and nitrogen, to survive, grow, and reproduce. They also produce a vast array of chemical compounds, which they use to cope with their environment, such as pigments, used by plants to absorb or detect light. This section explores how information may be gained on the physiological status of plants with respect to correlations between the NDVI and plant pigment content, plant water content, and plant nitrogen content.

6.4.1 NDVI as an indicator of pigment content

Among the most important molecules for plant function are the pigments, which allow plants to selectively absorb certain wavelengths of light to power chemical reactions. Chlorophyll selectively absorbs red and blue wavelengths to fuel photosynthesis; carotenoids help fuel photosynthesis by gathering wavelengths not readily absorbed by chlorophyll (Sims and Gamon 2002). When incident light energy exceeds that needed for photosynthesis, carotenoids dissipate excess energy, thus avoiding damage to the photosynthetic system (Demmig-Adams and Adams 1996). Anthocyanins help to protect leaves from excess light or from UV light (Barker et al. 1997; Mendez et al. 1999). These pigments may also serve as scavengers of reactive oxygen intermediates or as antifungal compounds (Coley and Kusar 1996; Yamasaki 1997). Because pigments are key to leaf function, variation in pigment content can be expected to provide information on the physiological state of leaves: chlorophyll content, for example, tends to decline when plants are under stress (Sims and Gamon 2002). A number of local, experimental studies have reported significant links between the NDVI and chlorophyll content, with the NDVI increasing as chlorophyll increases (Gitelson and Merzlyak 1994; Yoder and Pettigrew-Crosby 1995; Gamon et al. 1995; Kodani et al. 2002), although the link between these variables has not always been reported to be linear (Gitelson and Merzlyak 1997; Richardson et al. 2002). The NDVI has also been reported as insensitive to medium and high chlorophyll concentrations (Buschman and Nagel 1993; Vogelman et al. 1993; Gitelson and Merzlyak 1994; Lichtenthaler et al. 1996). No significant relationship has been established between the NDVI and carotenoid or anthocyanin content (e.g. Datt 1998).

Based on the accumulated results concerning the link between the NDVI and pigment content, one might conclude that using the NDVI to derive global information about changes in pigment content, such as chlorophyll, is problematic. First, the sensitivity of the NDVI to variation in chlorophyll content may depend on the band width and spectral coverage of the visible red band chosen to calculate the NDVI (the definition of this band slightly varies from one sensor to another; Yoder and Waring 1994). Ecosystems then tend to modify their photosynthetic capacity in relation to available resources through changes in leaf area, foliar chlorophyll content, and species and understorey composition, meaning that several factors, other than changes in chlorophyll content, may lead to a change in the NDVI (Danson and Curran 1993; Miller et al. 1997). Alternatives to the use of the NDVI to assess chlorophyll content have been recommended: studies have shown that changes in chlorophyll concentration can be related to shifts in the 'red edge', the inflection point that occurs in the rapid transition between the red and near-infrared reflectance (Horler et al. 1983; Curran et al. 1990). The 'green' NDVI, defined as NDVI green = (NIR − GREEN)/(NIR + GREEN) where GREEN represents the green wavelength near 550 nm, has been shown to be sensitive to a much wider range of chlorophyll concentration than the original NDVI (Gitelson et al. 1996).

6.4.2 NDVI as an indicator of plant water content

Vegetation water content is an important parameter in agricultural and forestry applications. Water stress is indeed one of the most frequent limitations to photosynthesis and primary productivity, and estimates of vegetation water content are of paramount importance for assessing drought risk and susceptibility to fire (Penuelas et al. 1993; Chuvieco et al. 2004; see also Chapter 5). This probably explains why various studies have tried to explore how remote sensing could help to monitor plant water content and detect water stress. Tucker (1979) suggested that the NDVI could be used to estimate leaf water content for grasses, which started a discussion as to whether or not the NDVI could be used to monitor plant water content. The idea is that the NDVI might supply information about leaf chlorophyll content, which is assumed to correlate strongly with leaf moisture content, and therefore with vegetation water content (Ceccato et al. 2001). Small scale, localized studies have reported

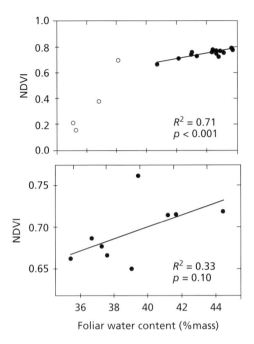

Figure 6.3 Regression of *Pinus edulis* (upper) and *Juniperus monosperma* (lower) foliar water content with the NDVI. (Reproduced from Stimson et al. 2005 with permission from Elsevier.) Reported relationships exclude trees subject to mortality (open symbols).

good links between water content and the NDVI: $R^2 = 0.71$ and 0.33 have been reported between the NDVI and foliar water conditions of *Pinus edulis* and *Juniperus monosperma*, respectively (Figure 6.3; Stimson et al. 2005). Foliage relative water content was positively, although not significantly, linked to the NDVI of chaparral vegetation in California ($R^2 = 0.35$, P < 0.10; Serrano et al. 2000), while the NDVI accounted for 48–54% of the variability in absolute water content of grassland–shrubland vegetation in Canada (Davidson et al. 2006). Successes have also been reported at larger scales: Cheng et al. (2006, 2008) demonstrated that NDVI values retrieved from MODIS could be successfully linked to equivalent water thickness (EWT) measurements (a variable correlated with canopy water content in all sites considered) in closed and open canopy crops ($R^2 = 0.50$), semi-arid shrublands ($R^2 = 0.70$ to 0.89), riparian zones ($R^2 = 0.72$), oak woodlands ($R^2 = 0.87$), and boreal conifer forests across the USA ($R^2 = 0.49$). Moreover Chen et al. (2005) showed that vegetation water content in Iowa, USA, could be positively related to NDVI data: for corn (Landsat: $R^2 = 0.72$; MODIS: $R^2 = 0.78$), and for soybeans (MODIS: $R^2 = 0.46$) (see Chapter 3 for more information on these sensors).

The accuracy with which the NDVI can estimate vegetation water content is limited (Ceccato et al. 2001, 2002), stemming from the facts that: (i) the link between chlorophyll content and vegetation water content is species specific; (ii) a decrease in chlorophyll does not imply a decrease in vegetation water content (as variation in chlorophyll content can also be caused by plant nutrient deficiency, disease, stress, toxicity, or phenological stage), and vice versa; (iii) the NDVI saturates at intermediate values of leaf area index (for more on NDVI caveats and limitations, see Chapter 3), meaning that the NDVI is not responsive to the full range of vegetation water content. Thus, NDVI-based approaches to monitor vegetation water content are only valid in regions where the correlations between leaf chlorophyll and water contents have been established (Ceccato et al. 2001). Alternatives to the NDVI for remotely monitoring vegetation water content have therefore been explored. In particular, numerous studies have highlighted that the near-infrared region and the shortwave infrared region are better suited for the monitoring of plant water content than the combination of the red and near-infrared regions (Tucker 1980; Carter 1991; Danson et al. 1992; Penuelas et al. 1997), leading to the suggestion that indices such as the Normalized Difference Water Index (NDWI), calculated as (NIR − SWIR)/(NIR + SWIR), where SWIR is the reflectance in the Short Wave InfraRed, are probably a better way to assess vegetation water content than the NDVI (Gao 1996; Cheng et al. 2006).

6.4.3 NDVI as an indicator of plant nitrogen content

Nitrogen is a critical element in the growth of all living organisms, because of its central role in all metabolic processes as well as in cellular structure and genetic coding. Plants therefore need it to grow, develop, and reproduce. Generally, the concentrations of nitrogen in plants reflect the supply of nitrogen in the root medium, and, as one might expect, yields increase as internal concentration of nitrogen in plants increases. Despite nitrogen being one of the most abundant elements on Earth, nitrogen deficiency is probably the most widespread nutritional problem affecting plants (Samborski et al. 2009). With herbivores also relying on nitrogen for their own growth, survival, and reproduction, nitrogen has therefore the potential to become a limiting nutrient for many herbivorous species (Mattson 1980).

The chlorophyll content of a plant is known to be a good indicator of nitrogen content (Evans 1983; Vos and Bom 1993; Blackmer and Schepers 1995). The link between nitrogen and chlorophyll content arises from the fact that nitrogen stress reduces the production of the chlorophyll that is involved in the production of the reduced compounds nicotinamide adenine dinucleotide phosphate and adenosine triphosphate. These compounds are used by an enzyme, the ribulose-1-5-biphosphate carboxylase, to drive dioxide carbon fixation. Reduced chlorophyll content can lead to increased reflectance of photosynthetically active light, resulting in nitrogen-stressed plants appearing yellow (Peňuelas et al. 1994; Gitelson et al. 2005). Since the NDVI is an indicator of photosynthetic activity, one might expect the NDVI to correlate with the amount of nitrogen in plants. Consequently, the link between the NDVI and plant nitrogen content has been extensively studied for crops, using handheld sensors (Samborski et al. 2009). For instance, Stone et al. (1996) reported that the NDVI was highly correlated with nitrogen uptake and grain yields of winter wheat. Tilling et al. (2007) reported a highly significant correlation ($R^2 = 0.42$, $P < 0.001$) between NDVI values and an index of nitrogen stress for wheat in Australia; and Clay et al. (2006) showed that the link between the NDVI and yield losses due to nitrogen stress was dependent on the phenological stage for corn. There are also studies linking the NDVI to crop nitrogen content using satellite-based information. Han et al. (2002) reported the existence of strong correlations between the NDVI derived from SPOT and readings from chlorophyll meters, which are commonly used to assess corn nitrogen status. Wright et al. (2004) reported significant positive correlations between NDVI data retrieved from QuickBird II and plant tissue nitrogen ($R^2 = 0.62$ and 0.56, for the 2002 and 2003 seasons respectively) for wheat in Idaho. Zhang et al. (2006) reported strongly positive correlations between IKONOS-based NDVI values and nitrogen concentration for rice ($R^2 = 0.84$ to 0.94, depending on the phenological stage). Recently Jia et al. (2012) showed that NDVI values collected by IKONOS correlated strongly with wheat nitrogen status parameters in China. Although several studies (such as those cited here) promote the use of satellite-based NDVI measurements to monitor nitrogen status in plants, a brief overview of the literature on this topic seems to suggest that (i) only very high-resolution satellites provide relevant information for precise nitrogen status monitoring; and (ii) hyperspectral information might be more appropriate for nitrogen monitoring than the NDVI.

6.5 Informing agriculture

As the world population continues to grow, measuring agricultural production at large spatial scales to improve agricultural productivity is becoming increasingly important to inform land use planning, and to prevent food insecurity and famines. In Sections 6.1–6.4, we have seen how the NDVI can supply relevant information on plant distribution, abundance, ecology and performance. It is therefore not surprising that NDVI variations may inform agricultural practices. This section now highlights how the NDVI can help (i) to remotely assess soil properties; and (ii) to remotely monitor and predict crop yields.

6.5.1 NDVI as a predictor of soil properties

Land use planning and land management can sometimes rely greatly on accessing reliable information about the spatial distribution of the physical and chemical soil properties affecting ecosystem processes. In some cases, the NDVI has been shown to be a significant predictor of soil properties, highlighting the potential role for this vegetation index to support agricultural development. Sumfleth and Duttmann (2008) showed that the distribution of carbon, nitrogen, and silt contents was quite closely related to the NDVI of vegetated surfaces in paddy soil landscapes in China. Also in China, regression analyses recently identified the NDVI as one of the best predictors for describing spatial patterns in soil

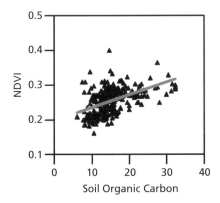

Figure 6.4 Regression of soil organic carbon (SOC, in g/kg) with the NDVI. (Reproduced from Zhang et al. 2012 with permission from John Wiley & Sons.) SOC estimates have been log-transformed. In all, 528 samples were considered; Pearson's correlation coefficient between the two variables was 0.55.

organic carbon (Zhang et al. 2012) (Figure 6.4). Similarly, work by Lobell et al. (2009) highlighted how the NDVI could be used to map soil salinity in the Red River Valley, North Dakota, although the authors reported that for this area, the EVI (see Chapter 2) outperformed the NDVI in terms of its ability to capture the spatial variability of this parameter (Lobell et al. 2009). In Iran, the NDVI was also reported to be significantly and positively correlated with soil organic matter (Ayoubi et al. 2011), an important factor related to soil quality, sustainability of agriculture, soil aggregate stability, and crop yield (Loveland and Webb 2003). Altogether, these studies thus highlight the potential for the NDVI to remotely assess soil properties over large areas and inform future agricultural needs and practices.

6.5.2 NDVI to monitor and predict crop yields

Crop yield is a key element for rural development and an indicator for national food security. It is also the ultimate indicator for describing agricultural response to changes in environmental and management conditions. Access to reliable measures of crop yield estimates across regions of the world combined with information on management procedures and environmental conditions can provide insights as to where and how productivity can be enhanced (Doraiswamy et al. 2003). Remote sensing-based approaches are important tools for monitoring yield production and help to forecast harvests. Satellite-based information allows the production of comparable, standardized measures of crop yields.

The partial convergence between the agricultural and remote-sensing agendas started in 1974, when LACIE—a joint effort of the NASA, the USDA, and the NOAA—began to apply satellite remote-sensing technology on experimental bases to forecast harvests in important wheat-producing areas (MacDonald 1979). Since then, several studies have explored the link between crop yields and the dynamics of vegetation indices such as the NDVI across a wide range of regions and crop types. In Burkina Faso, for example, Rasmussen (1992) demonstrated that an estimate of millet production of a given area by the end of the season could be made using the INDVI (see Chapter 3). Likewise, in the 1980s the Food and Agricultural Organization developed several procedures to compute the regional crop yields from the NDVI using low-resolution satellite images (Hielkema 1990). Doraiswamy and Cook (1995) demonstrated that accumulating the NDVI values for spring wheat only during the grain-fill period improved the estimates of potential crop yields in North Dakota, USA. More recently, Teal et al. (2006) showed that normalizing the NDVI for the number of growing degree days accounted for about 73% of variability in predicted grain yield. In Portugal, the NDVI was used to provide early predictions of wine yields (Cunha et al. 2010). Focusing on four test sites located in the main wine regions of Portugal (namely the Douro region, Vinhos Verdes, and Alentejo), the authors showed that NDVI measurements obtained from SPOT seventeen months before harvest enabled a reliable forecast of regional wine yield.

6.6 Conclusions

Figure 6.5 summarizes in map form the locations where NDVI dynamics in relation to plant ecology were assessed by studies cited in this book.

Based on currently available information on the relationship between the NDVI and plant species richness, one may conclude that: (i) plant species richness can be characterized and monitored using

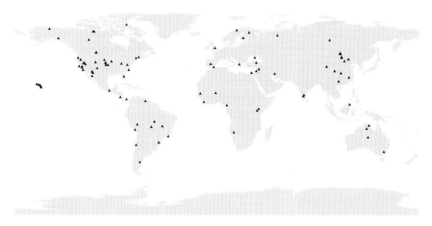

Figure 6.5 Spatial distribution of the study areas where the links between the dynamics of the NDVI and plant ecology were assessed. Each black triangle refers to the estimated barycentre or the published coordinates of a particular study area. Only the scientific publications cited in this book have been included.

the NDVI; (ii) spatial variability in the NDVI can be an important determinant of plant richness; and (iii) location, spatial scale, and spatial resolution are important factors determining the nature of the relationship between annual NDVI and plant species richness. Section 6.2 demonstrated the utility of the NDVI to monitor plant community composition, while highlighting its low potential for mapping the distribution of particular plant species (an exception being made for invasive plant species). Section 6.3 showed the great potential for the NDVI to inform patterns in plant growth and mortality, although the highest potential was observed when considering high- to very high-resolution satellite imagery. Section 6.4 demonstrated the high potential for the NDVI to support agricultural practices, especially in monitoring and predicting crop yields.

The successful applications of the NDVI in plant ecology should be balanced against several limitations. While reviewing the links between the NDVI and physiological attributes of plants, such as pigment concentration, water content, or nitrogen content, it appeared that the NDVI is not always the best tool to derive such information, although local successes can be reported in particular conditions. Greener areas are not necessarily those with the highest species richness of a given plant taxon or the highest density of individual plant species (Herrmann and Tappan 2013). These authors showed that, although a greening trend has been reported in central Senegal over recent decades, an impoverishment of the woody vegetation cover was observed on the ground, indicated by an overall reduction in woody species richness, a loss of large trees, an increasing dominance of shrubs, and a shift towards more arid-tolerant, Sahelian species. This example and the limitations discussed throughout this chapter highlight that, in many cases, NDVI-based analyses can offer a valuable starting point of investigation for plant ecologists. However, these analyses need to be informed with on-the-ground data to fully understand the processes leading to the reported patterns in NDVI spatio-temporal dynamics.

CHAPTER 7

NDVI and wildlife management

The most alive is the wildest.   Henry David Thoreau

Wildlife management can be defined as the application of ecological knowledge to populations of wild organisms in a manner that strikes a balance between the needs of those populations and the needs of people (Williams et al. 2002). The word 'management' comes from the Latin *manus*, meaning 'hand', and therefore the concept of management triggers the notion of 'human action' and 'intervention'. Wildlife is generally understood to comprise non-domesticated animals, although the definition of domestication and what constitutes 'non-domesticated animals' is not trivial (Mysterud 2010): wildlife, in this chapter, will refer mostly to non-captive animal populations. By bringing these two definitions together, we can conclude that wildlife management (a) refers to efforts directed toward wild animal populations; (b) is likely to require, among other things, a good understanding of the relationship between wild animal populations and their habitats; and (c) makes use of manipulations of habitats or populations to meet some specified human goals. In this regard, manipulations may include: the removal of individuals; the use of contraceptives; habitat restoration; or reintroductions.

In the face of current environmental change, making informed, appropriate wildlife management decisions are key to mitigate biodiversity loss and avoid unsustainable levels of resource offtake. In order to assess and anticipate the effects of these changes on wildlife, we do require, among other things, a good understanding of how trophic interactions are impacted by environmental change. As may be gathered from Chapters 3–5, the NDVI is associated with multiple opportunities to access such knowledge: this vegetation index enables researchers to combine climate, vegetation, and animal population data, and thus to quantify how changes in vegetation distribution, phenology, and productivity might affect upper trophic levels at various spatial scales. The NDVI can be used to assess primary productivity at multiple spatial and temporal resolutions (Chapter 3), and can help to assess temporal aspects of vegetation development and quality (Pettorelli et al. 2005b,c, 2007; Hamel et al. 2009). Such information can then be correlated with the distribution, behaviour, life history traits, and abundance of animal species (Table 7.1), in order to expand our knowledge on the nature and strength of factors shaping these parameters.

One benefit of exploring the link between the NDVI and wildlife is that such knowledge makes it possible to determine how much of the climatic effects on animal behaviour, performance, and dynamics are operating through the effects of climate on vegetation distribution, phenology, and productivity (i.e. through trophic interactions). Such insight might help to support efforts to manage and conserve wildlife efficiently in the face of climate change: to know the actual mechanisms and pathways by which environmental change affects herbivores could indeed enable better predictions of further changes. Focusing on ecosystems in northern latitudes, for example: if global warming mainly operates through decreasing snow depth then we are likely to see no further responses of herbivore populations to increased temperature once there is

Table 7.1 Examples of studies highlighting the usefulness of the NDVI for wildlife management.

Question	Group	Species	Reference
NDVI dynamics and space use	Invertebrates	Iberian spider	Jimenez-Valverde and Lobo 2006
	Invertebrates	Tick	Estrada-Peña 1999; Estrada-Peña and Thuiller 2008
	Invertebrates	Tsetse fly	Robinson et al. 1997
	Birds	Ptarmigans	Pedersen et al. 2007
	Birds	Ostrich	Verlinden and Masogo 1997
	Birds	Bustard and lark	Suarez-Seoane et al. 2002
	Amphibians	Golden-striped salamander	Arntzen and Alexandrino 2004
	Ungulates	Elk	Hebblewhite et al. 2008
	Ungulates	Elephant	Young et al. 2009
	Ungulates	Mongolian gazelle	Mueller et al. 2008; Olson et al. 2009
	Ungulates	Mule deer	Marshal et al. 2006
	Ungulates	Caribou/reindeer	Griffith et al. 2002; Hansen, B.B. et al. 2009a,b
	Ungulates	African buffalo	Ryan et al. 2006; Winnie et al. 2008
	Primates	Vervet monkey	Willems et al. 2009
	Primates	Grivet monkey	Zinner et al. 2002
	Carnivores	Lynx	Basille et al. 2009
	Carnivores	Leopard	Gavashelishvili and Lukarevskiy 2008
	Carnivores	Brown bear	Wiegand et al. 2008
NDVI dynamics and migration patterns	Birds	Marsh harriers	Klaassen et al. 2010
	Ungulates	Red deer	Pettorelli et al. 2005b; Bischof et al. 2012
	Ungulates	Roe deer	Andersen et al. 2004
	Ungulates	Wildebeest	Musiega and Kazadi 2004; Boone et al. 2006
	Ungulates	Saiga antelope	Singh et al. 2010
	Ungulates	Mongolian gazelle	Ito et al. 2006
	Ungulates	African buffalo	Cornelis et al. 2011
	Ungulates	Mule deer	Sawyer and Kauffman 2011
NDVI dynamics and population dynamics	Invertebrates	Mosquitoes	Britch et al. 2008
	Invertebrates	Locusts	Steinbauer 2011
	Invertebrates	Spiders	Shochat et al. 2004
	Amphibians	Salamander	Tanadini et al. 2012
	Birds	Pied flycatcher	Sanz et al. 2003
	Birds	Barn swallow	Saino et al. 2004
	Birds	White stork	Schaub et al. 2005
	Birds	Egyptian vulture	Grande et al. 2009
	Rodents	Great gerbil	Kausrud et al. 2007
	Rodents	Field vole	Bierman et al. 2006
	Rodents	*Akodon azarae*	Andreo et al. 2009a,b

continued

Table 7.1 Continued

Question	Group	Species	Reference
	Primates	Marmosets	Norris et al. 2011
	Ungulates	Roe deer	Pettorelli et al. 2006
	Ungulates	Mountain goat	Pettorelli et al. 2007
	Ungulates	Bighorn sheep	Pettorelli et al. 2007
	Ungulates	Ibex	Pettorelli et al. 2007
	Ungulates	Red deer	Pettorelli et al. 2005b; Mysterud et al. 2008
	Ungulates	Reindeer/caribou	Griffith et al. 2002; Pettorelli et al. 2005b; Couturier et al. 2009
	Ungulates	Moose	Herfindal et al. 2006a,b
	Ungulates	Elephant	Rasmussen et al. 2006; Chamaille-Jammes et al. 2007; Trimble et al. 2009
	Ungulates	African buffalo	Ryan 2006
	Ungulates	Chamois	Garel et al. 2011
	Ungulates	Soay sheep	Durant et al. 2005
	Ungulates	Domestic sheep	Texeira et al. 2008
	Carnivores	Brown bear	Wiegand et al. 2008
NDVI and species richness	Invertebrates	Beetles	Lassau and Hochuli 2008
	Invertebrates	Butterflies	Seto et al. 2004; Bailey et al. 2004; Levanoni et al. 2011
	Amphibians and reptiles		Qian et al. 2007
	Birds		Jorgensen and Nohr 1996; Hurlbert and Haskell 2003; Hawkins et al. 2003; Hawkins 2004; Foody 2004; Seto et al. 2004; Ding et al. 2006; Koh et al. 2006
	Mammals		Oindo and Skidmore 2002

no permanent snow cover established. However, if climate is influenced by plants, detailed knowledge about how plants respond to temperature will enable more robust predictions of future scenarios of environmental change (Pettorelli et al. 2005b).

This chapter reviews opportunities for the NDVI to support wildlife management. Section 7.1 explores the usefulness of this vegetation index for increasing our understanding of the factors influencing habitat use and selection, and deciphering patterns of movement at various temporal and spatial scales. Section 7.2 focuses on the links between NDVI variability and life-history traits, describing situations where patterns and trends in this vegetation index have been successfully linked to variability in animal body mass, survival, and recruitment. Section 7.3 examines the usefulness of the NDVI for predicting animal abundance, and Section 7.4 reviews knowledge on the links between NDVI variability and spatio-temporal patterns in species richness.

7.1 NDVI, habitat use, and animal movement

7.1.1 NDVI, habitat selection, and habitat use

Understanding the basis of habitat selection has important implications for explaining the distribution of organisms, as well as helping to differentiate among habitats of varying quality for effective management. Knowledge of those habitat characteristics essential for the viability of a species can also provide crucial information to wildlife managers dealing with reintroductions, translocations, and the development of new protected areas (Araujo and Williams 2000; Stamps and Swaisgood 2007). This sub-section highlights the NDVI's contribution to our understanding of the factors influencing habitat use and habitat selection (Box 7.1) by animals.

7.1.1.1 NDVI and home-range size

Most animals tend to use the same area over time, which has led ecologists to define and focus on the concept of home range (defined as the area used by an animal over a given time period; White and Garrott 1990; Fieberg and Börger 2012; Powell 2012). Determining the factors shaping inter- and intraspecific variation in home-range size and location has long been a subject of ecological interest. Theoretically, animals are expected to occupy the smallest possible areas within which they can acquire sufficient resources to enable reproduction and survival, while minimizing time and energy spent for territory defence and foraging (Maynard Smith 1974). Studies so far have revealed that the level of variability in home-range sizes between different species can be related to body-size-dependent metabolic requirements, declining rates of utilizable energy in the environment with increasing body weight, species overlap, biological time, and rate of home range use (McNab 1963; Harestad and Bunnell 1979; Damuth 1981; Lindstedt et al. 1986; Swihart et al. 1988). The level of variability in home-range sizes between different individuals of a given species has subsequently been related to sex, age, reproductive effort, availability of nesting sites and cover, and food availability (Mikesic and Drickamer 1992; Tufto et al. 1996; Asher et al. 2004; Begg et al. 2005; Getz et al. 2005; Said et al. 2005; Hayes et al. 2007).

Home range sizes and locations have both been successfully linked to NDVI dynamics for various terrestrial vertebrates inhabiting contrasting environments (Figure 7.1). In Eritrea, for example, the

---

Box 7.1 Habitat selection and habitat use

There can sometimes be a great deal of confusion between habitat selection and habitat use. Habitat selection generally refers to an individual's preference for a given habitat or a given landscape feature. Knowing which habitat features are selected by animals is expected to yield information on the environmental conditions needed for a species to reach its physiological optimum, under the assumption that these animals have been presented with the complete set of options available to them. Habitat use, on the other hand, refers to which features are used (not selected) by animals. The emergence of both concepts may be due partially to the nature and availability of data on animal distribution. For instance, animals equipped with devices constantly monitoring their location, associated with relevant maps of the area occupied by these animals, may enable ecologists to discern habitats that are not used from habitats that are used, and from habitats that are used and preferred. But in many cases, presence-only data (such as sighting locations) comprise the only type of data available to researchers: in these cases, it can be difficult to know whether habitats not associated with sightings are avoided, or whether animals do use these habitats but have not been spotted there. Therefore, habitat use is a term that tends to be associated with presence-only data, whereas habitat selection tends to be explored for animals equipped with monitoring devices (such as GPS collars). Exceptions, however, do exist, as selection can be very difficult to assess, even from animals equipped with monitoring devices (e.g. Thomas and Taylor 2006).

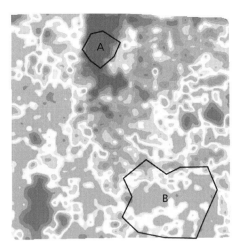

Figure 7.1 Schematic illustration of the potential effect of the NDVI on home range location and size of two animals (A and B) of a given species. Outlined polygons represent two theoretical home ranges, encompassing 95% of each individual's locations. The higher the NDVI value, the darker the pixel. The home ranges of animal A exhibit a significantly higher average NDVI value compared with the average of the survey area. The average NDVI value within animal B's home range is lower than the average NDVI value within animal A's home range; at the same time, animal B's home-range size is larger than that of animal A.

home ranges of the grivet monkey *Cercopithecus aethiops* exhibited a significantly higher average NDVI value compared with the average NDVI value of the survey area (Zinner et al. 2002). In Svalbard, Norway, a negative association between habitat productivity (as assessed by the NDVI) and home-range size was reported for reindeer *Rangifer tarandus* in Brøggerhalvøya (Hansen, B.B. et al. 2009a,b). In southern Africa, wet-season home-range size in the elephant *Loxodonta africana* was shown to correlate with seasonal vegetation productivity as estimated by the NDVI, whereas dry-season home-range sizes were best explained by heterogeneity in the distribution of vegetation productivity (Young et al. 2009). In the USA, home-range sizes of translocated wild turkeys *Meleagris gallapovo silvestris* were then related positively to within-home-range variability in the NDVI (Marable et al. 2012), whereas an inverse relationship between the NDVI and home-range size was reported for female bobcats *Lynx rufus* (Ferguson et al. 2009).

The positive link between the NDVI and home range locations and the negative link between the NDVI and home-range size are both likely to be underpinned by the NDVI being positively correlated with resource availability. For some herbivore species, the NDVI can indeed be expected to represent a good proxy of forage availability; for some carnivore species, areas with high NDVI values might represent areas preferred by their prey. In southern Norway, for example, roe deer *Capreolus capreolus* are the favoured prey of the lynx *Lynx lynx* (Basille et al. 2009). This prey species is found in areas with high primary productivity and therefore high NDVI values. Unsurprisingly, the preferred habitat of the lynx includes areas associated with high NDVI values (Basille et al. 2009).

7.1.1.2 Predicting animal distribution using the NDVI

The home range concept alone can sometimes be insufficient to understand how animals use their habitat (Box 7.2), as it does not explain how animals relate to certain habitat features (Bunnefeld et al. 2006; Reynolds-Hogland and Mitchell 2007). Therefore, several studies have focused on predicting animal presence (or occurrence) based on the distribution of habitat characteristics. In the course of such work, many have reported NDVI dynamics as being a strong predictor of animal presence.

In Europe, the distribution of the great bustard *Otis tarda*, the little bustard *Tetrax tetrax*, and the calandra lark *Melanocorhypha calandra* in Spain were all shown to be accurately predicted using, among other information, NDVI dynamics (Suarez-Seoane et al. 2002). In Portugal, Arntzen and Alexandrino (2004) reported that the southern form of the Golden-striped salamander *Chioglossa lusitanica* was preferentially found in areas with lower precipitation, air humidity, summer temperatures, and NDVI than the northern form. As previously discussed, Basille et al. (2009) demonstrated that the preferred habitat of the lynx in Norway includes high plant productivity areas (as indexed by the NDVI). The NDVI was also shown to be a good predictor of the presence of territorial rock ptarmigan *Lagopus muta* cocks in Svalbard (Pedersen et al. 2007).

In North America, parturient female caribou (caribou and reindeer refer to the same species) selected annual calving grounds with greater areas of high rates of greening (Griffith et al. 2002). Likewise, during most seasons, mule deer *Odocoileus hemionus eremicus* in the Sonoran Desert were found in areas with greater forage quality (as indexed by NDVI rates), when compared with random locations (Marshal et al. 2006). In the Walker River Basin of Nevada and California, hantavirus tended to be less detected in deer mice inhabiting areas characterized by low elevation, level topography, as well as low and relatively homogeneous vegetation productivity (as indexed using the NDVI) than in mice occupying sites with higher elevation, steeper topography, higher vegetation productivity and greater local variability in vegetation productivity (Boone et al. 1998).

In Asia, recent analyses on leopard *Panthera pardus* distribution suggested that animals in west and central Asia avoided areas with long-duration snow cover and low productivity (as indexed by the NDVI; Gavashelishvili and Lukarevskiy 2008). Spatio-temporal variation in NDVI dynamics was crucial to account for the observation of an unusual mass aggregation of Mongolian gazelle *Procapra gutturosa* (Olson et al. 2009): in 2008, thousands of gazelles were indeed observed attempting to move

> **Box 7.2 Defining habitat and habitat suitability**
>
> Understanding species–environment relationships is at the core of ecological research, and is fundamental to design management interventions for endangered species, as it allows identifying those resources and features that are relevant for their survival (Manly et al. 1993). The term habitat is central to this type of ecological research, being a descriptive, rather than a mechanistic, concept; in its most basic form it represents the biotic and abiotic variables at a given place and time (Kearney 2006). In a more elaborate definition, habitat is understood as the set of resources and conditions present in an area that enables occupancy—including survival and reproduction—by a given organism (Morrison et al. 1992). This definition is organism specific, and relates the presence of a species, population, or individual to an area's physical and biological characteristics. The spatial scale at which it is defined can be a function of several variables, such as the number of individuals considered or the size of the species considered. The term habitat can thus encompass several variables of differing importance according to the particular species, location, and spatio-temporal scale (e.g. something might be important for one species and not for another; something might be important for a species here but not there; or something might be important at that time and place for a given species, although it might become less important in the future).
>
> Habitat suitability can be understood as the capacity of a defined spatial unit to produce occupancy. This capacity is generally evaluated by modelling the link between environmental variables and species distributions. The outputs are a cartographic representation of either the probability of occurrence for a given species or its suitability gradient across the study region, in addition to a detailed depiction of the factors responsible for these patterns (Hirzel et al. 2002; Phillips et al. 2006). Various species distribution models (SDMs) have been developed for assessing habitat suitability: early modelling approaches include distance-based measures (e.g. DOMAIN) and envelope calculations (e.g. BIOCLIM), which are able to characterize sites based on their similarities to those where the species is known to be present (i.e. presence-only models; Guisan and Thuiller 2005). These were followed by the use of approaches requiring information about the species presence and absence across the study site (i.e. presence–absence models; Guisan and Thuiller 2005). Yet there are many challenges associated with collection of absence data, and their availability from existing databases is generally low. Due to these constraints, presence–absence methods were adapted to a presence-only framework (e.g. using pseudo-absences and use-versus-availability approaches), leading to new modelling techniques being developed (such as ecological niche factor analyses; Hirzel et al. 2002), which in some cases benefited from the incorporation of novel approaches such as machine-learning methods and artificial-neural networks (e.g. maximum entropy; Phillips et al. 2006).

from Mongolia into Russia. To understand the cause of this behaviour, the authors decided to map spatio-temporal changes in satellite-derived estimates of vegetation productivity over the previous decades. They discovered that a decade-long decline in available green biomass in Mongolia's steppes was likely to have triggered the gazelles' attempted mass emigrations.

In Africa, Verlinden and Masogo (1997) found a significant selection for higher NDVI signatures for ostrich *Struthio camelus* and wildebeest *Connochaetes taurinus*, the latter only when present in high numbers. The distribution of some subspecies of tsetse fly (e.g. *Glossina morsitans centralis*) in southern Africa is best correlated with the NDVI and the average maximum temperature (75% correct predictions; Robinson et al. 1997). Likewise, the NDVI was shown to be a good predictor variable of *Schistosoma* spp. prevalence in Ethiopia (Malone et al. 2001) and *Fasciolosis* spp. prevalence in East Africa (Malone et al. 1998), with greater prevalence in habitats associated with higher NDVI values.

Finally, in South America Estrada-Peña (1999) made a good estimation of the tick *Boophilus microplus* habitat suitability using two NDVI-based and four temperature variables. Human fasciolosis prevalence was linked to several NDVI-based metrics in the Bolivian Andes (Fuentes et al. 2001). Moreover, a stepwise multiple regression model of the log of schistosomiasis prevalence in Brazil showed that the best model included the NDVI as a significant variable, with the NDVI alone accounting for 25% of the prevalence variation within communities (Bavia et al. 2001).

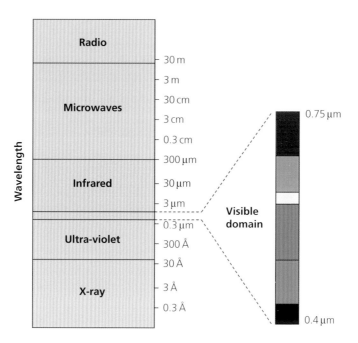

Plate 1 The electromagnetic spectrum. The human eye can perceive information only from a very small part of the electromagnetic spectrum, namely the visible spectrum (which ranges from ~400 nm to 750 nm). See also Figure 1.3.

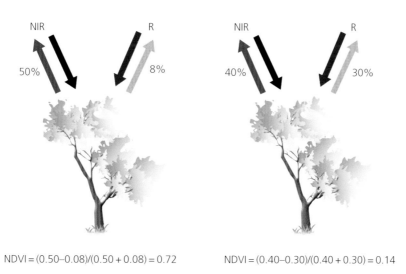

NDVI = (0.50−0.08)/(0.50 + 0.08) = 0.72   NDVI = (0.40−0.30)/(0.40 + 0.30) = 0.14

Plate 2 NDVI calculation for healthy (left) and senescent vegetation (right). Healthy vegetation absorbs incoming red light, while reflecting infrared radiation. Senescent vegetation reflects more visible light and less near-infrared light, leading to a reduced NDVI value (see also <http://earthobservatory.nasa.gov/Features/MeasuringVegetation/measuring_vegetation_2.php>). See also Figure 3.2.

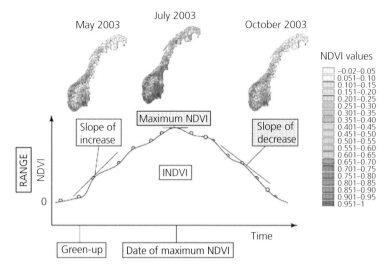

Plate 3 Presentation of the different indices (the slopes of increase (spring) and decrease (autumn), the maximum NDVI value, the Integrated NDVI (INDVI, i.e. the sum of NDVI values over a given period), the date when the maximum NDVI value occurs, the range of annual NDVI values, and the date of green-up (i.e. the beginning of the growing season) that may be derived from NDVI time-series over a year. Maps presenting NDVI values (ranging from 0 to 1) for Norway in May, July, and October 2003 are also shown. See also Figure 3.4. (Reproduced from Pettorelli et al. 2005b with permission from Elsevier).

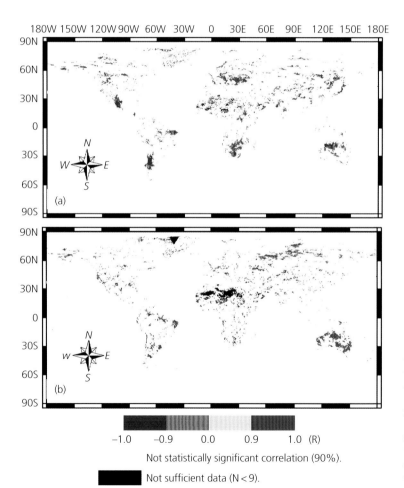

Plate 4 Global maps of correlation coefficients between annual averages of (a) NDVI and temperature and (b) NDVI and precipitation, for 1982–1990. Coloured areas indicate statistically significant associations (90% significance level). Grey areas are not statistically significant and black areas represent insufficient data ($n < 9$). See also Figure 4.1. (Reproduced from Ichii et al. 2002 with permission from Taylor & Francis.)

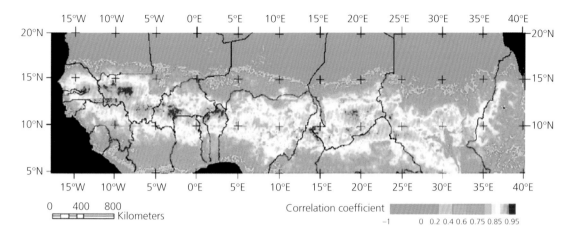

Plate 5 Linear correlations of monthly NDVI with three-monthly cumulative rainfall based on the Global Precipitation Climatology Project (GPCP) estimates for the period 1982–2003. The NDVI dataset used here is the GIMMS time-series. Both variables are highly correlated in the Sahel region. See also Figure 4.2. (Reproduced from Herrmann et al. 2005 with permission from Elsevier.)

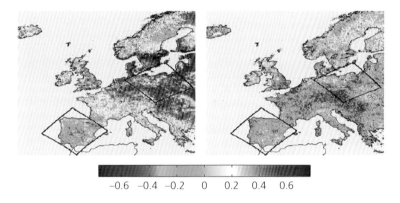

Plate 6 Point correlation fields of NAO versus NDVI in spring (left panel) and NAO versus NDVI in summer (right panel) for the period 1982–2002. The NDVI dataset used here is the GIMMS time-series. The NAO index used in this study is defined, on a monthly basis, as the difference between the normalized surface pressure at Gibraltar, in the southern tip of the Iberian Peninsula, and Stykkisholmur, in Iceland. The black frames identify north-eastern Europe and the Iberian Peninsula; the colour bar denotes the strength of the correlation: values above 0.42 and below −0.42 are associated with significant correlations between the NAO and the NDVI. See also Figure 4.3. (Reproduced from Gouveia et al. 2008 with permission from John Wiley & Sons.)

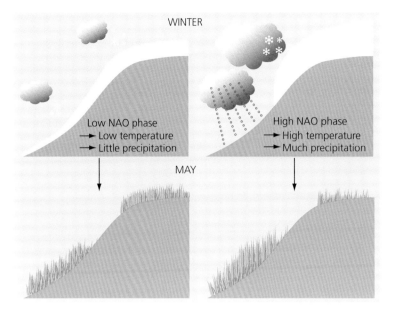

Plate 7 Linking the North Atlantic Oscillation (NAO) in winter and NDVI dynamics in Norway. At low altitudes, the vegetation was found to start earlier during high-winter NAO phases than low-winter NAO phases. At high altitudes, high-winter NAO phases mean more snow, which tends to delay the vegetation onset. See also Figure 4.4. (Adapted from the original figure in Pettorelli et al. 2005a with permission from The Royal Society.)

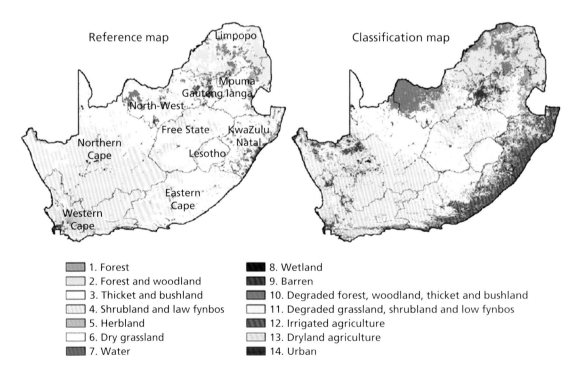

Plate 8 Reference and classification maps for South Africa. (Reproduced from Colditz et al. 2011 with permission from Elsevier.) The automated land cover classification methodology employed in this study makes use of MODIS time-series datasets. Input data are annual time-series of the spectral and emissive bands and the NDVI. Digitized national cartographies derived from Landsat 30 m data were employed as reference sets. Overall classification accuracy reached 80%. See also Figure 5.1.

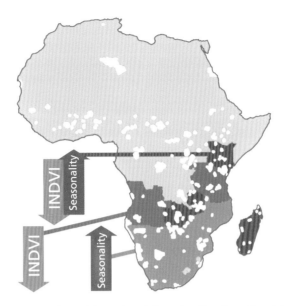

Plate 9 Map of Africa showing the 168 protected areas (white polygons). (Reproduced from Pettorelli et al., 2012 with permission from Elsevier.) Countries with protected areas displaying significant increase in NDVI seasonality are shown in orange. Countries with protected areas displaying significant decrease in the integrated NDVI (INDVI) are shown in dark green. Countries with protected areas displaying both a significant increase in NDVI seasonality and a significant decrease in INDVI during 1982–2008 are shown in purple. See also Figure 5.3.

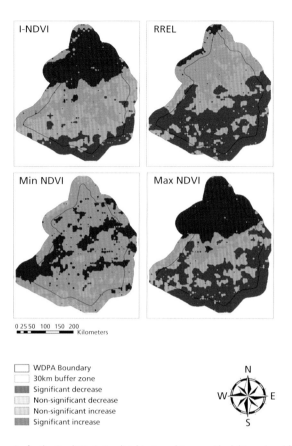

Plate 10 Trends in vegetation dynamics for the Ouadi Rimé–Ouadi Achim Faunal Reserve, Chad. (Reproduced from Freemantle et al. in press with permission from John Wiley & Sons.) The maps represent the spatio-temporal changes in the Integrated-annual NDVI (I-NDVI), Intra-annual Relative Range (RREL; an index of seasonality), and Maximum and Minimum NDVI (MAX and MIN) during the 1982–2008 period. See also Figure 8.2.

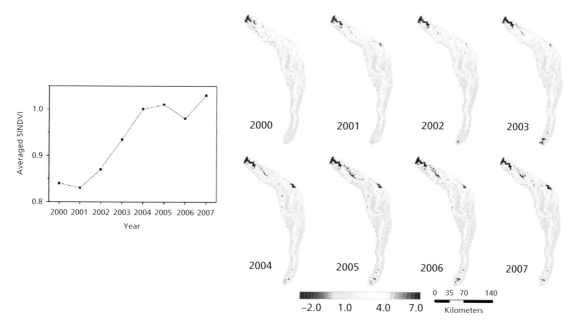

Plate 11 Time-series of the spatially averaged seasonally integrated NDVI (SINDVI) over the lower Tarim River (left) and spatial patterns of SINDVI from 2000 to 2007. (Reproduced from Sun et al. 2011 with permission from Elsevier.) SINDVI is NDVI integrated over the growing season. The Tarim River Basin is the largest continental river basin in Central Asia. Due to increasing water consumption and the construction of a reservoir in the 1970s, the groundwater table of the lower Tarim River (which is the study area considered here) fell substantially, leading to natural vegetation declining from 1972 to the late 1990s. In 2000, diversion of water from Bosten Lake to the lower Tarim River was implemented and eight intermittent water deliveries were put in place by the end of 2007. This figure allows assessing the effect of such interventions on the vegetation dynamics in the area. See also Figure 10.3.

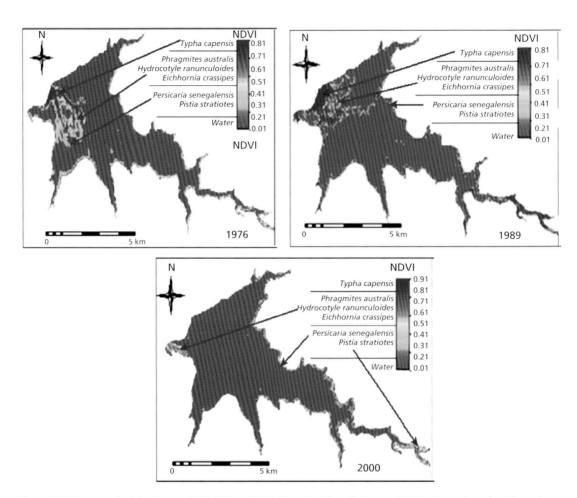

Plate 12 NDVI coverage for Lake Chivero in 1976, 1989, and 2000. (Reproduced from Shekede et al. 2008 with permission from Elsevier.) Information on the distribution of the main weed species is also shown. See also Figure 11.3.

All these studies support the argument that the NDVI can be successfully linked to animal distribution, irrespective of the set of environmental conditions or the taxa considered. They also demonstrate the great potential for this vegetation index to inform habitat suitability assessments.

7.1.2 Studying animal movement using the NDVI

Movement is the primary behavioural adaptation to spatio-temporal variability in resource availability. Establishing and deciphering patterns of movement at various temporal and spatial scales is essential for our understanding of the ecology and evolution of most animal species, as well as being a key determinant of our ability to manage their populations (Swingland and Greenwood 1987; Van Moorter et al. 2013). There are various types of movement, ranging from daily movements to dispersal and migration: these tend to be studied separately, thus referring to relatively disjointed literature. According to Dingle and Drake (2007), migration can evoke four different but overlapping concepts: (i) a type of locomotory activity that is notably persistent, undistracted, and linear; (ii) a relocation of the animal that is on a much greater scale, and involves movements of much longer duration than those arising from its normal daily activities; (iii) a seasonal to-and-from movement of populations between regions where conditions are alternately favourable or unfavourable (including one region in which breeding occurs); and (iv) movements leading to redistribution within a spatially extended population. Two broad types of dispersal, with different causes and consequences, are then usually considered: natal dispersal and breeding dispersal. Natal dispersal refers to the definitive movement of an individual from its birth location (or social group) to the place (or social group) where it will settle and first reproduce. Breeding dispersal, on the other hand, refers to the movement between two successive breeding areas or social groups (Howard 1960; Clobert et al. 2001). The high temporal resolution of the NDVI is particularly helpful in studying animal movements since data on vegetation productivity can be linked to simultaneous relocation data of individuals. Instead of traditional habitat selection analyses that use static habitat maps, dynamic landscape models based on the NDVI enable the examination of movement decisions of individuals or populations as resource availability changes in time. This section explores how the NDVI has furthered our understanding of daily movement, migration patterns and dispersal.

Migration is among the most studied type of movement (Dingle and Drake 2007), and as such, many of the studies linking the NDVI to movement patterns have actually focused on migrations (see also Box 7.3). For example, shifts in the relative distribution of the NDVI have been shown to explain movement to and from seasonal ranges in Mongolian gazelle (Ito et al. 2006) and saiga antelope *Saiga tatarica tatarica* (Singh et al. 2010). On

---

Box 7.3 NDVI and animal migration

In many cases, the reported positive association between NDVI variability and migration patterns may be underpinned by the species tracking high-quality forage. In seasonal environments, both plant crude protein content and digestibility peak early in the growing season, and then rapidly decline as the vegetation matures: higher forage quality is thus associated with early phenological stages where new green leaves dominate biomass (Crawley 1983). Newly emerged plants are particularly nutritious, and for species living in such seasonal environments (e.g. saiga antelope, wildebeest, red deer), NDVI dynamics could supply information on the spatio-temporal distribution of high-quality forage. This assumption is well supported by recent analyses in Canada, where the association was examined between NDVI indices and annual variability in the date of peak in faecal crude protein, which represents temporal variability in the availability of high-quality vegetation (Hamel et al. 2009). The authors reported that annual integrated NDVI in June was negatively correlated with dates of the peak in faecal crude protein, indicating that the NDVI can reliably measure yearly changes in the timing of the availability of high-quality vegetation. Likewise, data from the Bylot Island, Canada, showed that several NDVI metrics were significantly related to date of peak nitrogen concentration in this Arctic tundra ecosystem (Doiron et al. 2012). A positive exponential relationship between NDVI and above-ground biomass of plants was also reported for this system, being strongest early in the growing season.

Santa Cruz Island, Galapagos, adult giant tortoises *Chelonoidis nigra* of both sexes were recently reported to move up and down an altitudinal gradient in response to changes in vegetation dynamics, as indexed using the NDVI (Blake et al. 2013). The timing of downward movements was shown to occur, in this case, as lowland vegetation productivity peaked. In West Africa, both the launching and the orientation of the directional movement of buffalos (*Syncerus caffer*) was reported to match closely those of a large-scale NDVI gradient occurring within the study area during the first month of the early wet season (Cornelis et al. 2011). Likewise, in the Serengeti, migratory herds of ungulates have been shown to move over 'vast' geographical regions, following a rainfall gradient affecting plant quality in different seasons (McNaughton 1990). Building on these results, Musiega and Kazadi (2004) found that the great seasonal migration of herds of wildebeest in the Serengeti–Mara ecosystem was primarily driven by the availability of green vegetation, as indexed using the NDVI. Two years later, landscape models based on the NDVI were demonstrated to be key to explaining wildebeest migration in this ecosystem (Boone et al. 2006). In Norway, vegetation onset as indexed by the NDVI was shown to correlate strongly with the start of the altitudinal migration for red deer *Cervus elaphus* (Pettorelli et al. 2005b). Data from >250 Global Positioning System (GPS)-collared individuals helped to demonstrate how migratory individuals—and, to a lesser degree, residents—tracked phenological green-up through parts of the growing season by making small-scale adjustments in habitat use (Bischof et al. 2012). Congruent results were previously reported by Hebblewhite et al. (2008), who used the NDVI to trace habitat choices of single individual migrating red deer (or elk, as they are known in that part of the world) at a particular movement step. The novelty of the approach presented by Bischof et al., however, lies in the fact that they measured the speed of green-up in space, and correlated this information with the speed of red deer while migrating.

With respect to the NDVI and animal migrations, Sawyer and Kauffman (2011) published an interesting twist in terms of accounting for the environmental factors influencing movement decisions. The authors used fine-scale movement data collected from GPS collars to examine the ecological role of stopover use for migratory mule deer *Odocoileus hemionus* in western Wyoming, USA, where migration distances only range from 18 to 144 km. Although the individuals could easily complete migrations in several days, the authors reported that deer took an average of three weeks and spent 95% of that time in a series of stopover sites. Intrigued by this, they decided to test the general hypothesis that stopovers shape the migrations of land migrants, who seek to maximize energy intake, rather than speed, during their migrations. To do so, the authors characterized the forage quality of stopovers relative to movement corridors, and investigated whether mule deer used stopovers during time-periods when plant phenology was in an early state known to produce high-quality forage. The results matched their expectations: stopover sites did have higher forage quality than movement corridors. Forage quality of stopovers was then shown to increase with elevation and distance from the winter range, with mule deer use of stopovers corresponding to a narrow phenological range (Figure 7.2; Box 7.3). Being able to access relevant ecological metrics, such as those from the NDVI, enabled the demonstration that stopovers can play a key role in the migration strategy of large herbivores by allowing individuals to migrate in concert with plant phenology and maximize energy intake.

The links between NDVI variability and dispersal patterns have not been explored to the same extent as the links between NDVI variability and migration. However, there are exceptions. In Scandinavia, Andersen et al. (2004) correlated the expansion rate of roe deer populations from 1850 to 1980 with NDVI dynamics. The records used to assess expansion rate emanated from municipality wildlife boards, and helped demonstrate that expansion rates for this species were faster in marginal habitats. Similarly, there are relatively few studies on the relationships between NDVI variability and daily movements in animal species. Boettiger et al. (2011) focused on relating fine-scale movement pathways of African elephants to spatio-temporally structured landscape data, including vegetation productivity as indexed by the NDVI. Results showed that elephant responses to vegetation productivity indices were not uniform in time or space, indicating that elephant

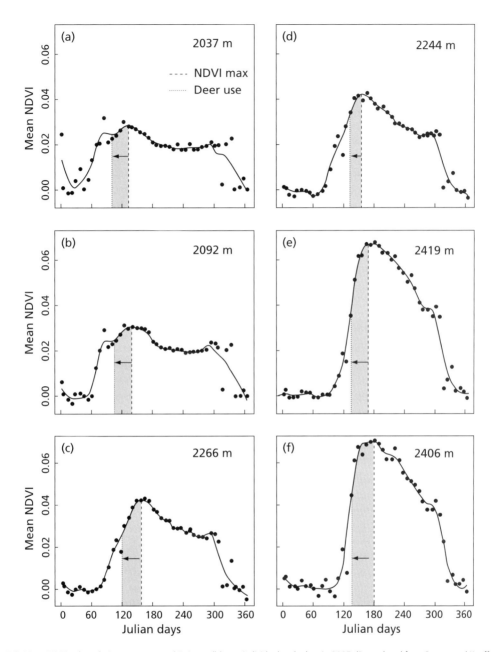

Figure 7.2 Mean NDVI values during stopover use (six in total) by an individual mule deer in 2005. (Reproduced from Sawyer and Kauffman 2011 with permission from John Wiley & Sons.) NDVI values (black spots) were taken every 8 days, and a smoothed curve (continuous dark line) was used to determine the date of maximum NDVI representing peak green-up (long hashed line). The median date on which the stopover was used by the deer (short hashed line) was subtracted from the date of maximum NDVI to characterize the phenological stage of vegetation (shaded area) for the time-period of stopover use. Panels (a)–(f) progress from winter range (a) to summer range (f). On average, deer occupied stopovers 44 ± 6 days (mean ± SD) before peak green-up. The mean elevation of the stopover is noted in the upper right corner of each panel.

foraging strategies are more complex than simply gravitation toward areas of high productivity. van Moorter et al. (2013) supply an interesting framework to assess the role of NDVI variability in determining daily movements: the authors developed an approach to simultaneously study movements at different spatio-temporal scales, based on the theory that the length and frequency of animal movements are determined by the interaction between temporal autocorrelation in resource availability and spatial autocorrelation in changes in resource availability. They tested their approach on moose *Alces alces*, and showed that frequent, smaller-scale movements were triggered by fast, small-scale changes in NDVI and snow depth, whereas infrequent, larger-scale movements matched slow, large-scale waves of change in these variables.

7.2 NDVI and life histories

During an animal's life, key events and periods shape its ability to produce the largest possible number of surviving offspring. Juvenile development, age of sexual maturity, number of offspring and level of parental investment, and senescence are part of these key events and periods, and depend on the physical and ecological environment that the animal experiences. The succession of these events and periods constitutes the animal's life history. This section discusses how the NDVI can supply information on the environment experienced by animals, thereby revealing how changes in the spatio-temporal dynamics of green biomass shape the outcomes of some of these key events and periods (Table 7.2).

7.2.1 NDVI and condition: body mass as an example

Body mass plays a fundamental role in shaping variation in life-history traits both at the interspecific (Peters 1983; Calder 1984) and the intraspecific (Sadleir 1969; Clutton-Brock 1991) level. In large herbivores, individual body mass has been shown to be a determinant of juvenile survival (Gaillard et al. 1997), adult survival (Bérubé et al. 1999), litter size (Hewison 1996), and age at first breeding (Albon et al. 1983). Understanding what causes variation in individual body mass can therefore help us to understand what shapes the population dynamics of many terrestrial vertebrates.

Primary productivity and vegetation phenology as indexed by the NDVI were shown to influence individual body mass in various ungulate populations. This is especially true for younger individuals. In Norway, the NDVI in spring was shown to be a strong predictor of juvenile body mass in reindeer (Pettorelli et al. 2005c). Birth and autumn body mass of migratory reindeer (also known as caribou) in Canada were influenced positively by habitat quality in June, as estimated by the NDVI (Couturier et al. 2009). The rate of change in primary production during green-up was negatively correlated with lamb body mass of bighorn sheep *Ovis canadensis* in Ram mountain, Canada, and with kid body mass of mountain goat *Oreamnos americanus* in Caw Ridge, Canada (Pettorelli et al. 2007). For moose, body mass during autumn was shown to be positively related to early access to fresh vegetation in spring, and to variables reflecting slow phenological development. Interestingly, the magnitude of the effects of environmental variation on body mass was larger in populations with small mean body mass or living at higher densities than in populations with large-sized individuals or those living at lower densities (Herfindal et al. 2006b). For alpine chamois *Rupicapra rupicapra*, body mass of both juvenile males and females was recently reported to decrease during years with late springs (−20%) and with increasing population abundance (−15%), with no interactive effect (Garel et al. 2011): such results suggested that body mass of juveniles could be used as an indicator of the relationship between chamois populations and their environment. As a general rule, correlation between NDVI-derived information and body mass thus seems to be highest (a) when considering young individuals; and (b) in highly seasonal environments (Table 7.3).

Sometimes it can be difficult to assess *a priori* which period might be critical in determining body mass of large herbivores. Vegetation conditions in spring and summer are generally regarded as decisive for the reproductive success and offspring condition of large herbivores inhabiting seasonal environments (see also Box 7.3), but objective ways to determine key periods during the growing season

Table 7.2 Studies highlighting links between NDVI dynamics and demographic parameters.

Parameter	NDVI metrics	Species	Location	References
Juvenile body mass	NDVI May	Reindeer	Norway	Pettorelli et al. 2005c
Body mass	NDVI May	Red deer	Norway	Pettorelli et al. 2005b
Body mass	NDVI June	Caribou	Canada	Couturier et al. 2009
Lamb body mass	Rate of change in primary production during green-up	Bighorn sheep	Canada	Pettorelli et al. 2007
Kid body mass	Rate of change in primary production during green-up	Mountain goat	Canada	Pettorelli et al. 2007
Juvenile body mass	NDVI spring	Chamois	France	Garel et al. 2011
Body mass	NDVI spring	Moose	Norway	Herfindal et al. 2006b
Juvenile Body mass	NDVI spring and NDVI autumn	Roe deer	France	Pettorelli et al. 2006
Recruitment	INDVI March–July	Soay sheep	UK	Durant et al. 2005
Number of lambs per ewe	Start of the growing season and NDVI mid autumn	Domestic sheep	Chile	Texeira et al. 2008
Conception rate	NDVI conditions during conception period	Elephant	Kenya	Wittemyer et al. 2007
Breeding success	NDVI spring	Pied flycatcher	Mediterranean region	Sanz et al. 2003
Clutch size	NDVI in winter quarters	Barn swallows	Trans-Saharan populations	Saino et al. 2004
Reproductive performance	NDVI seasonality and NDVI spatial variability	Brown bear	Spain	Wiegand et al. 2008
Annual survival rates	NDVI in winter quarters	White stork	Trans-Saharan populations	Schaub et al. 2005
Annual survival rates	NDVI in winter quarters	Egyptian vulture	Africa	Grande et al. 2009
Juvenile survival	NDVI during calving	Caribou	Canada	Griffith et al. 2002
Juvenile survival	Rate of change in primary production during green-up	Ibex	Italy	Pettorelli et al. 2007
Juvenile survival	Rate of change in primary production during green-up	Bighorn	Canada	Pettorelli et al. 2007
Anomalies in age structure	INDVI during gestation	Elephant	Africa	Trimble et al. 2009
Strength of density dependence	NDVI in March	Field vole	Britain	Bierman et al. 2006
Strength of density dependence	Coefficient of Variation of NDVI	Ungulates	Rocky Mountains, USA	Wang, G. et al. 2006
Population growth	NDVI in April	Great gerbil	Central Asia	Kausrud et al. 2007
Population growth	Length of the growing season	Roe deer	Norway	Melis et al. 2010

have not been defined, often due to limitations in plant data. Using the NDVI, Pettorelli et al. (2006) tackled this issue by determining how plant productivity from birth to autumn influenced the following winter body mass of roe deer fawns in two French populations, namely (i) a roe deer population inhabiting the low-productivity Chizé reserve in south-western France; and (ii) a roe deer population inhabiting the highly productive forest of Trois Fontaines in eastern France. Roe deer are income breeders (Andersen et al. 2000) and their growth pattern follows a monomolecular model (Portier et al. 2000), with a growth rate decreasing from birth onwards. The authors therefore expected a significant effect

Table 7.3 Expected correlation between the NDVI and large herbivore body mass.

	Young individuals	Adults
Highly seasonal ecosystems (alpine areas, high latitude)	Highest correlation; phenology and primary productivity both important	Lower correlation in general; possible better correlation in high-density populations and in less-productive environments; possible delay effects
Less seasonal ecosystems (temperate forests)	Lower correlation; primary productivity during first months of life important	Lowest correlation

of spring/summer conditions on winter body mass of roe deer fawns in both areas. However, contrasting results were obtained between sites: at Chizé, the authors found a strong correlation between plant productivity in April–May and the following winter body mass of roe deer fawns (Figure 7.3). Yet at Trois Fontaines, the only significant correlations reported between vegetation productivity at a fixed date and the body mass of fawns were negative. The relatively low variability in winter fawn body mass was thought to be responsible for the absence of NDVI effects at Trois Fontaines. Thus Pettorelli et al. used the NDVI to identify key periods during the growing season for body mass determination in large herbivores, and were able even to distinguish between different populations of the same species.

7.2.2 NDVI and reproduction

An individual's fitness refers to its ability to both survive and reproduce, assessed by its contribution to the gene pool of the next generation. An individual's reproductive output is thus key to its fitness, highlighting the importance of understanding the factors shaping reproductive success. Reproductive success can be influenced by several variables, such as (a) timing of reproduction; (b) number of offspring produced; and (c) offspring quality.

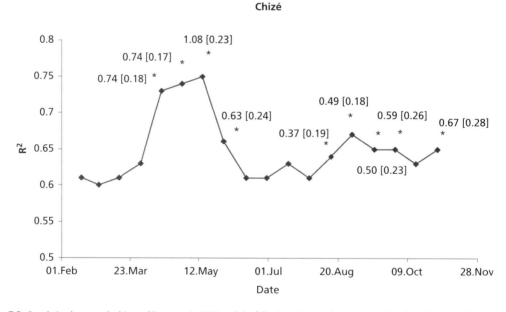

Figure 7.3 Correlation between the bi-monthly composite NDVI and the following winter body mass of roe deer fawns between February and October in the Chizé reserve, France. Slopes [SD] are given when the correlation is significant (<5%, marked with a star). (Reproduced from Pettorelli et al. 2006 with permission from John Wiley & Sons.)

Due partly to seasonality (and partly to stochasticity), most iteroparous organisms (i.e. organisms that can undergo many reproductive events throughout their lifetime) experience spatio-temporal variation in the accessibility and availability of resources, and such variability has the potential to influence reproductive success.

The links between resource availability dynamics and reproductive timing and success, for example, may be particularly strong in seasonal environments and for migrant species, as in both cases the life-history strategies of species have been selected to match the best environmental conditions (Box 7.4). This is well illustrated by the large herbivores inhabiting the strongly seasonal environments at northern latitudes. For those species, the timing of conception is expected partly to determine the timing of parturition (Berger 1992), and the following match with the phenology of the vegetation. Calving generally occurs in early summer each year to benefit from a longer growing season (Rutberg 1987). This enables females to minimize the costs of reproduction in the current offspring and reach a sufficient condition in autumn to enter the next reproductive cycle. This will also enable the offspring to benefit from the necessary care, thereby maximizing their chances for over-winter survival. Indeed, late-born offspring are more likely to die since they are smaller in autumn (Lindström 1999). However, females should not invest so as to jeopardize their own survival (Festa-Bianchet and Jorgenson 1998): prolonged investment in current-year offspring may lower female condition in autumn and hence her ability to ovulate sufficiently early to get a good start for the next calving season the following spring (Hogg et al. 1992; Langvatn et al. 2004). Therefore, the timing of events in a reproductive cycle (such as conception, calving, and weaning) may determine reproductive success.

Strong links between NDVI dynamics and reproduction may be expected (i) for animals living in seasonal environments (where NDVI dynamics are likely to correlate with forage availability dynamics); and (ii) for species where the NDVI is very likely to accurately index forage availability. Most of the successful studies linking the NDVI to animal reproduction belong to these categories. For example, the timing of the spring vegetation flush, identified from NDVI time-series, and the rutting and calving of red deer were all found to be later in Norway than in France (Loe et al. 2005). Birth occurrence and synchrony were reported to be driven by NDVI variation for buffalos in South Africa (Ryan et al. 2007), and in African elephants, the initiation of a female's reproductive bout was reported to be timed so that parturition occurred during the most likely periods of high primary productivity 22 months later (Wittemeyer et al. 2007; Figure 7.4).

Box 7.4 NDVI and the match–mismatch hypothesis

The match–mismatch hypothesis (Cushing 1990) is based on the prediction that a predator's recruitment will be high if the peak of its prey availability temporally matches the most energy-demanding period of the predator's breeding phenology. A mismatch, on the other hand, is expected to lead to poor recruitment. The hypothesis was first proposed for marine systems and suggested that the interannual variability in fish recruitment was a function of the timing of the production of their food (Hjort 1914). A similar hypothesis was later proposed for other systems (Nilsson 1998). NDVI data can supply information on the spatio-temporal dynamics of resource availability for herbivores, enabling the testing of such a hypothesis. Durant et al. (2005) did this, investigating how the timing of resource abundance influenced Soay sheep *Ovis aries* recruitment on Hirta Island, St Kilda, Scotland. Most Soay sheep offspring in this area are conceived in November and born in April. Consequently, food abundance for sheep was assessed using the INDVI (defined in Chapter 3) between 1 March and 21 July. A mismatch index was calculated, based on the average Julian date of lamb birth and the estimated date of the start of vegetation growth. The survival of Soay sheep lambs was significantly affected by the food abundance, taking into account the climate and density-dependence effects. Survival was not, however, affected by temporal mismatch between vegetation peak and the average birth date. Nevertheless, this work illustrates the potential for the NDVI in testing the match–mismatch hypothesis in various terrestrial systems.

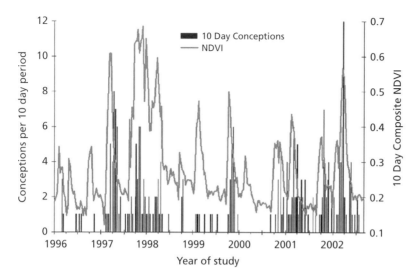

Figure 7.4 Mean 10-day composite NDVI values (grey line) and total number of conceptions per 10-day interval in elephants (black bars). (Reproduced from Wittemyer et al. 2007 with permission from John Wiley & Sons.)

In Chile, the most important controls of reproductive performance (as indexed using the number of lambs per ewe) of domestic sheep *Ovis aries* in the Patagonian steppe were foraging conditions during two critical periods (namely mating and late pregnancy), with reproductive performance showing a slight, yet consistent, improvement during years in which the growing season started earlier and had higher NDVI values in mid-autumn (Texeira et al. 2008). For the pied flycatcher *Ficedula hypoleuca*, the NDVI has been successfully linked to breeding success (Sanz et al. 2003); in barn swallows *Hirundo rustica*, it was related to clutch size and breeding date (Saino et al. 2004). Wiegand et al. (2008) recently showed that habitat quality of an endangered bear *Ursus arctos* population in Spain was linked to seasonal pulses in primary production captured by the NDVI; areas displaying strong NDVI seasonality and low spatial variability were associated with high reproductive performance for bear populations.

7.2.3 NDVI and survival

Understanding the factors influencing demographic parameters such as fecundity and survival is of both theoretical and practical importance, as these parameters shape population size variation. Assessing the factors limiting survival rate can yield insights for effective management, and help identify directions for future research (Williams et al. 2002). Environmental conditions have been shown previously to influence survival in various taxa, with the strength of the relationship between environmental conditions and survival being sometimes radically different according to the species considered as well as the individual characteristics (such as sex or age; Begon et al. 1996). In large ungulates, for example, prime-age survival of females tends to be highly resilient to changes in environmental conditions, while juvenile survival of males tends to be quite sensitive to environmental variability (Gaillard et al. 2000). There is a huge variety in the actual mechanisms by which animals can be affected by environmental variability. Changes in climatic conditions can directly and indirectly influence survival of individuals. Direct impacts of climatic conditions on life-history traits can be caused by increased energetic costs: during winter in northern latitudes, snow depth can positively correlate with body mass loss (Loison and Langvatn 1998), possibly due to extra energetic costs of movement in snow, thermoregulation, and reduced access to forage. If large enough, this loss of body mass can affect survival probability. Snow accumulation can also delay plant growth. On the other hand, a warm spring can speed up plant phenology. Changes in climatic conditions therefore lead to plant phenology being variable: yet, as already discussed, delayed phenology can affect condition and reproductive success.

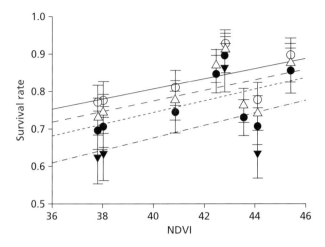

Figure 7.5 Relationship between NDVI index and survival estimates in Egyptian vultures. (Reproduced from Grande et al. 2009 with permission from John Wiley & Sons.) Black circles represent survival estimates for 1–2-year-old birds; white triangles are estimates for 3–4-year-old birds; black triangles for 5-year-old birds; white circles for >5-year-old birds. Standard errors and regression lines are also shown. Original NDVI values have been multiplied by 100, so as to range between 0 and 100.

Cases reporting direct correlations between NDVI measurements and survival are available mainly for birds and large herbivores, though emanating from studies carried out in various parts of the world. Variation in the primary production (as reflected by the NDVI) in the eastern Sahel, which is visited by migratory white stork *Ciconia ciconia* from October to November, was reported to contribute up to 88% to the temporal variation in annual survival rates (Schaub et al. 2005). NDVI values from the natal territories of Egyptian vulture *Neophron percnopterus* was positively correlated with juvenile survival, while annual survival rates for birds of all age classes were positively associated with the NDVI in their African wintering grounds (Grande et al. 2009; Figure 7.5). For caribou, the relative amount of forage available on the calving grounds, as indexed by the NDVI, was reported to be the best predictor of early calf survival (Griffith et al. 2002), while juvenile survival of bighorn sheep, ibex *Capra ibex*, and Soay sheep were all reported to correlate with vegetation dynamics in Canada, Italy and Scotland, respectively (Durant et al. 2005; Pettorelli et al. 2007).

7.3 NDVI and animal abundance

The amount of energy available in a system is thought to be one of the major determinants of species abundance and biomass (Blackburn and Gaston 2001; Carbone and Gittleman 2002). Although widely acknowledged, relatively few studies have actually tested this assumption, especially at large scales; possible reasons for this include the difficulties associated with measuring energy and monitoring abundance (Box 7.5). When it comes to measuring energy, satellite imagery and the NDVI facilitate standardized, quantified measures of energy availability for the whole globe, and several studies performed at various spatial scales and resolutions have now reported good correlations between NDVI-based measurements of energy availability and animal abundance.

In the Argentinian rangelands, livestock biomass and annual INDVI values were shown to be strongly and positively related (Oesterheld et al. 1998). A significant association between the NDVI and white-tailed deer density was reported in a temperate site in Mexico (Coronel Arellano et al. 2009); spatial variation in abundance of salamander *Salamandra salamandra* larvae in Switzerland was positively and significantly correlated with photosynthetic biomass as indexed by the NDVI (Tanadini et al. 2012); relative abundances of *Culicoides imicola* at 22 sites in Morocco were positively correlated with the average and minimum NDVI values (Baylis and Rawlings 1998); wolf spider Lycosidae abundance was reported to increase with productivity as indexed by the NDVI in Central Arizona (Shochat et al. 2004); and the density of the mosquito *Anopheles atroparvus* in Portugal was recently shown to be a function of temperature and the NDVI (Lourenco et al. 2011). Similar results were obtained for other mosquito species in the USA, where population changes were found to be strongly influenced by

> **Box 7.5 Assessing and monitoring abundance of animals**
>
> Abundance (i.e. the number of animals occupying a given area), as compared with species presence or species richness, can be difficult to assess and monitor. This is because researchers and wildlife practitioners only see all the animals in an area in rare cases. In situations where only a fraction of the animals present is seen, the issue becomes to estimate correctly the proportion of 'unseen' individuals. Several methods have been developed to assess this 'absolute' abundance, such as counts of the whole population, sample plots using aerial censuses in open areas, drive censuses, capture–mark–recapture, distance sampling, or hunting-related methods in closed areas. The choice of the method is generally a function of the area considered, the species under study, the structure of the landscape, the budget, and the number of people that could be involved in the monitoring (Rabe et al. 2002). This choice of methodology is associated with an important source of bias: survey methods can strongly affect abundance estimates (Bart et al. 2004; Walter and Hone 2003; Roberts and Schnell 2006). To overcome partly the difficulty of assessing and monitoring the number of animals occupying a given area, methods to monitor trends in population sizes have also been developed (Seber 1992). These methods aim at assessing and monitoring an index (relative abundance index) expected to correlate directly with 'absolute' density. Again, several methods have been developed, such as pellet group counts, or the monitoring of animal vocalizations, animal tracks, or animal body condition. For those kinds of approaches, the issue is to assess whether the index monitored truly correlates with 'absolute' density.

NDVI dynamics (Britch et al. 2008). Annual NDVI values were moreover reported to be positively correlated with African elephant densities in Zimbabwe (Chamaille-Jammes et al. 2007), as well as with large herbivore density estimates across Africa (Pettorelli et al. 2009).

Sometimes the NDVI can help to predict abundance based on the animal population's impact on the environment, rather than based on the expected positive link between energy and abundance. For example, ground measures of moth larval density were found to correlate with NDVI-based defoliation score in Fennoscandia (Jepsen et al. 2009; see also Chapter 5). By annually recording plant biomass in grazed control plots and in herbivore-free enclosures, Olofsson et al. (2012) showed that the regular interannual density fluctuations of voles and lemmings in a sub-Arctic ecosystem drove synchronous interannual fluctuations in vegetation biomass. Vole and lemming peaks were followed by reduced NDVI values the following year, thus it was possible to detect the effects of rodents on the vegetation using satellite imagery. In other cases, the NDVI can be used to assess the strength of the links in abundance fluctuations across populations. This approach was undertaken by Kausrud et al. (2007) to explore the role of climate in synchronizing the dynamics of great gerbils *Rhombomys opimus* in central Asia. Taking into consideration the NDVI in April increased the predictive power of their population model, suggesting that vegetation conditions in April may be expected to contribute to synchronizing abundances across space. The reported co-dependency between the NDVI in April, gerbil densities, and plague, together with predictions from historical climate records, led the authors to suggest that periods of relatively warm and/or moist conditions give rise to periods of high gerbil densities and large epizootics in otherwise dry areas.

Some of the studies exploring the link between energy availability and animal distribution have also highlighted factors that can affect the correlation between the NDVI and abundance. The absence of a relationship can indeed be telling (as later discussed in more depth in Chapter 9): the absence of a relationship between the NDVI and white-tailed deer abundance in a tropical area in Mexico, for example, revealed the importance of considering differences in anthropogenic pressures while exploring links between vegetation dynamics and animal abundance across populations (Coronel Arellano et al. 2009). Season was then shown to play a key role in determining the type of link between topi *Damaliscus lunatus* density and greenness in the Masai Mara, Kenya (Bro-Jørgensen et al. 2008).

7.4 NDVI and species richness

Species richness is a component of biodiversity (Box 7.6), and is defined as the number of different

> **Box 7.6 Species richness, species diversity, and biodiversity**
>
> Species richness is defined as the number of species present in a sample, community, or taxonomic group. Species richness is one component of species diversity, as species diversity is a concept that incorporates both species richness and species evenness (which refers to the relative abundance of species). Species diversity is thus one component of the broader concept of biodiversity, which can be defined as 'variation of life at all levels of biological organization' (Gaston and Spicer 2004): biodiversity (or biological diversity) indeed encompasses all the variety of life forms at all biological system levels (i.e. molecular, organismic, population, species, and ecosystem).

species in a given area. It represents a fundamental measure of community and regional diversity, and underpins many ecological models and conservation strategies (Gotelli and Colwell 2001). However, species richness is not constant worldwide, and many efforts have been made to understand the processes shaping the observed patterns in the spatial distribution of species richness. As for abundance, one theory states that the amount of energy available in a system should be a key determinant of species richness (Currie 1991; see also Chapter 6). More precisely, energy has been hypothesized to determine species richness through its effects on total species biomass or abundance. Higher energy availability may indeed allow more individual animals in total to persist in an area, which enables individual species to obtain higher population densities that reduce their risk of extinction, which consequently increases the species richness of the area (Hurlbert 2004, Evans et al. 2006).

Such an expectation has been very well supported over the years for many taxa: for instance, NDVI-based estimates of productivity in the USA were reported to explain up to 61% of avian species richness (Hurlbert and Haskell 2003). Likewise, Hawkins (2004) and Hawkins et al. (2003) showed that productivity indicators (in this case, the NDVI and actual evapotranspiration) both correlated with bird diversity data in North America. Seto et al. (2004) reported a positive correlation between the NDVI and bird species richness in western North America: it is interesting that the correlation was found to hold better at lower resolution. In a study carried out in the Great Basin, USA, Bailey et al. (2004) found positive linear relationships between the maximum NDVI and the number of functional guilds of birds and species richness of neotropical migrant birds (see also Figure 7.6). In East Asia, the average NDVI was the most important factor determining variation in bird species richness, with a linear positive correlation between the NDVI and bird species richness (Ding et al. 2006). Similarly, the NDVI was the most powerful predictor of bird species richness at the markedly smaller spatial scales

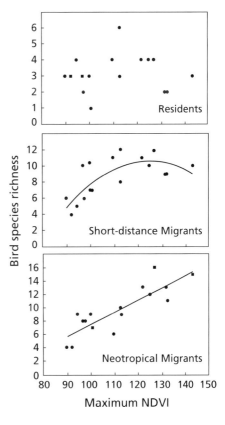

Figure 7.6 Relationship between species richness of birds and maximum NDVI at the canyon grain across the landscape, for residents, short-distance migrants, and neotropical migrants. (Reproduced from Bailey et al. 2004 with permission from John Wiley & Sons.) Solid lines denote significant relationships. Original NDVI values have been multiplied by 1000, so as to range between 0 and 1000.

of Jerusalem (where a hump-shaped relationship between NDVI and species richness was observed; Bino et al. 2008) and northern Taiwan (there, authors reported a linear positive correlation between the NDVI and bird species richness; Koh et al. 2006). These results, however, may need to be interpreted with caution. Lee et al. (2004) also found a hump-shaped relationship between the NDVI and bird species richness in Taiwan, but this became insignificant when effects of roads and elevation were accounted for.

Although most examples of studies linking the NDVI to species richness come from studies of birds, there have been some studies linking the NDVI to species richness of other groups. Seto et al. (2004) and Bailey et al. (2004) found strong correlations between the NDVI and butterfly species richness in the Great Basin of western North America. Lassau and Hochuli (2008) reported positive relationships between the NDVI and site-based beetle species richness and abundance, with the NDVI also being useful for predicting differences in beetle composition in open canopy forests. Qian et al. (2007) showed that temperature, precipitation, net primary productivity, minimum elevation, and range in elevation explain around 70% of the variance in amphibian and reptile species richness in China.

Occasionally, spatial heterogeneity in productivity can be more important than productivity *per se* in explaining species richness distribution. In Kenya, for example, higher yearly average NDVI values were correlated with lower species richness of mammals and plants, whereas standard deviation and coefficient of variation were correlated positively with species richness (Oindo and Skidmore 2002). In Israel, the variability in the NDVI, as measured both within and among adjacent elevation belts, was strongly and significantly correlated with butterfly species richness (Levanoni et al. 2011). In the Chihuahuan Desert of New Mexico, texture of NDVI (as opposed to mean NDVI) accounted for most of the variability in bird species richness, explaining up to 82.3% of the variability in this parameter (St-Louis et al. 2009). Thus, season and the spatial scale under consideration can also affect the strength and shape of the relationship between the NDVI and species richness. For example, bird richness patterns in North America (east of longitude 98°W) were correlated with average summer INDVI and annual INDVI values (Hawkins 2004). The average summer INDVI was positively linearly associated with richness, yet the relationship between bird richness and the annual INDVI was curvilinear. In sub-Saharan Africa, the relationship between avian species richness and the NDVI was reported to be significantly spatially variable and scale-dependent (Foody 2004).

7.5 Conclusions

The NDVI is a globally useful tool for terrestrial ecologists aiming to gain a better understanding of how vegetation dynamics and distribution affect diversity, life-history traits, movement patterns and population dynamics of animal populations (Figure 7.7).

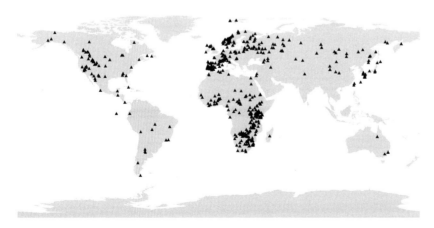

Figure 7.7 Spatial distribution of the study areas where the links between NDVI dynamics and wildlife were investigated. Each black triangle refers to the estimated barycentre or the published coordinates of a particular study area. Only the scientific publications cited in this book have been used for the production of this map.

The value of the NDVI for wildlife managers may be appreciated by reference to climate change and the variety of NDVI-based analyses cited in this chapter. Several climatic models aimed at predicting spatio-temporal changes in NDVI have been developed (see, e.g., Anyamba et al. 2006a; Funk and Brown 2006), and these may be used in conjunction with NDVI-based studies of wildlife habitat selection and distribution to estimate future range requirements and restrictions for many animal species. Such knowledge is especially important to protected area planning and reserve design; for species currently in protected areas at the edges of their range, range shift predictions are fundamental to assessing whether they will or will not be able to migrate and adapt in the face of global environmental change (Singh et al. 2010).

CHAPTER 8

NDVI for informing conservation biology

> For a century, environmentalism has divided itself into warring camps: conservationists versus preservationists. . . . The struggle pits those who would meddle with nature against those who would leave it be. . . . The only sensible way forward lies in a melding of the two philosophies. We must manage nature in order to leave it alone.
> **David Barron**

'What is conservation biology?' asked Michael Soulé in 1985, and this is a question worth answering before discussing the potential of any tool to inform this discipline. The term conservation biology was introduced as the title of a conference held at the University of California, San Diego, in La Jolla, California in 1978. According to one of the conference organizers, conservation biology can be defined as the application of science to conservation problems, addressing the biology of species, communities, and ecosystems that are perturbed, either directly or indirectly, by human activities or other agents (Soulé 1985). There are many definitions of conservation biology (e.g. Hunter 1996; Meffe and Groom 2006; van Dyke 2008), but Soulé's has the advantage of highlighting three important aspects of the discipline, namely that conservation biology: (i) is often a crisis discipline; (ii) has a strong applied nature; (iii) is a component of environmental and wildlife management.

We now turn to how the NDVI can be useful to conservation biology. Chapters 5–7 reviewed study cases in which the NDVI was successfully linked to the distribution and functioning of ecosystems, as well as to plant and animal ecology; we saw not only that the NDVI could inform theoretical and applied ecology, but that it could also yield insights into conservation biology. For instance, the NDVI could be applied to monitor ecosystem functioning and map habitat degradation (Chapter 5), or could be used to assess the relative impact of the direct and indirect effects of climate on animals (Chapter 7).

This chapter begins by discussing how the vegetation index can be used to inform the expansion and management of the current protected area network (Section 8.1), and then demonstrates how the NDVI can support the implementation of reintroduction programmes (Section 8.2). Landscape-scale connectivity among preserved patches is of paramount importance to buffer biodiversity against environmental change, highlighting the importance of taking pertinent, informed land-use management decisions. In this respect, the NDVI can be a useful tool in detecting and mapping wildlife corridors (Section 8.3). The NDVI offers great potential to better predict and mitigate the consequences of climate change on biodiversity, and Section 8.4 discusses recent examples of how the index may furnish realistic scenarios about climatic changes impacting on wildlife. The final section focuses on the NDVI in relation to invasion biology, that is, the detection, mapping, and monitoring of invasive species.

The Normalized Difference Vegetation Index. First Edition. Nathalie Pettorelli.
© Nathalie Pettorelli 2013. Published 2013 by Oxford University Press.

8.1 Supporting the management and expansion of protected areas

8.1.1 Protected areas as the cornerstone of global conservation efforts

Since the late nineteenth century the backbone of the conservation of biodiversity throughout the world has been the establishment of protected areas (Pressey 1996), which are defined by the World Conservation Union as 'an area of land and/or sea especially dedicated to the protection and maintenance of biological diversity, and of natural and associated cultural resources, and managed through legal or other effective means' (International Union for the Conservation of Nature (IUCN) 1994). These areas are renowned for their ability to act as refuges for species and ecological processes that cannot persist in intensely managed landscapes and seascapes, as well as for their capacity to enable natural evolution and future ecological restoration (Chape et al. 2008; Gaston et al. 2008; Dudley et al. 2010). Famous examples of protected areas include the Serengeti National Park in Tanzania, the Yosemite National Park in the USA, or the recently established Chagos marine reserve in the Indian Ocean, which contains the world's largest coral atoll. Aside from conserving biodiversity, protected areas can perform many other functions, such as protecting cultural heritage; maintaining vital ecosystem services; providing a range of socio-economic benefits; helping maintain microclimatic or climatic stability; or shielding human communities from natural disasters (Fiske 1992; IPCC 2007; Chape et al. 2008; Ezebilo and Mattsson 2010). Altogether, protected areas therefore constitute an important stock of natural, cultural, and social capital, yielding flows of economically valuable goods and services that benefit society, secure livelihoods, and contribute to the achievement of the Millennium Development Goals (Millennium Ecosystem Assessment Board 2005).

Protected areas can take many forms depending on their use and level of protection, ranging from conservation areas, national parks, game reserves, or forest reserves. The IUCN has classified protected areas into six categories. These are:

Ia: Strict Nature Reserves—managed mainly for science.
Ib: Wilderness Area—managed for wilderness protection.
II: National Park—managed mainly for ecosystem protection and recreation.
III: Natural Monument—managed mainly for conservation of specific features.
IV: Habitat/Species Management Area—managed mainly for conservation through specific intervention.
V: Protected Landscape/Seascape—managed mainly for landscape/seascape protection and recreation.
VI: Managed Resource Protected Area—managed mainly for sustainable use of natural ecosystem (IUCN 1994).

It is important to acknowledge that, although the IUCN has developed guidelines for the management of protected areas, individual countries have their own systems of protected area classification and management (IUCN 1994).

8.1.2 Identifying new protected areas

The creation of new protected areas is a key instrument in the battle to reduce biodiversity loss worldwide, and the Convention on Biological Diversity Programme of Work on Protected Areas is the accepted framework for creating comprehensive, effectively managed and sustainably funded national and regional protected area systems worldwide. Creating new protected areas ranks high in the 2010–2020 environmental agenda for many countries, with the parties of the CBD recently agreeing on specific goals in terms of protected area coverage. The Aichi Target 11 indeed states that 'By 2020, at least 17 per cent of terrestrial and inland water areas, and 10 per cent of coastal and marine areas, especially areas of particular importance for biodiversity and ecosystem services, are conserved through effectively and equitably managed, ecologically representative and well-connected systems of protected areas and other effective area-based conservation measures, and integrated into the wider landscapes and seascapes' (see also Chapter 10).

How do stakeholders decide where and how to set new protected areas? There are many factors that can be expected to influence the decision to establish a new protected area, such as the envisaged location, size, and shape; the level of biodiversity or endemism; the presence of threatened species; the establishment and management costs as well as the likely impact of climate change on some of these parameters (Brooks et al. 2006; McCarthy et al. 2006; Hannah et al. 2007; Gaston et al. 2008). Accessing reliable and comprehensive information about the distribution of biodiversity and the functioning of ecological systems at the scale required to inform the process of setting new protected areas can be extremely challenging, highlighting the importance of identifying practical, cost-effective tools to guide decision-making.

In some situations, the NDVI can be used to derive relevant information for the setting of new protected areas. Krishnaswamy et al. (2009) introduced a new multi-date NDVI-based Mahalanobis distance measure to index tree biodiversity and ecosystem services for the Western Ghats. This measure successfully quantified habitat and forest variability, with low values corresponding to moister, denser, more evergreen forest habitats with high evapotranspiration, and high carbon storage; higher values, on the other hand, corresponded to more open, dry deciduous, and scrub habitats with low evapotranspiration and lower carbon storage. This approach thus enabled the description of forest type and ecosystem services over large landscapes to be captured by a single continuous numerical scale. Such a tool could help stakeholders prioritize areas of high conservation value, thereby supporting the development of new protected areas in this region, and possibly elsewhere.

Another example is that of Singh and Milner-Gulland (2011), who discussed how to best develop a set of protected areas in Kazakhstan for the benefit of a threatened migratory species, namely the Saiga antelope *Saiga tatarica*. This species is found in central Asia, and has experienced a 95% reduction in population size over the last two decades (Milner-Gulland et al. 2001). The authors investigated the factors influencing the species' migration patterns in the considered area, and concluded that Saiga distribution in spring was determined by an intermediate range of temperature and intermediate primary productivity (as indexed by the NDVI), by the availability of areas at intermediate distance from water and away from human settlements. This enabled them to derive a habitat suitability map for the country and explore the current match between suitable habitats for Saiga and protected area distribution. They then explored the potential effect of climate change on temperature and primary productivity in the region, using recent predictions for the area (IPCC 2007; de Beurs et al. 2009; Zhao and Running 2010). From these predictions they derived spatially explicit information about how climate changes might affect the distribution of suitable habitats for the species. Thus they were able to assess the fit of the existing protected area network for Saiga antelope in the light of the predicted impact of climate change on the distribution of their suitable habitats, and to make recommendations for adapting the development of the network accordingly. The framework presented by Singh and Milner-Gulland illustrates well the capacity for the NDVI to support the identification of new protected areas, and is a step forward in terms of designing protected area networks that are robust to future changes in distributions and densities of key target species.

8.1.3 Monitoring protected area effectiveness

Over the last few decades, the number of protected areas worldwide has increased rapidly (Coad et al. 2009), yet this has not been followed by a reduction in the rate of biodiversity loss (Millennium Ecosystem Assessment Board 2005; Secretariat of the Convention on Biological Diversity 2010). Such an absence of correlation illustrates the point made by Chape et al. (2005), that is, that the number and extent of protected areas do not supply information on a key determinant for meeting global biodiversity targets, namely the 'effectiveness' of protected areas. Since protected areas are associated with one of the most significant resource allocations on the planet (Balmford et al. 2003b), the monitoring of their effectiveness (Box 8.1) is of key importance for making relevant management decisions in the face of future environmental change (Gaston et al. 2008).

Assessing protected area effectiveness relies on the evaluation of a series of criteria represented by carefully selected indicators (quantitative and qualitative) against agreed objectives or standards (Box 8.1). But what are, and what should be,

Box 8.1 Monitoring protected area effectiveness

Protected area monitoring is defined as 'collecting information on indicators repeatedly over time to discover trends in the status of the protected area and the activities and processes of management' (Hockings et al. 2006). The evaluation of effectiveness is generally achieved by the assessment of a series of criteria represented by carefully selected indicators (quantitative and qualitative) against agreed objectives or standards. Accordingly, Salzer and Salafsky (2003) distinguished between 'status assessment' and 'effectiveness measurement,' whereby status assessment indicates the existing condition of biodiversity at a particular point in time or over various points in time, whereas effectiveness measurement indicates whether conservation interventions are having their intended effect, i.e. links goals and objectives with activities, management processes and indicators used to measure progress toward achieving conservation goals and objectives (Stem et al. 2005).

The evaluation of protected area management effectiveness is generally undertaken for reasons such as: (i) promoting better protected area management; (ii) guiding project planning, resource allocation, and priority setting; (iii) maintaining accountability and transparency; and (iv) increasing community awareness, involvement and support (Chape et al. 2008). It can include evaluation of protected area design, adequacy of management systems and processes, and delivery of protected area objectives (Hockings et al. 2004, 2006). Unsurprisingly, effectiveness at addressing priorities in protected areas has been shown to be linked with availability of monitoring data (Timko and Innes 2009), while good monitoring was shown to correlate with overall effectiveness (Dudley et al. 2004).

Table 8.1 Chronological presentation of the definitions proposed to characterize the concept of ecological integrity.

Definitions	References
The sum of physical, chemical and biological integrity	Karr and Dudley (1981)
The capacity to support and maintain a balanced, integrated, adaptive biological system having the full range of elements and processes expected in the natural habitat of a region	Karr and Chu (1995)
A condition that is determined to be characteristic of its natural region and likely to persist, including abiotic components and the composition and abundance of native species and biological communities, rates of change and supporting processes	Parks Canada Agency (2000)
The ability of an ecological system to support and maintain a community of organisms that has species composition, diversity, and functional organization comparable to those of natural habitats within a region	Parrish et al. (2003)
Measures of representation and maintenance of key biodiversity features	Gaston et al. (2006, 2008)
A measure of the composition, structure, and function of an ecosystem in relation to the system's natural or historical range of variation, as well as perturbations caused by natural or anthropogenic agents of change	Tierney et al. (2009)

these agreed objectives and standards? Although protected areas have often been established with many goals in mind, most conservationists are likely to argue that the primary objective of existing protected areas should be to maintain ecological integrity (Table 8.1; Ervin 2003a,b; Dudley 2008). Ecological integrity assessment can involve quantifying changes in ecological processes and functioning (Parks Canada Agency 2005), as well as the evaluation of the threats and pressures faced by protected areas (Parrish et al. 2003; Parks Canada Agency 2005; Stem et al. 2005).

The idea that remote sensing information can represent a great addition to the monitoring toolkit for protected areas is not new, with various authors having recommended the use of satellite data for protected area monitoring and effectiveness assessment (e.g. Gillespie et al. 2008; Alcaraz-Segura et al. 2009; Nemani et al. 2009; Wiens et al. 2009; Kinyanjui 2011; Nagendra et al. 2013). We saw in Chapter 5 how the NDVI could be used to track the impact of global environmental change on vegetation dynamics in the African and Spanish protected area network. Making use of the NDVI time series to detect significant anomalies in vegetation dynamics and quantify changes in ecological processes and functioning was an approach also undertaken by the European Union, which funded the Assessment of African Protected Areas project <http://bioval.jrc.ec.eu/PA/> (Hartley et al. 2007).

There are, however, other relevant NDVI-based initiatives. For instance, Seiferling et al. (2012) investigated the links between protected area isolation, protected area status, and land use intensity. The study was motivated by the observations that (i) species extirpations and extinctions are occurring inside protected area borders (DeFries et al. 2005); and (ii) there seems to be a correlation between the failure to prevent species extirpations in protected areas and the degree of contrast between the patterns of vegetation cover in the protected area and in the matrix that surrounds it (Parks and Harcourt 2002; Newmark 2008). If this is correct, then measuring isolation (i.e. this degree of contrast between vegetation cover patterns inside and outside protected areas) could inform the assessment of protected area effectiveness, and identifying protected areas with a high level of isolation could help prioritize conservation efforts. To quantify isolation, the authors used the NDVI as a measure of the vegetation cover across the sampled landscapes and applied the contagion metric (for more information on this metric, see O'Neill et al. 1988; Li and Reynolds 1993; Proulx and Fahrig 2010) to NDVI values as a measure of the pattern of land-cover heterogeneity. Under such settings, higher values of contagion indicate less heterogeneous land cover. Only protected areas of IUCN categories I–V between 50 000 and 70 000 ha were considered. Protected areas with 10–20 km buffers that overlapped coastlines or included large water bodies were eliminated, as well as those that were part of reserve networks or that shared borders with other protected areas. Vegetation cover was consistently more heterogeneous outside protected areas than inside their borders, for the vast majority of the 114 protected areas considered. Protected areas characterized as more uniformly green inside their border were reported to have higher isolation values (Figure 8.1). The study also revealed a negative link between human activity and isolation, with protected areas in which low levels of human activity are permitted being more isolated than areas in which high levels are permitted. More research is probably needed to confirm this unexpected pattern, as well as some form of quantification between this NDVI-based metric and the likelihood of species extirpation at the protected area scale. Nevertheless, the study promotes an

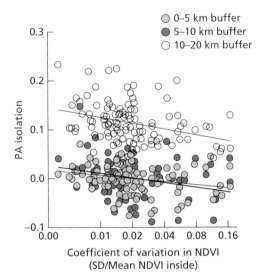

Figure 8.1 Relation between protected area (PA) isolation, measured as the difference between mean contagion inside and in three non-overlapping buffers outside each protected area (i.e. 0–5, 5–10 and 10–20 km buffers), and the coefficient of variation of the NDVI (CV-NDVI). (Reproduced from Seiferling et al. 2012 with permission from John Wiley & Sons.) Isolation measured as the difference between mean contagion inside the protected area and in the 10–20 km buffer is strongly positive and always higher than when measured as the difference between mean contagion inside the protected area and in the 0–5 km or 5–10 km buffers. Protected areas with a lower CV-NDVI ratio (i.e. more uniformly green) inside their border tended to have higher isolation values.

interesting concept for the development of NDVI-based tools for the assessment of protected area effectiveness.

8.2 NDVI for support of habitat assessments for reintroductions

Global biodiversity is under increasing threat from anthropogenic impacts, leading to an unprecedented rate of species loss (Chapter 1). A range of actions is available to relevant stakeholders to slow the rate of populations and species extinctions; these may include the creation and effective management of protected areas (Section 8.1.3), or the strengthening of the current legal framework supporting biodiversity conservation (Chapter 10). Another possible action consists in reintroducing species in areas where they have been extirpated or become extinct (Box 8.2). Reintroduction can support biodiversity

> **Box 8.2 From reintroductions to reinforcement and assisted translocations: some definitions**
>
> **Reintroduction:** This term describes the intentional movement and release of an organism inside its indigenous range from which it has disappeared (IUCN 2012). A reintroduction thus aims to re-establish a viable population of the focal species within its indigenous range. Re-establishment is sometimes used as a synonym of reintroduction, yet the term implies that the reintroduction has been successful.
>
> **Reinforcement:** This describes the intentional movement and release of an organism into an existing population of conspecifics. Reinforcement aims to enhance population viability, for instance by increasing population size, by increasing genetic diversity, or by increasing the representation of specific demographic groups or stages (IUCN 2012). Augmentation, supplementation, re-stocking, and enhancement (for plants only) are accepted synonyms of reinforcement.
>
> **Translocation:** The term refers to the human-mediated movement of living organisms from one area, with release in another. Translocations may move living organisms from the wild or from captive origins; they can be accidental (e.g. stowaways) or intentional. Intentional translocations can address a variety of motivations, including reduction in population size; for welfare, political, commercial, or recreational interests; or for conservation objectives (IUCN 2012).
>
> **Conservation translocation:** This refers to the intentional movement and release of a living organism where the primary objective is a conservation benefit: this will usually comprise improving the conservation status of the focal species locally or globally, and/or restoring natural ecosystem functions or processes. Conservation translocations can entail releases either within or outside the species' indigenous range. The indigenous range of a species is the known or inferred distribution generated from historical (written or verbal) records, or physical evidence of the species' occurrence. Where direct evidence is inadequate to confirm previous occupancy, the existence of suitable habitat within ecologically appropriate proximity to proven range may be taken as adequate evidence of previous occupation (IUCN 2012).
>
> **Conservation introduction:** This is the intentional movement and release of an organism outside its indigenous range (IUCN 2012). Two types of conservation introduction are recognized: assisted colonization and ecological replacement.
>
> **Assisted colonization:** This refers to the intentional movement and release of an organism outside its indigenous range to avoid extinction of populations of the focal species. This is carried out primarily where protection from current or likely future threats in current range is deemed less feasible than at alternative sites. The term includes a wide spectrum of operations, from those involving the movement of organisms into areas that are both far from current range and separated by non-habitat areas, to those involving small range extensions into contiguous areas (IUCN 2012). Benign introduction, assisted migration and managed relocation are accepted synonyms of assisted colonization.
>
> **Ecological replacement:** The term refers to the intentional movement and release of an organism outside its indigenous range to perform a specific ecological function.
> This is used to re-establish an ecological function lost through extinction, and will often involve the most suitable existing sub-species, or a close relative of the extinct species within the same genus (IUCN 2012). Taxon substitution, ecological substitutes/proxies/surrogates, and subspecific substitution are accepted synonyms of ecological replacement.

conservation efforts in several ways: first, it increases the number of populations for a given species, therefore reducing the overall extinction risk. Second, re-establishment can sometimes help to restore the functionality of the recipient ecosystem, by helping it return to its historic equilibrium—for example the re-establishment of gray wolves *Canis lupus* in Yellowstone National Park. The species was extirpated from the ecosystem in 1926, leading to growing populations of large herbivores. Increased herbivore pressure resulted in a loss of biodiversity through overgrazing, impacting vegetation structure, productivity, species composition, and habitat quality for other fauna (Estes et al. 2001; Ripple and Beschta 2012a). Gray wolf packs were reintroduced to Yellowstone National Park from 1995, and the growing wolf population soon started to impact the resident elk *Cervus elaphus* populations, both through direct predation and its effects on the species' life histories. The presence of wolves was indeed shown to alter elk movements, browsing patterns, and foraging behaviour, indirectly affecting the reproductive physiology and the demography of elk (Ripple et al. 2001; Creel et al. 2007). This in turn rapidly led to increased species composition in vegetation communities and increased sizes of various vertebrate populations, such as bison *Bison* spp., beaver *Castor Canadensis*, or coyote *Canis latrans* (Ripple et al. 2001; Ripple and Beschta 2004; Smith and Bangs 2009).

Successfully reintroducing a species in a given area is a difficult task, and the increasing number of reintroductions and translocations over past decades led to the establishment of the IUCN/SSC Species Survival Commission's Reintroduction Specialist Group in the 1990s. One of the first tasks for this group was to update the IUCN Position Statement on the Translocation of Living Organisms developed in 1987 (IUCN 1987). This task was motivated by the general consensus that more detailed guidelines were needed to provide a more comprehensive coverage of the various factors involved in reintroduction exercises. In 1998, these new IUCN guidelines were published, in the hope that these would help ensure that reintroductions achieve their intended conservation benefit, and do not cause adverse side-effects of greater impact (IUCN 1998). These guidelines were subsequently updated in 2012 (IUCN 2012).

The 1998 and 2012 guidelines provide a comprehensive set of recommendations intended to act as a guide for procedures useful to reintroduction projects. These include a set of pre-project activities to be carried out before reintroducing individuals in an area. One such activity relates to the importance of carrying out a detailed habitat assessment of the proposed site for reintroduction. As stated in the 1998 guidelines (p. 8), this is indeed necessary to ensure that 'the area [has] sufficient carrying capacity to sustain growth of the reintroduced population and support a viable [self-sustaining] population in the long run.' The practicalities of habitat assessments can be logistically challenging when dealing with large-bodied animals requiring wide spaces for their survival, such as large carnivores or megaherbivores. A typical cheetah *Acinonyx jubatus* home-range size, for example, reaches 800–1000 km^2 (Durant et al. 2010), which means that areas of sufficient carrying capacity to support a viable population would average thousands of square kilometres in size. Field-based assessments of potential reintroduction areas can therefore quickly become labour intensive and time-consuming. In this respect, the combined use of Earth observation data and geographic information systems can offer a standardized, low-cost approach for evaluating potential habitats for reintroductions at various spatial scales. For example, available NDVI timeseries can supply data on long-term trends in habitat degradation, or on temporal changes in primary productivity dynamics.

Up to now, few publications have made use of the NDVI to inform habitat assessment prior to reintroduction implementation. One of the exceptions is the recent work by Freemantle et al. (in press), who assessed the Ouadi Rimé–Ouadi Achim Faunal Reserve (OROAFR) in Chad as a potential reintroduction site for the Scimitar-horned oryx *Oryx dammah*. The species is currently classified as 'extinct in the wild' by the IUCN. The OROAFR was, in recent history, one of the stronghold areas for the species (Newby 1980), and recent habitat suitability modelling work has identified this faunal reserve as a good candidate for reintroduction. Using the NDVI data collected by the NOAA satellites and processed by the GIMMS group, Freemantle et al. showed that: (i) average annual primary productivity in the OROAFR has been increasing over the past three decades; (ii) this average trend was mainly driven by increasing primary productivity in the south of the reserve; (iii) the northern part of the reserve was associated with decreasing annual primary productivity over the period 1982–2008; and (iv) opposite trends are currently leading to an increased contrast in primary productivity dynamics between the north and the south of the reserve (see also Figure 8.2). These results therefore highlighted that changes in the spatial dynamics of primary productivity are occurring in OROAFR, changes that need to be taken into consideration when planning a possible reintroduction of oryx in the area: the sub-desert transition zone is indeed preferred by oryx, and this habitat is currently narrowing. This implies a potential reduction of favourable habitat for the species, which could have detrimental effects on the success of establishing a self-sustaining reintroduced population. The example illustrates how long-term, freely available remote-sensing information can support

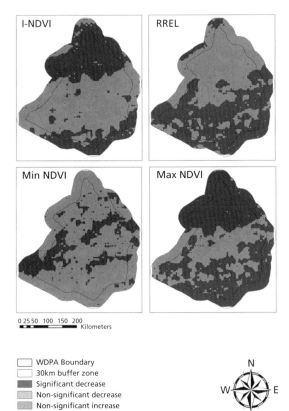

Figure 8.2 Trends in vegetation dynamics for the Ouadi Rimé–Ouadi Achim Faunal Reserve, Chad. (Reproduced from Freemantle et al. in press with permission from John Wiley & Sons.) The maps represent the spatio-temporal changes in the Integrated-annual NDVI (I-NDVI), Intra-annual Relative Range (RREL; an index of seasonality), and Maximum and Minimum NDVI (MAX and MIN) during the 1982–2008 period. See also Plate 10.

reintroduction efforts, by helping assess the state of, and trend in, suitable habitat availability.

8.3 Identifying corridors

8.3.1 What are corridors?

As acknowledged by the Millennium Ecosystem Assessment Board (2005; see also Chapter 1), habitat loss and fragmentation are among the main threats to biodiversity. These threats operate in multiple ways, and one pathway of concern is how habitat loss and fragmentation can lead to the isolation of small populations, which may worry wildlife managers. One concern is linked to the loss of evolutionary adaptability to environmental changes. In small populations, random fluctuation in gene frequencies (also referred to as random genetic drift) indeed tends to reduce genetic variation (Lande 1988). A second concern is linked to rapid population loss in historically large populations, which can lead to inbreeding depression and an increased likelihood of recessive deleterious mutations becoming fixed (Lande 1988). A third concern is then linked to what is known as the 'Allee effect,' which refers to loss of the benefits associated with the presence of conspecifics, and the resulting positive correlation between population density and per capita population growth rate in small populations (Stephens et al. 1999). A fourth concern is linked to the increasing importance of demographic stochasticity in determining extinction risk in small populations. Large populations are indeed assumed to be primarily influenced by one type of random demographic factor, namely environmental stochasticity; small populations, on the other hand, become vulnerable to both environmental stochasticity and demographic stochasticity, which increases their susceptibility to extinction (Lande 1988). Altogether, smaller populations are known to have higher extinction rates (Pimm et al. 1988).

The existence of a link between habitat loss and fragmentation and the appearance of isolated, small populations has motivated conservation biologists to discuss the actions that are needed to increase the effective size of local populations. One of the possible strategies has been the recommendation that corridors be included in conservation plans to increase the connectivity of otherwise isolated patches (Meffe and Carroll 1994). Corridors can be defined as strips of native vegetation or habitat connecting otherwise isolated remnants (Hobbs 1992), and are thought to facilitate movement between connected patches of habitat, thus increasing gene flow, promoting re-establishment of locally extinct populations, and increasing species diversity within otherwise isolated areas.

8.3.2 NDVI to help identify corridors

Because several NDVI-based metrics have successfully been linked to habitat suitability for many species (see examples in Chapter 7), and because the NDVI can help in some cases to identify strips of 'native' vegetation in a heterogeneous landscape (see examples in Chapters 5 and 6), this vegetation index is a valuable tool for identifying corridors at large spatial scales. For instance, Pittiglio et al. (2012) investigated the environmental and anthropogenic factors related directly and indirectly to the transit corridors of elephants *Loxodonta africana* in Tanzania. Elephants represent an ideal species for modelling transit corridors in savannah ecosystems, and the study system, the Tarangire–Manyara ecosystem, hosts the largest population of savannah elephant in northern Tanzania. To identify corridors, the authors started by identifying those variables that were positively correlated with elephant presence in the ecosystem over the period 1995–2004.

They found that in both seasons, elephant presence was: (a) negatively associated with distance from permanent drinking-water sources, distance index from protected and semi-protected areas, slope, and their topographic position index; and (b) positively associated with distance from villages, minor roads, monthly average NDVI, closed woody vegetation, and cultivated areas. These results were then used to identify transit corridors in the ecosystem. Most of the transit corridors identified matched migration routes described in the 1960s, suggesting that corridors in the area are temporally stable.

Another interesting example comes from the study by Gavashelishvili and Lukarevskiy (2008), who modelled the habitat of leopard *Panthera pardus* in west and central Asia, and analysed the connectivity between different known populations in

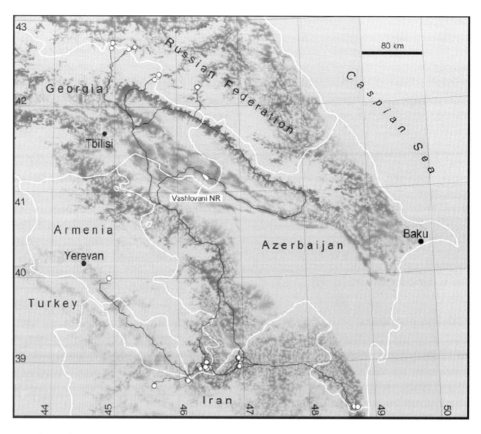

Figure 8.3 Map showing least-cost paths (black lines) computed between known leopard populations (hollow points) in the Caucasus. (Reproduced from Gavashelishvili and Lukarevskiy 2008 with permission from John Wiley & Sons.) The paths follow higher probability values of leopard habitat (darker areas).

the Caucasus. The species is classified as endangered in the region, and the idea of this work was to support conservation efforts with relevant information on leopard habitat requirements, to enable more effective management strategies to be implemented. The authors compiled information on the species distribution over the period 2001–2005 in Armenia, Azerbaijan, Georgia, Iran, Russian Federation, Turkey, and Turkmenistan. These presence/absence data were then used in conjunction with logistic regressions and information on habitat variables related to climate, terrain, land cover, and human disturbance to develop a spatially explicit model of leopard habitat suitability. Annual vegetation productivity (as indexed using the NDVI), days of snow cover per year, and distance to urban areas were the variables best explaining the distribution of leopard in west and central Asia. Having identified the model that had the best predictive power, the authors used the associated predictions to define least-cost paths among the known leopard populations. The least-cost path algorithm used in this case sought to link these populations by routes that followed higher probability values of leopard habitat (Figure 8.3). The approach allowed the identification of two major paths connecting the Greater Caucasus with Karabagh Mountains, which is part of the Lesser Caucasus. The study shows how remote sensing and vegetation indices such as the NDVI, used in combination with the latest developments in spatial analyses and modelling, can support conservation efforts worldwide by improving access to relevant information at relevant scales for implementation.

8.4 Predicting climate change effects on wildlife

Chapter 4 discussed the existing links between climatic conditions and NDVI dynamics; Chapter 7 reviewed examples of successful links established between NDVI and wildlife. From these bases, prediction of future changes in climatic conditions may enable the development of a set of realistic scenarios as to how such changes will impact NDVI and therefore wildlife.

Thus far, few correlative studies between the NDVI and wildlife have made this extra step of qualitatively linking climate change predictions with the NDVI and wildlife. Part of the explanation for such an absence of predictive approaches can be attributed to the relative paucity of studies addressing the links between changes in climate and changes in NDVI dynamics (but see, e.g., De La Maza et al. 2009; Larsen et al. 2011). Another element is linked to the constraints associated with predicting the effects of climate change on wildlife (Berteaux et al. 2006). There are indeed uncertainties associated with the climatic scenarios, which creates further uncertainties associated with predicting the impact of climate change on NDVI dynamics and wildlife. Global warming may also result in future climatic conditions being out of the range of the previous conditions observed in the area considered, so that forecasting without extrapolation may be impossible. Primary productivity represents only one parameter in the set of parameters defining the link between an animal and its habitat. The strength of the link between this parameter and the species under study can be expected to be temporally variable, as other variables influencing habitat suitability change. Therefore, although climate change might not significantly affect NDVI dynamics in certain areas (for example, leading to the prediction of species X inhabiting an area Y might not be severely impacted by future changes in climatic conditions), its effects on variables such as plant community composition might be more important for defining habitat suitability for a given species.

However, there are exceptions. For instance, Singh and Milner-Gulland (2011; see Section 8.1.2) developed a set of predictions regarding the impact of climate change on NDVI dynamics in central Asia to assess how such changes might affect Saiga antelope distributions in the future. The authors were then in a good position to discuss the fit of the existing protected area network for Saiga in light of predicted changes in climatic conditions, and to make recommendations regarding the management of the protected area network in the region.

Hu and Jiang (2011) have also developed a habitat suitability model to explore how the distribution of Przewalski's gazelle *Procapra przewalskii*, an endangered ungulate whose historical and current ranges encompass Mongolia and China, may be impacted by future climate change. The authors considered nineteen environmental predictors across four types of data, namely (i) climatic data; (ii) habitats, which included a land-cover layer and the NDVI for April, May, July, and August; (iii) some human influence index; and (iv) information on topography and elevation. Using a set of occurrence data collected over the period 2002–2008, they then modelled the current suitability of habitat for the species. The best-fit model was thereafter used to investigate the impact of climate change on Przewalski's gazelle habitat suitability, by replacing the current values of the retained climate predictors with the values for these predictors extracted from three global climate models under two greenhouse gas emission scenarios selected for the years 2020, 2050, and 2080. The authors in this case did not evaluate how changes in climatic conditions might impact the NDVI, and subsequently habitat suitability for gazelles. Instead, they used extrapolations of past trends in all non-climatic variables as estimators of the future. This approach might yield a conservative method of predicting changes in NDVI and its impact on wildlife on the medium term, and may promote the emergence of more studies on how climate change will affect wildlife.

8.5 NDVI and invasive species

8.5.1 The pertinence of NDVI to tracking invasive species

The human-mediated introduction of plants and animals to areas beyond their natural geographic range limits is a major driver of biodiversity loss worldwide (Vitousek et al. 1996; Millennium Ecosystem Assessment Board 2005). It has been so pervasive that almost every environment worldwide now houses these 'alien' species (Box 8.3). Invasive

---

Box 8.3 From introduced species to invasive species and invasive alien species—definitions

Introduced species: An introduced species (which can also be referred to as an alien species, an exotic species, a non-indigenous species or a non-native species) is a species living outside its native distributional range, which has arrived there by human activity, either deliberate or accidental. Examples of introduced plants include peaches *Prunus persica* (which originated from China, but can now be found in many parts of the world), tomatoes *Solanum lycopersicum* (which are native to the Andes) or tobacco *Nicotiana tabacum* (which is native to the Americas, but is now commonly found across Europe). Examples of introduced animals include the grey squirrel *Sciurus carolinensis* in Europe, the red fox *Vulpes vulpes* in Australia, or the Common Brushtail Possum *Trichosurus vulpecula* in New Zealand (Owen and Norton 1995; Kinnear et al. 2002; Genovesi 2005; Davis 2009).

Invasive alien species: This is an introduced species whose introduction causes, or is likely to cause, economic or environmental harm, or harm to human, animal, or plant health. Famous invasive alien species include the black rat *Rattus rattus*, which originated from tropical Asia, but spread in just about every region of the world, heavily impacting bird, reptile, and other small vertebrate communities where introduced (see, e.g., Jones et al. 2008). Another example of a famous invasion is that of Guam by the Brown Tree snake *Boiga irregularis*: the species is thought to have been introduced accidentally in imported cargo (Colvin et al. 2005) and was first detected on the island in the 1950s. In addition to the dramatic ecological consequences of its spread, the invasion of the island by the snake has been associated with non-negligible societal and financial consequences: the species causes frequent power cuts by climbing on electrical wires and wrapping itself around the power lines.

Invasive species: Although invasive species are typically exotic species that have been introduced by humans in new areas, some native species can, under certain circumstances, increase in number and range, and become invasive, threatening biological diversity. These species are sometimes referred to as natural invaders, for example the genus *Prosopis*, a highly invasive group of plants comprising 44 species of trees and shrubs. Advancements in invasions for species in this genus have been reported to occur as 'bursts', in response to highly favourable but irregular climatic events such as periods of exceptional rainfall and floods (see Wise et al. 2012). A distinction can thus be made between invasive species and invasive alien species, whereby invasive species refers to any species (native or introduced) causing, or likely to cause, economic or environmental harm, or harm to human, animal, or plant health.

species eradication is extremely challenging: preventing new invasions is therefore essential, being probably the most efficient way to minimize their impact on global biodiversity, human health, and the success of human economic and agricultural enterprises. Yet the prevention of new invasions requires that the factors allowing certain species to establish and spread into foreign environments be understood, so that situations where invasion risk is high can be identified (Davis 2009).

Invasion success is a sequential process (Figure 8.4), requiring a species to be transported outside its native range; to be introduced (released or escaped) into the novel environment; to establish a viable exotic population; and finally to spread away from the location of establishment (Blackburn et al. 2011). The first two stages are largely driven by the availability of species for transport and introduction, but whether or not a species establishes and spreads depends on a complex interaction between the characteristics of the location of introduction, the characteristics of the species introduced, and details of the introduction event (Duncan et al. 2003).

Many species are tightly linked to NDVI variability (Chapters 6 and 7), and studies have revealed that the match between a species' native environment and the environment the introduced individuals experience at the new location can explain both establishment success and the extent of subsequent spread (Duncan et al. 2001). Primary productivity dynamics in sites of introduction may thus be an important parameter when it comes to determining the establishment success of an introduced species. Likewise, for invasive plant species with phenology distinct from the phenology of native species, the NDVI might help to detect the presence of an invasive species and track the evolution of its distribution.

8.5.2 The NDVI to track invasive species distribution

Knowing the exact patterns in the spatial distribution of invasive species is a prerequisite to the design of any successful management action aiming to mitigate their impacts on biodiversity and economies. With regard to tracking invasive plant species distribution using the NDVI, the rationale is relatively simple: if the ecology and/or the leaf phenology of the invasive plant are distinct from those of the native species, then temporal opportunities exist to reliably detect the presence of these invasive plants at large spatial scales.

Several studies have now successfully implemented methods based on such a rationale: for example, the Amur honeysuckle *Lonicera maackii* is an Asian shrub that invades North American forests, expands leaves earlier and retains leaves later than native woody species. Using the spectral information collected by Landsat 5 TM and Landsat 7 ETM+, Wilfong et al. (2009) explored the possibility of predicting Amur honeysuckle cover from vegetation indices in woodlots in south-western Ohio and eastern Indiana. They evaluated the predictive

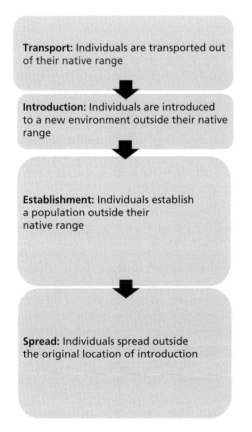

Figure 8.4 Invasion process. This can be divided into a series of stages (transport, introduction, establishment, and spread), each associated with barriers that need to be overcome for a species or population to pass on to the next stage. This figure is a simplified representation of the framework introduced by Blackburn et al. (2011).

abilities of six spectral vegetation indices and concluded that the NDVI was the best-performing predictor variable. With a fit of up to 75%, the authors concluded that, with refinement, this approach could help to map current and past understory invasion by Amur honeysuckle. Tamarisk *Tamarix* spp. is another Asian tree/shrub species that is invading riparian zones in the USA, being described as a 'high-priority' invasive species (Morisette et al. 2006, Evangelista et al. 2009). As for Amur honeysuckle, the NDVI was key to map areas likely to be invaded by tamarisk in continental USA. Using thousands of field-sampling points collected and co-ordinated through the USGS, and nationwide environmental data layers derived from MODIS, Morisette et al. (2006) were able to generate a map for tamarisk that is 90% accurate. In this case, the usefulness of the NDVI was highlighted by exploratory data analysis, which revealed that locations with known tamarisk presence showed much less absolute difference between the range in the NDVI and the range in the EVI (see Section 2.2.3) than did areas without tamarisk. Known tamarisk locations tended to fall along the line where the seasonal variability in the NDVI was equal to the seasonal variability in the EVI, whereas locations without tamarisk tended to fall off this line. As the difference between the EVI and NDVI stems from an adjustment for the atmosphere and soil background (Chapter 2), the authors hypothesized that the trend for tamarisk to grow along the one-to-one line was due to the soil. Tamarisk is a species that tends to spread quickly and is thick enough to cover most soil; once present, such a species therefore reduces or blocks any signal from the soil. Conversely, non-tamarisk locations in riparian areas could be expected to have either bright sandy or dark wet soils. Such differences in cover between locations with and without tamarisk would show up as differences in the range in the EVI and the NDVI, explaining the reported patterns.

Case studies outside the USA do exist: one of the first applications of the NDVI to map invasive plant species originates from India, where Venugopal (1998) used NDVI data derived from SPOT to monitor the infestation of *Eichhornia crassipes* (water hyacinths) in Bangalore. More recently, Hoyos et al. (2010) used the NDVI to map glossy privet *Ligustrum lucidum* invasion in Argentina. Glossy privets are the most widespread invasive trees in central Argentina, and were originally imported from China as an ornamental. The trees of this species grow fast, and can reach heights of up to 17 m. Glossy privets can thrive under both shaded and full-sun conditions, produce abundant seed that are dispersed by birds, and can propagate vegetatively. They are thus aggressive invaders, with traits allowing them to colonize relatively undisturbed native forests and outcompete most of the native vegetation. To inform the delimitation of the areas invaded by glossy privets in the region of interest (namely the Sierras Chicas of Cordoba), the authors used the spectral information collected by Landsat TM and ETM+. Glossy privet-dominated stands are known to differ substantially in structural and physical characteristics from native forest stands, with invaded canopies being more dense and close, resulting in higher absorption in the visible domain and higher reflection in the near-infrared domain. Because of this, the authors decided to calculate the NDVI for each scene, to guide the visual interpretation of the images. They then combined this information with the use of a support vector machine to map the distribution of the invaders. The reported reliability of the approach is particularly impressive in this example, as the method was associated with an overall accuracy of 89%.

8.5.3 The NDVI as a predictor of invasion

The invasion of ecosystems by non-native species has been recognized as a major driver of biodiversity loss worldwide, and, as stated earlier, extirpation is a challenging and costly process. The prevention of further invasions is thus a key component of the successful protection of the remaining native biodiversity: yet to successfully prevent these new invasions, one needs, among other things, to identify and actively protect areas at high risk of future invasion (Hobbs and Humphries 1995).

Previous work has demonstrated that remotely sensed data can improve data-collection capacity and increase the accuracy of predictive spatial models of invasion (Cohen and Goward 2004). The NDVI has enabled identification of areas at high risk of future invasion (Bradley and Mustard 2006).

Previous work by Peterson (2005) showed that successful remote detection of cheatgrass *Bromus tectorum*, an invasive plant species found in the Great Basin, USA, could be achieved using the NDVI, due to the species' early growth relative to native shrubs and grasses. Using this reliable detection method, Bradley and Mustard examined the factors positively correlated with temporal changes in cheatgrass distribution over the period 1973 to 2001. Their results highlighted strongly how land use can increase the risk of cheatgrass invasion. In 2001, cheatgrass was indeed 20% more likely to be found within 3 km of cultivation, 13% more likely to be found within 700 m of a road, and 15% more likely to be found within 1 km of a power line. Having established the links between invasion probability and landscape features, the authors were able to create a risk map of future cheatgrass invasion to support future land management decisions.

So far, all the studies illustrating the use of the NDVI to track and predict invasions have focused on plants, yet the NDVI can also be used to predict the spread and potential geographical distribution of invasive animal species. Roura-Pascual et al. (2004), for example, used the NDVI to explore the potential distributional expansion of Argentine ants *Linepithema humile* with warming climates. Native to central South America, Argentine ants are now found in many Mediterranean and subtropical climates around the world, and the species has been designated as one of the world's worst invasive alien species. To predict changes in the ants' distribution associated with expected changes in climatic conditions, the authors assembled a dataset of more than 1000 ant occurrences, and, by using it in combination with relevant environmental information and ecological niche models, predicted the species' overall range, including both the native distributional area and invaded areas worldwide. When a highly predictive model had been developed (i.e. once the authors possessed a model able to accurately predict the current range of the species), the best model was used in conjunction with future climate change scenarios to explore the potential for the species to expand its distributional range with climate change. The best model included information on primary productivity as indexed by the NDVI, and was associated with an overall accuracy varying between 87% and 89%. Nevertheless, as NDVI projections associated with climate change scenarios are not currently available, the authors had to use a sub-optimal model (based on climate only) to assess potential expansions in the distribution range of the species.

8.6 Conclusions

Conservation science aims to support efforts to maintain biological diversity and the delivery of ecosystem services. Conservation science and the effective implementation of conservation actions mostly require reliable, continuous, long-term data to increase our scientific knowledge, but also to assess the outcomes of conservation actions. Such data may be rare—for example, monitoring efforts generally constrained by the short time-frame of many funding streams. In this context, remote sensing is a source of information that conservation

Table 8.2 Studies having successfully used the NDVI to support species conservation efforts.

Aim	Focal species	Location	Reference
Inform the setting of new protected areas	Saiga antelope	Kazakhstan	Singh and Milner-Gulland 2011
Habitat assessment to inform reintroductions	Scimitar oryx	Chad	Freemantle et al. in press
Identify transit corridors	African elephant	Tanzania	Pittiglio et al. 2012
Assess population connectivity	Leopard	West and Central Asia	Gavashelishvili and Lukarevskiy 2008
Assess the likely impact of climate change on the distribution of an endangered species	Przewalski's gazelle	Mongolia and China	Hu and Jiang 2011
Track invasive species distribution	Amur honeysuckle	USA	Wilfong et al. 2009
Identify areas at risk of invasion	Argentine ants	Worldwide	Roura-Pascual et al. 2004

scientists and conservation practitioners cannot afford to disregard. The examples in this chapter underline the potential for the NDVI to inform conservation science and support conservation efforts worldwide (Table 8.2). From identifying new protected areas, to informing reintroduction planning, and to predicting potential invasions, the NDVI provides key information for improving the design and implementation of mitigation and adaptation tools to reduce the current rate of biodiversity loss. Much more could be achieved by developing quantitative predictions on the effects of changes in climatic conditions on NDVI distribution and dynamics under the current set of climate change scenarios. This much-needed information is essential for increasing our ability to efficiently counteract the impact of global environmental change on biological diversity.

CHAPTER 9

NDVI falls down: exploring situations where it does not work

> Science, my lad, is made up of mistakes, but they are mistakes which it is useful to make, because they lead little by little to the truth.
> **Jules Verne**

So far, I have reviewed many examples where the NDVI has been successfully used to support environmental and wildlife management. Chapters 5–8 are replete with studies in which the NDVI has been successfully linked to the distribution and functioning of ecosystems, as well as to the distribution and performance of plants and animals. Yet there are also several studies challenging the usefulness of this vegetation index for ecological research (Table 9.1), and for environmental management in general. Such an accumulation of studies could be seen as a deterrent to using the NDVI—hence this chapter, which is dedicated to these failures.

Failures to detect a significant effect of a parameter on a variable can be as important as detecting a significant impact. This chapter reviews known cases in which the NDVI could not be successfully linked to the parameter of interest and relates this information to the mechanisms shaping potential limitations. This may enable us to identify situations where its usefulness might be reduced.

9.1 Diet matters

9.1.1 Herbivores

As we saw in Chapter 7, the link between the NDVI and herbivores relies partially on the assumption that the NDVI correctly indexes food availability for these species. Yet, depending on the diet, the season, and the location, the reliability of this assumption may vary. For example, Pettorelli et al. (2009) assessed the correlation between the NDVI and ungulate abundance in Africa, finding that density estimates were more strongly related to NDVI variation for some species than for others. Because the interaction between species identity and NDVI and the interaction between diet and NDVI were not significant, the case was not discussed further. Yet it is possible to imagine why some species would display a weaker link with the NDVI than others: browsers are indeed very selective species while grazers tend to be more generalist, processing a larger amount of resources of more variable quality. Therefore, a high NDVI value might not systematically be associated with high forage availability for highly selective species that depend on particular food resources, whereas it would be more likely to reflect better forage conditions for less selective species (Figure 9.1). Simply put, it can be easily hypothesized that the NDVI might have tended to better index forage availability for grazers than for browsers for the African national parks considered in this study.

Another example of the importance of the diet, and, in this case, of the season in determining the link between the NDVI and herbivores, comes from Kenya, where Bhola et al. (2012) examined the link between NDVI variability and distribution of ten mammalian herbivore species with different body

Table 9.1 Examples where no significant relationship between the NDVI and plant or animal parameters could be detected.

Putative link	Result	Taxa/species	References
Environmental heterogeneity as a predictor of species richness	No measure of heterogeneity was a better predictor of local richness than mean pH within plots	Plants	Costanza et al. 2011
Environmental heterogeneity as a predictor of species richness	No direct effect of NDVI on species richness	Plants	Pau et al. 2012
Environmental productivity as a predictor of species richness	Summer temperature was consistently the strongest predictor of observed species richness, by comparison with the NDVI	Birds	Evans et al. 2008
Plant phenology as a determinant of the amount of fat deposed	Climatic variables describing spring conditions performed better than plant phenology variables	Raccoon	Melis et al. 2010
Primary productivity as a predictor of dietary composition and trophic diversity	Diet was not related to habitat productivity here but proved to be most significantly related to mean temperature	Martens	Zhou et al. 2011
Primary production as a predictor of abundance	Winter severity is driving population dynamics	Soay sheep	Berryman and Lima 2006
Primary production as a predictor of density	The link between animal density and primary production was a function of the NDVI spatial resolution	Topi	Bro-Jørgensen et al. 2008
Primary production as a predictor of secondary productivity	NDVI-I was not related with secondary productivity, whereas December and annual maximum NDVI were positively related with secondary productivity	Sheep	Posse and Cingolani 2004
Primary production as a predictor of body mass	Not all populations examined displayed a link between body mass and NDVI	Red deer	Martinez-Jauregui et al. 2009;
		Roe deer	Pettorelli et al. 2006
Primary production as a predictor of body mass	No link between NDVI and body mass	Wild boar	Mysterud et al. 2007
Primary production as a predictor of distribution	Negligible effect of NDVI in predicting species presence	Iberian spider	Jimenez-Valverde and Lobo 2006
Primary production as a predictor of distribution	Climatic variables generally better predictors of tick distributions than NDVI	African ticks	Cumming 2002
Primary production as a predictor of distribution	NDVI not a significant predictor of occurrence	Leopard, African civet, Caracal, Marsh mongoose, Gambian mongoose, side-striped jackal	Burton et al. 2012
Primary production as a predictor of distribution	Presence not significantly related to NDVI	Gemsbok, eland, wildebeest, springbok; birds of the genus *Grallaria*, giraffes	Verlinden and Masogo 1997; Parra et al. 2004; Bhola et al. 2012
Primary production as a predictor of distribution	NDVI was not a predictor of range size or chosen patch	African buffalo	Ryan et al. 2006
Primary production as a predictor of distribution	NDVI patterns were linked to herd movements at large spatial scales but space use at smaller scales could not be predicted by the vegetation index	African buffalo	Cornélis et al. 2011

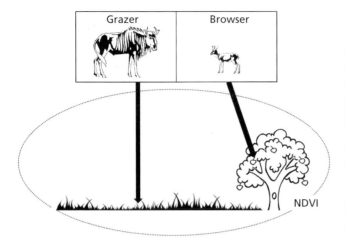

Figure 9.1 The link between the NDVI, resource availability, and diet. The NDVI is determined by the abundance of green biomass, which encompasses grasses and leaves. The abundance of fruits, consumed by duikers (the browser species in this figure), is not directly related to NDVI values (as more fruits does not mean higher photosynthetic capacity and higher NDVI values). In shrublands, the NDVI is thus likely to provide a more reliable index of resource availability for grazers (such as the wildebeest shown in this figure) than for browsers such as duikers.

sizes and feeding styles. In the Mara reserve, the authors expected that in the late wet season, grass would be tall and dense, low in crude protein and high in lignin, and hence of low digestibility. Furthermore, the tall grass was expected to heighten predation risk for herbivores. Thus they assumed that high NDVI values would be associated with taller, more mature, and less nutritious grasses, leading to the prediction of a negative association between the NDVI and hotspots of small and medium herbivores but a positive association with hotspots of large herbivores in the reserve. By contrast, during the dry season vegetation quality and quantity are known to be lower in the region, leading to the expectation that herbivore density would increase with increasing NDVI during this season. As expected, during the wet season hotspots of two species of the small (Thomson's gazelle *Eudorcas thomsonii*, impala *Aepyceros melampus*) and three of the medium (topi *Damaliscus lunatus*, wildebeest *Connochaetes taurinus* and zebra *Equus quagga*) herbivores occurred in areas of low NDVI, while the distribution of hotspots of the large herbivores (buffalo *Syncerus caffer*, elephant *Loxodonta africana*) peaked in areas of high NDVI values. Contrary to expectations, however, hotspots of Grant's gazelle *Nanger granti* and hartebeest *Alcelaphus buselaphus* were unrelated to the NDVI. Also in contrast to expectations, Thomson's gazelles and impalas concentrated in areas of low NDVI values during the dry season, but Grant's gazelles, wildebeest, buffalo, and zebra were all reported to occur in areas of high NDVI. No correlation was established between hotspots of both topi and hartebeest and the NDVI during the dry season, a result interpreted by the authors as topi and hartebeest being likely to feed on short and dry grasses that are not reflected by the NDVI, or displacement from open habitats by migrants. Likewise, giraffe *Giraffa camelopardalis* and elephant, both of which browse on woody plants, showed no relationship with dry season NDVI and concentrated in riparian woodlands. Bhola et al.'s study confirms how diet can strongly influence the link between NDVI variability and wildlife distribution, and how such an effect can be modulated by factors such as season or scale.

9.1.2 Carnivores and omnivores

The relationship between the NDVI and animal distribution has only recently been assessed for carnivorous and omnivorous animals, and in many situations this led to successful applications of the vegetation index. Wiegand et al. (2008) reported that brown bears *Ursus arctos* selected areas with specific NDVI characteristics, and Herfindal et al. (2005) showed that the productivity of a given study site had a negative relationship with lynx *Lynx lynx* home-range size. Basille et al. (2009) moreover demonstrated that the preferred habitat of the lynx includes high plant productivity (as indexed by the NDVI) areas. Recent work on vervet monkey *Chlorocebus pygerythrus* home range location suggests that the animals prefer areas with elevated productivity and reduced seasonality as indexed by the NDVI, with monthly NDVI values correlating with field measurements of leaf cover and food availability for this species (Willems et al. 2009). The potential for satellite-based indices of primary productivity to explain inter-population variation in home-range sizes of carnivore species was investigated by Nilsen et al. (2005), who found that the explanatory power of the considered index varied between 16% and 71% (n = 12 species considered).

Despite these success stories, NDVI usefulness has been challenged with respect to predicting the distribution and abundance of carnivores and omnivores. The hypothesis behind the expected link between the NDVI and omnivore parameters (such as body mass, survival, or abundance) is that the NDVI can supply information on resource availability for omnivores, either directly through its link with primary productivity, or indirectly through the expected preference by potential prey for areas with higher productivity (Figure 9.2). Since NDVI is expected to correlate positively with both prey abundance and vegetal resource availability, one could assume that omnivores should respond quite strongly to NDVI variability. Yet habitat productivity was not significantly related to wild boar *Sus scrofa* numbers in Eurasia (Melis et al. 2006). Similarly, in Poland, the NDVI was not successfully linked to wild boar annual body mass (Mysterud et al. 2007). In Finland, climatic variables describing spring conditions performed better than plant phenology variables in explaining the variation in raccoon dog *Nyctereutes procyonoides* fat deposition (Melis et al. 2010). These studies all suggest that NDVI-based information, effective in explaining life-history traits in herbivores, might not necessarily reflect variation in food abundance and quality for omnivore species. The number of studies successfully linking the NDVI to mammalian omnivores is currently lower than the studies failing to report such a link. However, there are exceptions (Chapter 7), indicating that the optimum conditions for the NDVI to yield relevant information on resource availability for omnivores need to be further researched. In this respect, a recent review by Bojarska and Selva (2012) points out that brown bear populations in locations with relatively low productivity (as assessed using the NDVI) consumed significantly more vertebrates, fewer invertebrates, and less mast (mast referring here to foods that are produced in natural habitats from trees, shrubs, and other plants) than populations inhabiting more productive habitats. This may suggest that the temporal link between NDVI dynamics and performance in omnivores might be mediated by the average level of productivity of the location considered.

If the distribution and performance of herbivores can be linked to NDVI dynamics, and if areas most favourable to herbivores can be identified, then we can assume that these areas might also be favoured by carnivores. As we saw earlier, these assumptions have been validated for some species and some locations, but there are exceptions. For instance, habitat productivity was not reported to significantly affect red fox *Vulpes vulpes* numbers in Eurasia (Barton and Zalewski 2007). While analysing the links between vegetation density, the presence of drinking water, and lion *Panthera leo* kills at the Klaserie Private Nature Reserve, South Africa, de Boer et al. (2010) highlighted that the presence of water, rather than vegetation density in riverine areas (as indexed using the NDVI), increased predation risk for

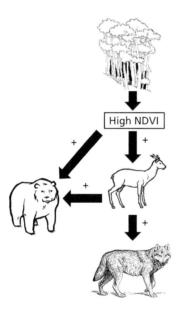

Figure 9.2 The NDVI and its expected effects on herbivores, omnivores, and carnivores. Large herbivores are expected to perform better in more productive areas. Resource availability, both in terms of prey and vegetal biomass, is therefore expected to be higher in more productive areas, which should benefit omnivores and carnivores.

water-dependent prey species, such as African buffalo, kudu *Tragelaphus strepsiceros*, and wildebeest. A global investigation of the biogeographical patterns in dietary composition and trophic diversity across the genus *Martes* in relation to geographical range and environmental variables failed to relate marten diet and habitat productivity (Zhou et al. 2011). Such an absence of correlation was interpreted as NDVI variability being unable to reflect marten food supply accurately. Zhou et al.'s results were also coherent with the outcome of a previous analysis carried out by Nilsen et al. (2005), which failed to identify a relationship between home-range size of the American marten *Martes Americana* and habitat productivity as indexed by the NDVI. The number of studies on links between the NDVI and carnivore parameters is far smaller than the number on the NDVI and herbivores; therefore much is still to be uncovered to develop a good understanding of the situations where the NDVI can be used to derive reliable information about carnivore distribution and ecology.

9.1.3 Parasites

Many studies have reported a link between NDVI-based metrics and the distribution and abundance of ectoparasites (such as ticks and mosquitoes) and endoparasites (such as flatworms; see Chapters 5 and 7). Yet not all studies agree on the importance of vegetation dynamics as assessed by the NDVI in shaping the distribution and dynamics of these parasites. For example, in the Nile Delta, the NDVI did not significantly correlate with schistosoma prevalence in the 41 sites surveyed by Malone et al. (1997). Multiple regression approaches showed that the NDVI was not significantly associated with anopheline mosquito densities in Eritrea (Shililu et al. 2003). When comparing the abilities of climatic variables, soil physical attributes, and monthly NDVI to define the geographical distribution and environmental niche of the spinose ear tick *Otobius megnini* in both tropical and neotropical regions, Estrada-Pena et al. (2010) reported that the best predictive values were obtained from ground-derived climate, while air-derived features ranked second; the remaining environmental information had poor discriminatory abilities. Benavides et al. (2012) found that home range productivity as assessed by the NDVI failed to explain gastrointestinal parasite richness in a terrestrial subtropical vertebrate (chacma baboon, *Papio ursinus*), whereas longer daily travel distances were associated with higher host parasite richness. This lack of association was thought to support the idea that movement patterns within a relatively stable home range, rather than variation in the home range area itself, determine parasite exposure and subsequent infection. What can be learned from the studies failing to report a link between NDVI-based metrics and the distribution of parasites? Host behaviour, parasite ecology, the level of spatial heterogeneity in vegetation distribution, and the scale at which the study is carried out seem to be of paramount importance in shaping the relationship between the NDVI and parasites. Apparently the NDVI is not always key in predicting disease outbreaks, but the large number of successful studies suggests that expecting this variable to shape disease outbreak risk is probably a good bet.

9.2 Scale matters

In ecology, scale is known to be a key determinant of the patterns and processes observed, as well as being key in shaping data availability and data collection protocols (Wiens 1989; Levin 1992; Cumming et al. 2006). Unsurprisingly, one of the challenges associated with coupling plant or animal data with NDVI data is to define the timing as well as the size and shape of the area relevant to the plant or animal variable considered. Uncertainties when estimating these parameters can potentially weaken the link between datasets, ultimately leading to NDVI-based metrics being poorly related to plant or animal variables (Figure 9.3). This section discusses how the spatial and temporal resolutions of NDVI data, and the potential resulting spatio-temporal mismatches between NDVI data and wildlife data, can lead to the index being poorly related to plant and animal distribution and ecology. In this context, spatial mismatches refer to situations where there is no appropriate alignment between (i) the spatial resolution or scale of the NDVI data; and (ii) the spatial resolution or scale of the biological data. Temporal mismatches refer to situations where there is no appropriate alignment between (i) the temporal

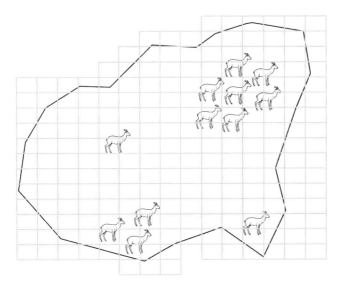

Figure 9.3 Linking animal data with NDVI data. Imagine a protected area (delimited here with a black line) where ungulate numbers are estimated each year. Each grey rectangle represents an NDVI pixel. Imagine that only count data are available—with no information on the spatial distribution of the ungulates. To link the NDVI to population counts, one would need to assess the average NDVI profile of the whole protected area and use this information to explore a potential link between yearly abundance and primary production dynamics as estimated by the NDVI. Yet one can imagine that ungulates do not use the whole protected area (as represented here), meaning that only a fraction of these pixels are associated with relevant information for ungulate dynamics.

resolution or scale of the NDVI data; and (ii) the temporal resolution or scale of the biological data.

9.2.1 Spatio-temporal mismatch and NDVI–wildlife links

Available NDVI datasets have fixed temporal and spatial resolutions (Chapters 1–3). For instance, many NDVI datasets yield NDVI information at a 10–16-day temporal resolution, and at a 250 m to 8 km spatial resolution. Yet biological systems function at scales and paces which are not systematically in line with these resolutions, and thus spatio-temporal resolutions and scales that might be aimed for when collecting biological data are rarely in line with available spatio-temporal resolutions of remote sensing imagery. These misalignments can seriously impair the links that may exist between NDVI variability and wildlife.

An illustration of this issue comes from Africa, where the relationship between NDVI variability and elephant population density in protected areas was recently explored (Duffy and Pettorelli 2012). In this example, elephant data come from the most recent African Elephant Status Report (Blanc et al. 2007), which includes estimates of elephant population numbers, survey locations, survey area sizes and types, as well as a reliability rating for each population estimate. Elephant numbers are provided for each protected area, but in each case the report contains no information about: (i) whether the whole protected area or only a fraction of it was surveyed (the total elephant number may therefore be a derived estimate); (ii) whether this elephant population uses the whole protected area or not; and, if it does not, which part it preferentially uses. Thus, the authors were forced to assess the spatially averaged NDVI profile of each protected area and use this information to explore a potential link between elephant density and primary production as estimated by the NDVI. In accordance with their expectations, the authors found a significant relationship between primary productivity and savannah elephant density estimates (Figure 9.4). Although significant, this association was weak, and several reasons for this were put forward. It was acknowledged that other factors, such as soil nutrient status, poaching and hunting rates, artificial fencing, water availability, habitat structure, and agricultural practices were not taken into account, despite being known to affect elephant densities. One important conclusion drawn by the authors was that the strength of the relationship reported between the NDVI and elephant density might have been impacted by the nature of the data available, which is likely to have caused some spatio-temporal mismatch between the measure of primary productivity and elephant densities. Available protected area shapefiles, for example,

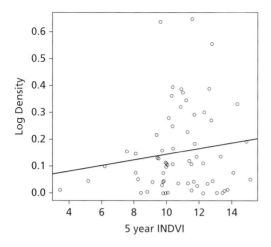

Figure 9.4 Population densities of African savannah elephants as a function of the five-year average annual INDVI. (Reproduced from Duffy and Pettorelli 2012 with permission from John Wiley & Sons.) The line represents the estimated $\log_{10}$ values of the population densities for the range of primary productivity considered, under a simple linear model linking INDVI and the $\log_{10}$ values of elephant population density.

were likely to have been different in shape and size compared with the actual survey areas listed in the African Elephant Status Report. In addition, as the survey data consisted of one-off population estimates, the season in which the count was taken could have influenced the distribution of elephants (see, e.g., Marshal et al. 2011). All in all, this example illustrates how data availability can constrain the ability to match plant or animal data with the most relevant NDVI data, and thus might be a potential factor in some of the poor correlations reported.

Mismatches between the resolution at which wildlife data have been collected and the resolution at which NDVI data have been collected may well influence the NDVI–wildlife links when the study area considered is highly heterogeneous. Various studies have examined the links between climatic conditions, NDVI variability, and bird richness at multiple spatial scales and resolutions (Table 9.2). These studies have so far failed to agree on which environmental variables (e.g. temperature, rainfall, NDVI) best predict bird species richness. One non-exclusive explanation for this inconsistency may be that, depending on the spatial scales and resolutions considered, environmental variables differ in their ability to identify patterns in environmental productivity relevant to the level of spatial variability captured by the species richness dataset under study. In most situations described in Table 9.2, variables such as temperature, NDVI, rainfall, and

Table 9.2 Studies on the links between climatic conditions, NDVI variability, and bird richness, at multiple spatial scales and resolutions.

Location	Environmental variables considered	Best predictor of species richness variability	References
Britain	Summer temperature, NDVI	Summer temperature	Evans et al. 2008
North and Central America as far south as Nicaragua	Temperature, precipitation, NDVI, elevation	Temperature	H-Acevedo and Currie 2003
Sub-Saharan Africa	Temperature, precipitation, NDVI	NDVI	Foody 2004
Taiwan	Temperature, precipitation, NDVI, elevation, number of land-cover types, urbanization, population density, naturalness, distance to road	NDVI	Koh et al. 2006
Worldwide	Elevation, habitat diversity, NDVI, temperature, population density, agricultural area, ice coverage	Elevation	Davies et al. 2007
East Asia	Land area, NDVI, elevation range, population density	NDVI	Ding et al. 2006
North America	Temperature, NDVI	The structure of winter assemblages responds more strongly to temperature, whereas breeding assemblages tended to respond more strongly to NDVI	Evans et al. 2006

species richness data have been collected at different spatial resolutions. Everything being equal, one might expect the predictive variables collected at a spatial resolution closest to the spatial resolution of the species richness dataset to perform better than predictive variables collected at coarser resolutions: such an assumption can be expected to be particularly true for highly heterogeneous study areas.

9.2.2 Not all resolutions lead to the same outcome

The importance of the spatial scale of analysis (Dungan et al. 2002) has received only limited attention in ecological studies using the NDVI, yet this factor has considerable potential to affect the strength of correlation between NDVI fluctuations and the plant or animal parameter considered. The fact that the spatial resolution of a given analysis can affect whether or not patterns are detected has been illustrated by Costanza et al. (2011), who investigated how measures of heterogeneity from abiotic environmental variables, vegetation productivity data, and land-cover classifications related to plant species richness in North and South Carolina, USA. The authors examined this link at four different spatial scales, ranging from within plots to across regions. They found that, whatever the scale considered, no measure of heterogeneity was a better predictor of local richness than mean pH within plots. However, they also showed that each of the three classes of heterogeneity measures they used had a distinct scale at which it performed better than the others.

The importance of spatial scale in ecological studies has also been shown by Bro-Jørgensen et al. (2008), who investigated how the distribution of the two sexes in a lek-breeding population of topi antelopes relates to resource abundance. Höglund and Alatalo (1995) found that in lekking species such as topis, females were attracted to clusters of resource-poor male display territories for mating purposes, yet the relative strength of this attractive force (as opposed to predation risk and resource distribution) in determining the overall spatial distribution of the two sexes is generally unknown. Focusing on the savannah-dwelling topi antelopes, the authors therefore decided to use the NDVI to test two alternative hypotheses about the mechanisms shaping the observed patterns in the spatial distribution of the sexes. The first hypothesis stated that topis distribute themselves to maximize resource abundance in their vicinity, leading to the prediction of a positive correlation between the NDVI and topi density. The second hypothesis stated that mating and/or antipredator benefits of resource-poor areas such as leks outweigh the benefits of resource maximization, leading to the prediction of a negative correlation between the NDVI and topi density. Because topi have a synchronized rut coinciding with the wet season, the authors decided to investigate patterns before and during the rut separately.

To the surprise of the authors, both hypotheses were supported by their results: in the dry season preceding the rut, topi density was found to correlate positively with the NDVI, although the pattern was less pronounced in males than in females. Resources are relatively scarce during the dry season, and these results showed that topi tend to prefer pastures where green grass is widely abundant. The observation that the correlation between the NDVI and topi abundance was less pronounced for males than for females suggested that the need for territorial attendance prevents males from tracking resources as freely as females do. During the rut, which occurs in the wet season, both male and female densities were shown to correlate negatively with the NDVI (Bro-Jørgensen et al. 2008). At that time of the year, resources are generally plentiful and the negative correlation between the NDVI and topi abundance suggested that distribution during the rut is mostly determined by the benefits of aggregating on relatively resource-poor leks for mating, and possibly antipredator, purposes. The results from this study therefore improved our understanding of the processes governing lekking species distribution, by conveying sequential information on the role of overall resource availability in determining topi distribution in the Mara ecosystem.

However, the most interesting aspect of the work presented by Bro-Jørgensen et al. (in relation to this sub-section of the book) concerns the link between the NDVI and topi distribution using NDVI data collected at different spatial resolutions (ranging from 250 m to 8 km). This enabled them to show that in the dry season preceding the rut, topi density correlated positively with the NDVI at the large scale but not at the fine scale. During the rut, on the other hand, both male and female densities correlated negatively with NDVI at the fine scale. At the large scale, no correlation between density and NDVI was found during the rut in either sex. This kind of outcome illustrates how the spatial resolution of the analysis can affect whether or not patterns are detected. The negative correlation between the NDVI and density during the rut was indeed only apparent at the finer scales. The likely reason is that leks are too localized in space to be reflected in NDVI values based on large areas. On the other hand, the positive correlation between topi density before the rut and the NDVI was apparent only at the larger scales. This finding suggests that while topi at this time prefer ranges where food is generally abundant, other factors affect their distribution at the fine scale. This example represents a strong case for advising any researcher conducting analyses of animal distributions to do so at multiple spatial scales, as this affects not only whether patterns are detected but it may also shed light on the causal factors behind observed distribution patterns.

There are also examples from the world of plants. Harrison et al. (2006) assessed the relationship between productivity, as indexed using the NDVI, and beta diversity in herbaceous plants at 105 widely distributed sites on serpentine soil in California. One measure of beta diversity (the species dissimilarity between paired 500 m^2 plots on adjacent north and south slopes) was positively related to productivity as measured by the NDVI. However, this effect was not strong enough to transform the neutral relationship of NDVI with alpha (1 m^2) diversity to a positive relationship of productivity with gamma (1000 m^2) diversity. These studies illustrate that ecological processes occurring across multiple scales influence animal and plant parameters differently, and more broadly suggest that correlations between the NDVI and animal and plant parameters should be expected to be scale dependent (i.e. an absence of correlation between a parameter and the NDVI might just indicate that the spatial resolution chosen for the NDVI data is not adequate).

9.2.3 Responses might be spatially or temporally inconsistent

Regional and global analyses aiming to link NDVI-based metrics to a given parameter (e.g. abundance, body mass, species richness) generally assume the existence of a spatially consistent link between the metrics and the parameter considered. Yet there are reasons to expect spatial inconsistency (Nielsen et al. 2012) that may arise from spatial variation in factors limiting population growth or from differences in the links between NDVI metrics and variability in ecologically relevant local resource availability. Spatial inconsistency in the link between the NDVI and biological parameters has already been reported in the literature. Nielsen et al. (2012) reported positive effects of early summer NDVI values on body mass of lambs *Ovis aries* in autumn in all but one Norwegian study area, Hardangervidda West. NDVI-based metrics did not influence body mass variability in all populations of red deer *Cervus elaphus* examined by Martinez-Jauregui et al. (2009).

Two issues are associated with the existence of spatial or temporal inconsistencies. First, in many cases we still lack the ecological understanding required to detail the processes shaping these inconsistencies; second, if such an understanding is not sought, then the presence of inconsistencies might be used as an argument to invalidate the use of the NDVI as a tool to enhance ecological knowledge. Thankfully, recent work such as that by Bhola et al. (2012) is clearly heading towards dissecting the processes shaping the links between NDVI-based metrics and wildlife parameters. Bhola et al. compared the link between animal distribution and NDVI variability in the dry and wet seasons for 10 herbivore species inhabiting (a) the Mara reserve and (b) group ranches bordering the reserve. They reported clear differences in the links between primary productivity and animal distribution across seasons and between herbivores with differing diets (see also Section 9.1.1). However, they also reported spatial differences for a given species and season. Because the authors had a good knowledge of the ecology of the species considered, and because analyses were performed at a scale where the authors had a good understanding of the factors driving vegetation dynamics and ecosystem functioning, these differences were expected. Namely, the authors strongly suspected that in the late wet season, high NDVI tended to indicate the presence of less-nutritious grasses in the reserve. In group ranches, they knew that grass tended to be kept in a short, active growth stage by livestock grazing, thereby increasing its quality but decreasing its biomass. Hence in ranches high NDVI values were expected to be associated with higher vegetation quality. During the dry season, on the other hand, vegetation quality and quantity were known to be lower in the region. However, the authors also knew that grass height is generally taller inside than outside the reserve because of the absence of livestock grazing in the reserve. They therefore expected herbivore hotspots to positively correlate with the NDVI in the reserve during this period of short food supply, but to show the opposite pattern in the group ranches because of competition with livestock for forage (Bhola et al. 2012). This study illustrates beautifully how a good understanding of the system can lead to opposite expectations in terms of the link between the NDVI and wildlife, and to spatio-temporal inconsistencies in the reported patterns.

9.3 Location matters

9.3.1 Working in densely vegetated areas

As discussed in Chapter 3, the ability of the NDVI to index primary productivity in densely vegetated areas tends to be challenged, as (i) the NDVI is known to saturate in these areas; and (ii) cloud cover tends to be heavy in these regions. This means that NDVI fluctuations in rainforests are less likely to supply reliable information on the level of variation in leaf area index and productivity. Therefore, studies linking the NDVI to wildlife in areas where the correlation between the NDVI and primary productivity is challenged are likely to report non-significant results. For instance, Parra et al. (2004) aimed to evaluate the relative merit of three alternative environmental datasets in the Ecuadorian Andes—climatic data, NDVI, and elevation data—to model the distribution of six bird species of the genus *Grallaria*. Models including climate variables performed relatively well across most measures, whereas models using the NDVI performed poorly.

Elevation-based models were relatively good at predicting most sites of expected occurrence but showed a high overprediction error. Combinations of datasets improved the performance of the ecological niche models, but not significantly.

The difficulty of linking NDVI-based metrics to wildlife in densely vegetated areas is also apparent from the studies by Duffy and Pettorelli (2012) on elephants (Section 9.2.1) and by Pettorelli et al. (2006) on roe deer *Capreolus capreolus* in France (Chapters 3 and 7). In the former, the authors examined the correlation between NDVI variability and African elephant densities for savannah and forest elephants. Although they found a positive link between savannah elephant density and the NDVI at this continental scale, no link between NDVI variability and animal density was discovered for forest elephants (Duffy and Pettorelli 2012). Multiple reasons were discussed as to why no significant correlation was found for forest elephants: one was that surveyors' visibility could have been impaired by the amount of dense vegetation in the forested areas that this sub-species inhabits, leading to reduced visibility of elephants from afar and biased population estimates in these areas. Another possible explanation was linked to differences in the behaviour of elephants found in forested and non-forested habitats. Given that individuals found in these contrasting habitats have been shown to differ genetically and morphologically, one could indeed hypothesize that their spatial use patterns might be different. A third possible cause was linked to the nature of the classification of forest/non-forest environments, which was rather crude, and the nature of the sample size, which was small ($n = 13$ populations). A fourth possibility might involve known limitations associated with the use of the NDVI in dense forests, that is, the saturation of the relationship between the NDVI and vegetation productivity in these highly productive areas.

This point is reinforced by the study on roe deer in France (Pettorelli et al. 2006; see also Figure 9.5). The authors used the NDVI to assess how plant productivity from birth to autumn influences the following winter body mass of roe deer fawns. Two populations were followed, the first inhabiting the poorly productive Chizé reserve in south-western France with an oceanic climate, and the second from Trois

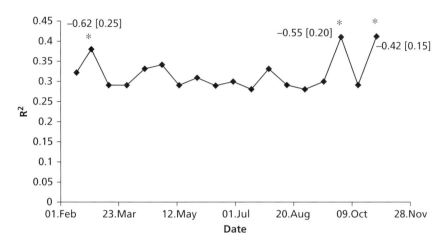

Figure 9.5 Correlation between the bimonthly composite NDVI and the following winter body mass of roe deer fawns between February to October in Trois Fontaines, France. (Reproduced from Pettorelli et al. 2006 with permission from John Wiley & Sons.) Slopes [SE] are given when the correlation is significant (these dates are indicated by a star).

Fontaines, a highly productive forest with a continental climate in the east of France (see also Chapter 7). The results were contrasting, with a strong effect of plant productivity in spring (April–May) in the Chizé population, but no effect in the Trois Fontaines population. The higher level of variability in winter fawn body mass in Chizé in comparison with Trois Fontaines was considered as a possible explanation. However, it was also acknowledged that these results might indicate a limitation in the use of the NDVI in forested areas; the majority of forest mammalian herbivores indeed rely predominantly on the forest floor vegetation (Gebert and Verheyden-Tixier 2001; Gilliam 2007). In highly productive areas such as Trois Fontaines, the relationship between the NDVI and plant productivity on the ground can be expected to be weaker, leading to an absence of correlation between NDVI variability and deer life-history traits (Pettorelli et al. 2006).

9.3.2 Working in seasonal forested environments

As discussed in the Section 9.3.1, only a fraction of the radiation reflected by forests originates from the ground layer, which can compromise the usefulness of the NDVI for wildlife managers operating in such areas. However, in seasonal forested environments, the contribution of the ground vegetation layer and the tree canopy to the amount of radiation reflected by the vegetation and received by the sensors on board satellites is temporally variable. The timing of tree and shrub leaf development somehow defines the moment when canopy reflectance starts to dominate the received signal. At the same time, developing tree leaves can negatively affect the growth of ground vegetation, which can then cause a decrease in its biomass. In such situations, the expected high NDVI values at the height of the summer season are thus likely to correlate negatively with, or not describe accurately, forage conditions on the ground. All of this therefore suggests that in the case of indexing forage availability for animal species living in seasonal forested habitats and strictly depending on the ground vegetation, NDVI variability in transitional seasons is the information most likely to provide reliable information about the forage conditions for these individuals. This expectation is well founded by many studies reporting a link between spring NDVI values and life-history parameters of animals living in seasonal deciduous forested areas (see Chapter 7).

9.4 Confounding effects

Correlative approaches without the existence of an assumed causal framework can lead to erroneous results, and NDVI-based analyses can sometimes be at risk of such drawbacks. Pau et al. (2012) used structural equation models combined with 10 years of satellite data to disentangle the effects of precipitation, structure, and NDVI-estimated productivity on woody plant species richness in Hawaiian dry forests. The NDVI has been shown to increase with tree species richness by many studies (Chapter 6), explaining between 30% and 60% of the variance in species richness (Bawa et al. 2002; Feeley et al. 2005; Gillespie 2005; Cayuela et al. 2006; Foody and Cutler 2006; Gillespie et al. 2009). Pau et al. therefore expected a positive correlation between NDVI variability and tree species richness in their study area. Accordingly, simple linear regressions showed that mean NDVI and CV NDVI can predict 30–40% of the variation in species richness in this area. However, when the relationship between NDVI and species richness was simultaneously considered with other confounding biotic and abiotic effects (such as precipitation), there was no direct relationship between the NDVI and species richness.

This outcome boils down to the fact that the mechanisms producing the previously reported links between NDVI and plant species richness are not well established, leading to the report of correlative, yet not causal, links. Many of the studies listed above cite the species–energy hypothesis when using the NDVI to predict plant species richness, whereby species are assumed to partition available energy (as indexed using the NDVI) so that energy is a limiting resource to the number of species in an area (see Chapter 6 for more details). However, few authors question the use of the NDVI as a correct proxy of available energy for plants. The NDVI is known to increase with precipitation in this part of the world (Asner et al. 2005), and precipitation is a strong limiting factor in the structural development of tropical dry forests (Bullock et al. 1995). Forest structure has also been shown to affect the NDVI (e.g. Feeley et al. 2005), so that the relationship between the NDVI and species richness is likely to be an indirect result of precipitation and forest structure—and this is what the study by Pau et al. established. In other words, they demonstrated that: (i) the link between plant richness and the NDVI is correlative, but not causal; (ii) the NDVI is not a relevant proxy of energy availability for plants. Thus, poor understanding of the mechanisms producing a given link between the NDVI and wildlife may lead to inconsistent results among study cases, and emphasizes the need to consider possible underlying factors to improve predictive power.

9.5 Conclusions

Factors shaping the successful use of the NDVI in ecology and wildlife management include spatiotemporal match between wildlife and environmental data, scale, resolution, location, and species attributes. A good understanding of the systems considered is key to develop scientifically sound local expectations about the links between the NDVI and wildlife; this chapter also reiterates the fundamental importance of on-the-ground data collection to complement and make sense of remote sensing information. The studies detailed here offer valuable guidance to ecologists dealing with NDVI data. If you have strong reasons to expect a link yet cannot find it, have you tried to consider NDVI data collected at different spatial resolutions? Have you tried to aggregate NDVI data over a different timeframe, such as the season instead of the year? Have you tried to increase the precision of your definition of the study area? If your species is using a clear selection of land-cover types, and if your study area is heterogeneous, have you tried to focus your analyses on those NDVI values associated with these specific pixels? These options might still not work, but their consideration might reduce the probability of not detecting something that might be there.

CHAPTER 10

Ecosystem services, NDVI, and national reporting: matches and mismatches

> I have been impressed with the urgency of doing. Knowing is not enough; we must apply. Being willing is not enough; we must do.
> **Leonardo da Vinci**

Science has a long tradition of worshiping individual geniuses haunted by their own quest to solve the mysteries that fascinated them. As a twenty-first century ecologist, however, it is difficult to divorce oneself from the challenges facing society and to conduct scientific activities purely based on one's own interests. Faced with a growing world population, loss of species, and increasingly dangerous concentrations of greenhouse gases, world leaders do need the support of ecologists and conservation biologists to make informed decisions on issues related to environmental and wildlife management. In other words, present-day ecologists cannot afford to restrict themselves merely to generating the science necessary to plug the current knowledge gaps; given the current context, they should also devote themselves to making science user-useful while meeting user needs (Mooney and Mace 2009).

Now 'with great power comes great responsibility' said Uncle Ben, and that is particularly true for ecologists who embark upon the quest of making their science useful to society. The prime responsibility is to produce scientifically sound outputs, if these are to inform future policies. The less obvious responsibility is to become informed about policy- and decision-making processes locally and globally to ensure that scientific outputs meet user needs and are presented in a way that minimizes the risks of them being ignored or misused. As Johns (2010) states 'It is not enough to say how the biological world works and that we think it is good to sustain it. The point is to be heard.'

There are many national, regional, and international instruments in the form of resolutions, declarations, platforms and conventions that deal directly or indirectly with the subject of the preservation of biological diversity: these instruments are generally associated with aims, targets, and goals of reducing the rate of biodiversity loss. There are several ways to address the rate of biodiversity loss: one may focus on reducing the level of a given threat to biodiversity (e.g. by cutting greenhouse gas emissions); one may also focus on increasing the number and effectiveness of management actions that promote biodiversity restoration (e.g. by rehabilitating abandoned government-leased fishponds to mangroves). There is, however, little point in having targets and goals that cannot be tracked: if one commits to achieving 'A', for example, one needs to be able to know when 'A' is achieved or how far one is from achieving 'A'. Assessing progress toward a given target therefore requires information on past and current states and trends in the entity under consideration. Yet biodiversity is a multidimensional concept, and this can lead to a number of difficulties when trying to measure and monitor biodiversity. Moreover, there are many interacting drivers of biodiversity loss, rendering the whole task of reporting on progress toward a reduction in the rate of biodiversity loss extremely challenging (Balmford et al. 2005; Mace and Baillie 2007; Baillie et al. 2008).

Remote sensing technologies provide a wealth of opportunities to monitor the different constituents of the planet's natural environment, allowing access to standardized, global, continuous data directly relevant to the monitoring of, and pressures on, biodiversity (Pereira and Cooper 2006; Scholes et al. 2008). The NDVI, in particular, represents a useful indicator of ecological change that can help to measure past ecological impacts and provide early warning signs of impending change, thereby enhancing our ability to manage and solve associated problems. Specifically, the concept of ecological indicator can be defined as measures allowing the assessment and monitoring of the state of the environment over time (Cairns et al. 1993; Morellet et al. 2007): these should be 'easily measured, be sensitive to stresses on the system, respond to stress in a predictable manner, be anticipatory, predict changes that can be averted by management actions, be integrative, have a known response to disturbances, anthropogenic stresses, and changes over time, and have low variability in response' (Dale and Beyeler 2001). The NDVI can supply information on ecosystem state, changes in ecosystem functioning as well as changes in biodiversity (Chapters 5–7). The NDVI is also easily measured, sensitive to stresses (e.g. drought, insect infestation, deforestation, grazing pressure, and land-use conversion), and integrative. Therefore in many situations NDVI-based metrics can meet the criteria associated with the definition of ecological indicators. This chapter introduces the main international platforms and conventions relevant to the preservation of biodiversity and ecosystem services, and discusses the importance of other regional and national agreements. We then explore how the NDVI data may support existing environmental policies, as well as inform future policies. Finally, the limitations and constraints associated with the NDVI within this context are discussed.

10.1 International conventions and platforms on the preservation of biodiversity and ecosystem services

10.1.1 Convention on Biological Diversity

In 1988 the United Nations Environment Programme (Box 10.1) convened a Working Group of Experts on Biological Diversity to explore the need for an international Convention on Biological Diversity (CBD). Its work culminated in the drafting of the CBD, which was opened for signature in June 1992 at the Rio 'Earth Summit' and has now been ratified by 193 countries (as of 12 March 2013).

The CBD is central to international and national efforts to preserve biodiversity, being key to the development of the protected area network. The Convention is a non-binding international treaty for the conservation and sustainable use of biodiversity and the equitable sharing of the benefits from utilization of genetic resources. It seeks to address all notable threats to biodiversity and ecosystem services, including those from climate change, through

---

Box 10.1 The United Nations Environment Programme (UNEP)

UNEP is an international institution that coordinates the United Nations environmental activities, headquartered in Nairobi, Kenya. The primary goal is to assist developing countries in implementing environmentally sound policies and practices. UNEP was founded in June 1972, as a result of the United Nations Conference on the Human Environment. UNEP is a key actor when it comes to the formulation and implementation of international guidelines, treaties, and conventions relevant to the preservation of biodiversity and ecosystem services. Together with the World Meteorological Organization, it established the Intergovernmental Panel on Climate Change (IPCC) in 1988 (see Box 1.1, for more information on the IPCC). UNEP is also one of several implementing agencies for the Global Environment Facility (GEF). The Programme is a key member of the Biodiversity Indicators Partnership (BIP), which brings together a host of international organizations working on biodiversity indicator development, to provide the best available knowledge on biodiversity trends. More information can be found at www.unep.org.

(i) the provision of scientific assessments; (ii) the development of tools, incentives, and processes; (iii) the implementation of technologies and good practices; (iv) the full and active involvement of relevant governmental and non-governmental stakeholders <http://www.cbd.int>.

In 2002 the Conference of the Parties (COP; Parties in this case referring to nation states), the governing body of the CBD, developed a Strategic Plan to guide the conservation and sustainable use of biodiversity at national, regional, and global levels. The mission statement agreed during this sixth COP was to achieve, by 2010, 'a significant reduction of the current rate of biodiversity loss at the global, regional, and national level as a contribution to poverty alleviation and to the benefit of all life on Earth.' This statement became known as the 2010 Biodiversity Target. To inform the tenth COP in 2010 in Nagoya, Japan, national reports were submitted to the CBD. The aim was to evaluate the state of the biodiversity in each of the signatory countries, and determine whether the 2010 Biodiversity Target was met. Sadly, it quickly became clear that nations failed to meet the Target (Gordon et al. 2010), leading to the development of a revised and updated Strategic Plan for Biodiversity 2011–2020, and the formulation of 20 headline targets known as the Aichi Targets (Table 10.1). At the recent COP in October 2012 in Hyderabad, India, developed countries agreed to double funding to support efforts in developing states towards meeting these internationally agreed Biodiversity Targets, and the main goals of the Strategic Plan for Biodiversity 2011–2020 (CBD 2012).

10.1.2 Convention on Migratory Species

The Convention on the Conservation of Migratory Species of Wild Animals (also known as CMS or Bonn Convention <http://www.cms.int>) was adopted in June 1979, and aims to conserve terrestrial, aquatic, and avian migratory species throughout their range. Like the CBD, the CMS is an intergovernmental treaty, concluded under the aegis of UNEP. Since the Convention's entry into force, its membership has grown steadily to include 118 Parties (as of 12 March 2013). Like the CBD, the decision-making organ of CMS is the COP (see Figure 10.1). A Standing Committee provides policy and administrative guidance between the regular meetings of the COP. A Scientific Council, consisting of experts appointed by individual member states and by the COP, gives advice on technical and scientific matters. Migratory species of concern to the CMS are categorized under two headings; those threatened with extinction are listed in Appendix I of the Convention. Besides establishing obligations for each State joining the Convention, CMS promotes concerted action among the Range States of many of these species. Migratory species that need or would significantly benefit from international co-operation are listed in Appendix II of the Convention.

CMS acts as a kind of framework Convention: the agreements negotiated under the CMS umbrella may range from legally binding treaties to less formal instruments, such as Memoranda of Understanding, and can be adapted to the requirements of particular regions.

Examples of CMS agreements include the African–Eurasian Waterbird Agreement (AEWA; <http://www.unep-aewa.org>), which represents the largest agreement developed so far under CMS auspices. AEWA is an intergovernmental treaty dedicated to the conservation of migratory waterbirds and their habitats across Africa, Europe, the Middle East, Central Asia, Greenland, and the Canadian Archipelago. Administered by UNEP, AEWA brings together countries and the wider international conservation community in an effort to establish coordinated conservation and management of migratory waterbirds (such as divers, grebes, pelicans, cormorants, herons, storks, rails, ibises, spoonbills, flamingos, ducks, swans, geese, cranes, waders, gulls, terns, tropic birds, auks, and frigate birds) throughout their entire migratory range. Sixty-eight countries and the European Union have so far become a Contracting Party to AEWA; three main bodies structure the agreement: the Meeting of the Parties (MOP), which is the governing body of AEWA, the Standing Committee, and the Technical Committee, each responsible for steering the operations between sessions of the MOP and for supplying scientific advice.

Agreements under the aegis of CMS also include the West African Elephant *Africana Loxodonta* Memorandum of Understanding, which provides an international framework for Range State governments,

Table 10.1 The Aichi targets, as defined by the Convention on Biological Diversity.[a]

Target	Description
1	By 2020, at the latest, people are aware of the values of biodiversity and the steps they can take to conserve and use it sustainably.
2	By 2020, at the latest, biodiversity values have been integrated into national and local development and poverty reduction strategies and planning processes, and are being incorporated into national accounting, as appropriate, and reporting systems.
3	By 2020, at the latest, incentives (including subsidies) harmful to biodiversity are eliminated, phased out or reformed in order to minimize or avoid negative impacts, and positive incentives for the conservation and sustainable use of biodiversity are developed and applied, consistent and in harmony with the Convention and other relevant international obligations, taking into account national socio-economic conditions.
4	By 2020, at the latest, governments, businesses, and stakeholders at all levels have taken steps to achieve or have implemented plans for sustainable production and consumption, and have kept the impacts of use of natural resources well within safe ecological limits.
5	By 2020, the rate of loss of all natural habitats, including forests, is at least halved and where feasible brought close to zero, and degradation and fragmentation is significantly reduced.
6	By 2020, all fish and invertebrate stocks and aquatic plants are managed and harvested sustainably, legally and applying ecosystem-based approaches, so that overfishing is avoided, recovery plans and measures are in place for all depleted species, fisheries have no significant adverse impacts on threatened species and vulnerable ecosystems, and the impacts of fisheries on stocks, species and ecosystems are within safe ecological limits.
7	By 2020 areas under agriculture, aquaculture and forestry are managed sustainably, ensuring conservation of biodiversity.
8	By 2020, pollution, including from excess nutrients, has been brought to levels that are not detrimental to ecosystem function and biodiversity.
9	By 2020, invasive alien species and pathways are identified and prioritized, priority species are controlled or eradicated, and measures are in place to manage pathways to prevent their introduction and establishment.
10	By 2015, the multiple anthropogenic pressures on coral reefs and other vulnerable ecosystems impacted by climate change or ocean acidification are minimized, so as to maintain their integrity and functioning.
11	By 2020, at least 17% of terrestrial and inland water areas, and 10% of coastal and marine areas, especially areas of particular importance for biodiversity and ecosystem services, are conserved through effectively and equitably managed, ecologically representative and well-connected systems of protected areas and other effective area-based conservation measures, and integrated into the wider landscapes and seascapes.
12	By 2020, the extinction of known threatened species has been prevented and their conservation status, particularly of those most in decline, has been improved and sustained.
13	By 2020, the genetic diversity of cultivated plants and farmed and domesticated animals and of wild relatives, including other socio-economically as well as culturally valuable species, is maintained, and strategies have been developed and implemented for minimizing genetic erosion and safeguarding their genetic diversity.
14	By 2020, ecosystems that provide essential services, including services related to water, and contribute to health, livelihoods, and well-being, are restored and safeguarded, taking into account the needs of women, indigenous and local communities, and the poor and vulnerable.
15	By 2020, ecosystem resilience and the contribution of biodiversity to carbon stocks has been enhanced, through conservation and restoration, including restoration of at least 15% of degraded ecosystems, thereby contributing to climate change mitigation and adaptation and to combating desertification.
16	By 2015, the Nagoya Protocol on Access to Genetic Resources and the Fair and Equitable Sharing of Benefits Arising from their Utilization is in force and operational, consistent with national legislation.
17	By 2015 each Party has developed, adopted as a policy instrument, and has commenced implementing an effective, participatory and updated national biodiversity strategy and action plan.
18	By 2020, the traditional knowledge, innovations and practices of indigenous and local communities relevant for the conservation and sustainable use of biodiversity, and their customary use of biological resources, are respected, subject to national legislation and relevant international obligations, and fully integrated and reflected in the implementation of the Convention with the full and effective participation of indigenous and local communities, at all relevant levels.
19	By 2020, knowledge, the science base, and technologies relating to biodiversity, its values, functioning, status and trends, and the consequences of its loss, are improved, widely shared and transferred, and applied.
20	By 2020, at the latest, the mobilization of financial resources for effectively implementing the Strategic Plan for Biodiversity 2011–2020 from all sources, and in accordance with the consolidated and agreed process in the Strategy for Resource Mobilization, should increase substantially from the current levels. This target will be subject to changes contingent to resource needs assessments to be developed and reported by Parties (countries).

[a] For more information on these targets, see <http://www.cbd.int/decision/cop/?id>.

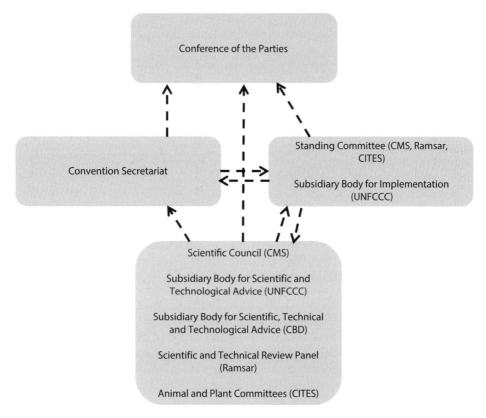

Figure 10.1 Common bodies structuring major international Conventions (in this case referring to the Ramsar Convention, the Convention on Biological Diversity (CBD), the Convention on International Trade in Endangered Species of Wild Fauna and Flora (CITES), the United Nations Framework Convention on Climate Change (UNFCCC), and the Convention on Migratory Species (CMS)). In all cases, the Conference of the Parties (COP) is the decision-making organ. Each Convention possesses its own COP. A scientific committee generally informs the COP and the Secretariat. In some cases (such as for the CMS, Ramsar Convention, CITES, and UNFCCC), another body (referred to as the Standing Committee or the Subsidiary body for implementation) exists, to facilitate the implementation of the Convention. The links and hierarchy between these bodies is variable from one convention to another (dashed arrows).

scientists, and conservation groups to collaborate in the conservation of elephant populations and their habitats. The aim of this Memorandum of Understanding is to provide an intergovernmental structure to help monitor and co-ordinate conservation activities relevant to elephants in West Africa, where many of the most viable populations span the national boundaries of two or more countries. The African Elephant Specialist Group of the IUCN Species Survival Commission is the co-ordinator of the Memorandum; it supplies technical assistance to catalyse transboundary conservation activities and implement existing national conservation strategies. The group also helps to prepare meetings and enables information sharing.

10.1.3 Ramsar Convention on wetlands

The Convention on Wetlands of International Importance, generally referred to as the Ramsar Convention, is an intergovernmental treaty aiming at 'the conservation and wise use of all wetlands through local and national actions and international cooperation, as a contribution towards achieving sustainable development throughout the world.' The Ramsar Convention is the only global environmental treaty that deals with a particular type of habitat, and was adopted in the Iranian city of Ramsar in 1971. The Convention uses a broad definition of the types of wetlands included in its mission, ranging from lakes and rivers, swamps and marshes, wet grasslands and peatlands, oases, estuaries, deltas and tidal flats, near-shore marine areas, mangroves and coral reefs, as well as human-made sites such as fish ponds, rice paddies, reservoirs, and salt pans. Unlike the other global environmental conventions (such as the CBD), the Ramsar Convention is not affiliated with the United Nations system of Multilateral Environmental Agreements (MEA), but it works very closely with the other MEAs and is a full partner among the 'biodiversity-related cluster' of treaties and agreements. For example, in January 1996, the secretariats of the Ramsar Convention and the CBD signed a first Memorandum of Cooperation, and in November of that year the third COP of the CBD invited the Ramsar Convention 'to cooperate as a lead partner' in implementing CBD activities related to wetlands. The COP of both Conventions have also called for increased communication and cooperation between their subsidiary scientific bodies, the CBD's Subsidiary Body for Scientific, Technical, and Technological Advice (SBSTTA) and the Ramsar Scientific and Technical Review Panel, and members of both these bodies regularly participate in the work and meetings of one another (Figure 10.1). One hundred and sixty-five contracting parties are currently associated with the Ramsar Convention (as of 12 March 2013; visit <http://www.ramsar.org/> for an update on this number). Becoming a contracting party of the Ramsar Convention signals a commitment to support the 'three pillars' of the Convention: (i) ensuring the conservation and wise use of wetlands designated as wetlands of international importance; (ii) including as far as possible the wise use of all wetlands in national environmental planning; (iii) consulting with other parties about implementation of the Convention, especially in regard to transboundary wetlands, shared water systems, and shared species. Upon joining the Ramsar Convention, each contracting party is required to designate at least one wetland site for inclusion in the list of wetlands of international importance.

10.1.4 World Heritage Convention

The Convention concerning the Protection of the World Cultural and Natural Heritage is an international treaty adopted by the United Nations Educational, Scientific and Cultural Organization (UNESCO; <http://whc.unesco.org>) in 1972. It seeks to encourage 'the identification, protection, and preservation of cultural and natural heritage around the world considered being of outstanding value to humanity'. The stated objectives of the World Heritage Convention are: to encourage countries to ensure the protection of their natural and cultural heritage; to encourage States that are Party to the Convention to nominate sites within their national territory for inclusion on the World Heritage List; to encourage Parties to establish management plans and set up reporting systems on the state of conservation of their World Heritage sites; to help Parties to safeguard World Heritage properties by providing technical assistance and professional training; to provide emergency assistance for World Heritage sites in immediate danger; to support Parties' public awareness-building activities for World Heritage conservation; to encourage participation of the local population in the preservation of their cultural and natural heritage; and to encourage international co-operation in the conservation of our world's cultural and natural heritage. One hundred and ninety State Parties have so far ratified the World Heritage Convention (as of 19 September 2012). The World Heritage Committee is the main body in charge of the implementation of the Convention. It is assisted by a Secretariat appointed by the Director-General of UNESCO, and advised by the International Centre for the Study of the Preservation and Restoration of Cultural Property (ICCROM), the International Council on Monuments and Sites (ICOMOS), and the IUCN.

10.1.5 Convention on International Trade in Endangered Species of Wild Fauna and Flora (CITES)

CITES is an international agreement to which Parties (countries) adhere voluntarily, which aims to ensure that international trade in specimens of wild animals and plants does not threaten the survival of these species. The Convention was drafted as a result of a resolution adopted in 1963 at a meeting of members of the IUCN, yet the text was only agreed at a meeting of representatives of 80 Parties in 1973 and

CITES only entered into force in 1975. CITES is a legally binding Convention—that is, member States have to implement the Convention, yet the Convention does not take the place of national laws. Rather it provides a framework to be respected by each Party, which has to adopt its own domestic legislation to ensure that CITES is implemented at the national level. One hundred and seventy-seven state parties have so far ratified CITES (as of 1 March 2013; visit <http://www.cites.org for an update on this number>). The species covered by CITES are listed in three Appendices, according to the degree of protection they need. Appendix I includes species threatened with extinction, and trade in specimens of these species is permitted only in exceptional circumstances. Appendix II includes species not necessarily threatened with extinction, but in which trade must be controlled in order to avoid utilization incompatible with their survival. Appendix III includes species that are protected in at least one country, which has asked other CITES Parties for assistance in controlling the trade.

10.1.6 Intergovernmental Platform on Biodiversity and Ecosystem Services (IPBES)

In June 2010, a number of governments recommended the establishment of IPBES. This platform—which mirrors the Intergovernmental Panel on Climate Change (IPCC; see Box 1.1)—was established in April 2012, after years of international negotiations, as an independent intergovernmental body open to all member countries of the United Nations <http://www.ipbes.net>. The idea behind the creation of the IPBES is to provide a bridge between the wealth of scientific knowledge and worldwide government action required to halt and reverse the declines in biodiversity and ecosystem services (Larigauderie and Mooney 2010; Reyers et al. 2010). IPBES is expected to respond to requests for scientific information related to biodiversity and ecosystem services from Governments, relevant multilateral environmental agreements and United Nations bodies, as well as other relevant stakeholders. Governments have agreed that the four main functions of IPBES will be to (i) identify and prioritize key scientific information needed for policy-makers and to catalyse efforts to generate new knowledge; (ii) perform regular and timely assessments of knowledge on biodiversity and ecosystem services and their inter-linkages; (iii) support policy formulation and implementation by identifying policy-relevant tools and methodologies; (iv) prioritize key capacity-building needs to improve the science–policy interface, and to supply and call for financial and other support for the highest-priority needs related directly to its activities. The first meeting of the Platform's Plenary (IPBES-1) was held in Bonn, Germany, from 21 to 26 January 2013; the event was hosted by the Government of Germany. The meeting led to an agreement on rules of procedures for the platform and for the meetings of the platform. It also led to the election of the Bureau and Multidisciplinary Expert Panel members, and to an agreement on the next steps by which the IPBES work programme can become operational as soon as possible.

10.1.7 United Nations Framework Convention on Climate Change (UNFCCC)

The UNFCCC's text was adopted at the United Nations Headquarters, New York, on 9 May 1992, and entered into force on 21 March 1994. The aim is to set an overall framework for intergovernmental efforts to tackle the challenge posed by climate change. It enjoys near universal membership, with 195 countries having ratified the Convention (194 States and one regional economic integration organization; as of 12 March 2013). Under the Convention, governments: (i) gather and share information on greenhouse gas emissions, national policies and best practices; (ii) launch national strategies for addressing greenhouse gas emissions and adapting to expected impacts, including the provision of financial and technological support to developing countries; and (iii) co-operate in preparing for adaptation to the impacts of climate change <http://unfccc.int>. Again, the COP is the 'supreme body' of the Convention, that is, its highest decision-making authority.

An international convention about climate change is relevant in several respects to scientists dealing with biodiversity and ecosystem services management and conservation. The first link is built on the observation that climate change influences biodiversity and other ecosystem services in both direct and indirect ways (Walther et al. 2002). Corals, for example, are affected through ocean acidification and sea-surface temperature rise leading to coral bleaching (Hoegh-Guldberg 2007), and this has secondary effects on marine animal communities and human communities that depend upon reef ecosystems (Millennium Ecosystem Assessment Board 2005; IPCC 2007; Munday et al. 2008). Climate change also influences timing of flowering and reproduction by plants (Chmielewski and Rotzer 2002; Willis et al. 2008) and these changes have trophic cascading effects on the faunal communities that depend on plants. Birds, with their complicated and intricately timed migratory patterns are currently among the most strongly influenced by changes in climate (Devictor et al. 2008; La Sorte et al. 2009), although strong responses to phenological changes have also been reported for arctic and alpine mammals (e.g. Réale et al. 2003; Kausrud et al. 2008). Any policy aimed at reducing greenhouse gas emissions, therefore, has consequences for the conservation of biodiversity and ecosystem services.

The second link is derived from the fact that certain anthropogenic activities detrimental to biodiversity are also associated with substantial greenhouse gas emissions. Deforestation and forest degradation are good examples of such activities, contributing to greenhouse gas emissions in two different ways: (i) through the release of carbon stores to the atmosphere; (ii) through the removal of sequestration capacity, reducing the ability of forest to act as a carbon sink. These paired effects are especially dramatic in native tropical forests, where deforestation rates are highest and where a large share of the world's terrestrial carbon is held (Chomitz 2007; Laurance 2007). It has been claimed that the forestry sector represents upwards of 50% of greenhouse gas emission mitigation potential (IPCC 2007). Given the contribution of deforestation and forest degradation to greenhouse gas emissions and the important potential for a reduction in greenhouse gas emissions through better forest management practices, policy-makers involved with the UNFCCC have recently expressed increased attention in the forest sector. Yet this sector has also been of high interest to policy-makers involved with the preservation of biodiversity and ecosystem services, as forests may contain high levels of biodiversity; provide wildlife corridors; support livelihoods; contribute to the robustness of other ecosystems by providing regulating services such as modulating local climates and maintaining populations of species that are key to ecosystem function; deliver essential ecosystem services important for human wellbeing (Millennium Ecosystem Assessment 2005; Malhi et al. 2008).

The third link is through what is known as 'ecosystem-based adaptation' (e.g. <www.EBAflagship.org>), which is defined by the CBD as 'the use of biodiversity and ecosystem services to help people to adapt to the adverse effects of climate change' (Secretariat of the Convention on Biological Diversity 2009). As detailed in Decision X/33 on Climate Change and Biodiversity, this definition also includes the 'sustainable management, conservation and restoration of ecosystems, as part of an overall adaptation strategy that takes into account the multiple social, economic and cultural co-benefits for local communities.' Ecosystem-based adaptation relies on the idea that healthy ecosystems can help to build resilience against the impacts of climate change, something that has been recognized by the UNFCCC in decision 1/CP.16, which invited Parties to enhance action on adaptation by 'building resilience of socioecological systems, including through economic diversification and sustainable management of natural resources.' The interest of the UNFCCC for healthy ecosystems able to help build resilience against climate change impacts can stimulate additional financial and societal support for addressing the current biodiversity crisis, leading to possible win–win situations in many parts of the world.

10.1.8 Regional and national agreements

Up to now, we have focused our attention on the main international platforms and agreements relevant to the conservation of biodiversity and ecosystem services. Yet, in some areas of the world, national or regional treaties and agreements might also be key to achieving a reduction in biodiversity loss and in ecosystem services preservation. Most African countries, for example, have ratified the Ramsar Convention, the Convention on International Trade in Endangered Species of Wild Fauna and Flora, the Convention on the Conservation of Migratory Species of Wild Animals, the Convention Concerning the Protection of the World Cultural and Natural Heritage, the UNFCCC, and the CBD. These global multilateral environmental agreements are then complemented by sub-regional and regional agreements, such as the African Convention on the Conservation of Nature and Natural Resources, which was ratified in Algiers in 1968 by forty states. This regional Convention has specific provisions on ecosystems and species, stating that 'the Contracting States shall ensure conservation, wise use and development of faunal resources and their environment' (African Union 2011). Likewise, the Lusaka Agreement on Co-operative Enforcement Operations Directed at Illegal Trade in Wild Fauna and Flora (1994) is a regional agreement originally ratified by eight eastern and southern African countries in 1992. The agreement resembles CITES, with a mission to support the member states and collaborating partners in reducing and ultimately eliminating illegal trade in wild fauna and flora through: (i) the facilitation of co-operative activities in undertaking law enforcement operations; (ii) the investigations on violations of national wildlife laws; (iii) the dissemination and exchange of information on illegal trade activities; and (iv) capacity building including the promotion of awareness <http://www.lusakaagreement.org>.

This section has not listed all relevant national and regional agreements that could be relevant to biodiversity and ecosystem services conservation. Rather, the aim has been to illustrate that international conventions are not the only legal instrument impacting environmental and wildlife management. These national and regional instruments should not be forgotten when assessing the legal context in which scientists and their research can inform the management of wildlife and ecosystems.

10.2 NDVI and the legal sphere

Each of the agreements cited in Section 10.1 is associated with specific goals, targets, and aims (e.g. the current CBD's targets are listed in Table 10.1). Becoming a signatory Party to these conventions thus creates obligations for countries, which need to be implemented when the agreements enter into force. For example, if the CBD's target is to achieve 'a significant reduction of the current rate of biodiversity loss at the global, regional, and national level' (a target that was known as the 2010 target), each Party to this specific Convention is expected to commit to help reach this target. Yet to determine how current conservation efforts can be improved and to guide new strategies, it is crucial that progress toward the targets set by these agreements and beyond is monitored (Pereira and Cooper 2006). Satellite images have already facilitated the decision-making of a number of international environmental agreements. Most notably, imagery from the total ozone mapping spectrometer satellite showing the variable extent and depletion of the ozone layer helped establish consensus for the Montreal Protocol on Substances that Deplete the Ozone Layer (de Sherbinin et al. 2002).

Because the NDVI is a global, continuous, standardized remote sensing product that provides long-term, easily accessible, sensitive, and integrative information, it represents an appealing variable in the context of biodiversity monitoring. Clearly, there are limitations and constraints to its use, many of which have been discussed at length in Chapters 3–9. However, in many cases there is also a clear match between information needs and NDVI access and capabilities. This section illustrates how existing and/or new NDVI-based monitoring frameworks could help to monitor progress toward targets set by multilateral international agreements.

10.2.1 NDVI and management of the global protected area network

Parties of the CBD are currently aiming to set at least 17% of terrestrial and 10% of coastal and marine areas as effective protected areas (Target 11, see Section 10.1). As discussed in Chapter 8, the NDVI can be used to: (i) supply relevant information for the setting of new terrestrial protected areas; and (ii) monitor protected area effectiveness. The NDVI therefore has the potential to support the implementation of the current CBD targets related to the management of the protected area network.

One example of this potential is that of the Digital Observatory of Protected Areas (DOPA; Dubois et al. 2011), which is a platform aiming to support the assessment, monitoring, and forecasting of the state and pressure of protected areas at the global scale; it has been created as a component of the observation network established by the intergovernmental Group on Earth Observations Biodiversity Observation Network (GEO BON; see also Box 11.1). GEO BON is an initiative that aims to co-ordinate activities relating to the Societal Benefit Area on Biodiversity of the Global Earth Observation System of Systems (GEOSS). Some 100 governmental, intergovernmental, and non-governmental organizations are collaborating through GEO BON to organize and improve terrestrial, freshwater, and marine biodiversity observations globally and make their biodiversity data, information and forecasts more readily accessible to policy-makers, managers, experts and other users. GEO BON has also been recognized by the Parties to the CBD. Within DOPA, the eStation web service <http://estation.jrc.ec.europa.eu/> is automatically processing environmental parameters derived from remotely sensed data, including the NDVI, to compute environmental trends and detect anomalies (Figure 10.2), for each protected area (Clerici et al. in press). Data on protected areas, species distributions, and socio-economic indicators are then combined with this remote sensing-based information to generate the necessary environmental indicators, maps, and alerts. DOPA is still in its infancy, but this monitoring platform shows how the NDVI can support the implementation of international and regional agreements.

Averaged temporal trends in NDVI dynamics are not the only type of information that can be used to guide protected area management; Table 10.2 includes examples of national and international studies that have carried out NDVI-based approaches to support protected area monitoring. Valuable information can be derived from changes in the spatial patterns of NDVI dynamics, an approach yet to realize its full potential.

10.2.2 NDVI for monitoring habitat restoration targets

Member states from the European Union are set to restore at least 15% of degraded ecosystems (Target 2, <http://ec.europa.eu/environment/nature/index_en.htm>). Yet how do we identify which areas to restore? And how do we assess whether restoration programmes have been successful? NDVI-based methodologies such as that developed by Prince et al. (2009) (see Chapter 5) might be a cost-effective solution to this monitoring challenge. One can indeed envisage estimating, for all terrestrial ecosystems, the potential and actual production in homogeneous land capability classes. The annual monitoring of the potential and actual production for each spatial unit could yield, at least for some parts of the world, reliable information about changes in degradation rates while identifying large areas experiencing high levels of degradation. Other studies employing the NDVI to assess damage and evaluate ecological restoration success through vegetation patterns include: Sun et al. (2011), using the NDVI to assess the efficiency of water diversion efforts in restoring the lower Tarim River area, a seriously degraded riparian ecosystem in China (Figure 10.3); Spruce et al. (2009), analysing MODIS NDVI time-series data to aid visual assessments of forest damage and recovery from hurricanes in coastal Louisiana and Mississippi; and Tuxen et al. (2008), who implemented a semi-automated technique using colour infrared aerial photography and the NDVI to document vegetation colonization in a restoring salt marsh and to assess the potential for remote sensing to support post-restoration monitoring of tidal-marsh ecosystems. Although the global implementation of these types of approaches still represents technical and logistical challenges, they could provide standardized information to support decision-making in many regions.

10.2.3 NDVI and the Ramsar Convention on wetlands

Wetlands are transition areas between aquatic and terrestrial systems, and among the most biologically productive ecosystems on Earth: they generate myriad benefits to ecosystems and society such as habitat for bird, fish, and other wildlife, playing key roles in biogeochemical hydrological cycles, regulating water quality, reducing shoreline erosion, offering flood protection, moderating climate, and supporting numerous economic activities such as hunting, fishing, and recreation (Keddy 2010). However, wetlands are being threatened or destroyed at alarming rates, with estimates suggesting that, of the world's total

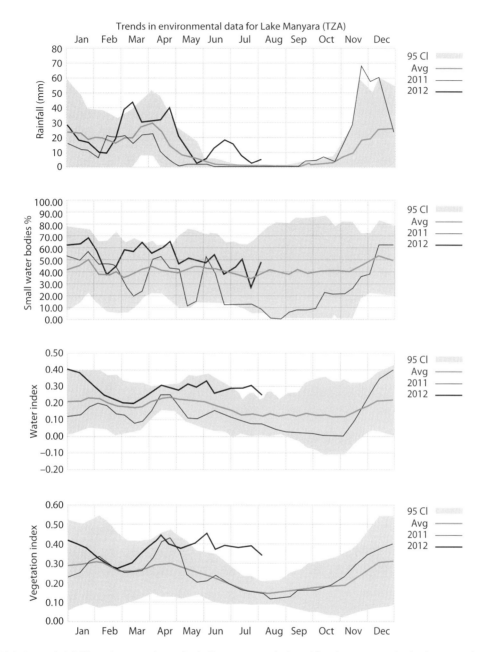

Figure 10.2 Excess of rainfall (in mm), NDVI, and Normalized Difference Water Index (NDWI) for Lake Manyara National Park, Tanzania. These remote sensing based observations have been collected for each 10-day period and are contrasted against 10-year averages (thick grey line). The associated 95% confidence intervals for these averages are also shown (grey areas). Courtesy of Gregoire Dubois, European Commission, Joint Research Centre.

wetlands extant at the turn of the twentieth century, half have been destroyed (Dugan 1993). The NDVI can be a useful, but not optimal, tool to help monitor wetland across the globe. In Vietnam, mean, maximum, minimum, standard deviation, and range of NDVI values from wetland sites were used to monitor vegetation health and heterogeneity, and to supply indirect information on waterbird abundance (Seto and Fragkias 2007). In Malaysia, vegetation indices such as the NDVI were reported to be useful

Table 10.2 Examples of national and international studies that have carried out NDVI-based approaches to support protected area monitoring.

Aim	Protected areas considered	Metrics	References
Track annual changes in average plant production	1015 protected areas of ≥500 km²	Annual Integrated NDVI (INDVI)	Tang et al. 2011
Track spatio-temporal changes in plant production dynamics	Spanish national parks	Annual INDVI; annual maximum (MAX) and minimum (MIN) NDVI; annual relative range (RREL); date of maximum and minimum NDVI values	Alcaraz-Segura et al. 2009
Track the impact of climate change on primary production in Africa	All African protected areas of category I and II	Annual INDVI; MAX; MIN; RREL	Pettorelli et al. 2012
Measure protected area isolation	114 very large protected areas worldwide	NDVI-based contagion metrics	Seiferling et al. 2012
Track temporal changes in plant production dynamics	All African protected areas	NDVI time-eries	Hartley et al. 2007
Track temporal changes in vegetation dynamics	Argentinean protected areas	Annual INDVI; MAX; MIN; RREL	Garbulsky and Paruelo 2004
Track temporal changes in vegetation dynamics across land-cover types	Tanzanian protected areas	Monthly NDVI	Pelkey et al. 2000, 2003

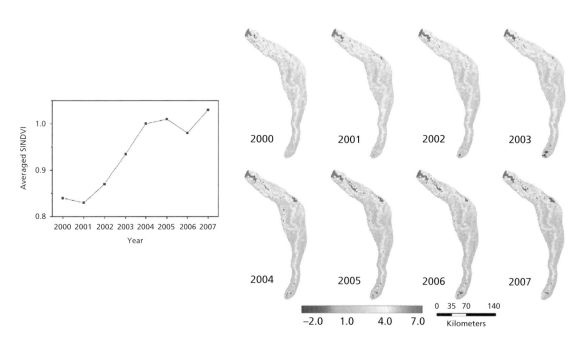

Figure 10.3 Time-series of the spatially averaged seasonally integrated NDVI (SINDVI) over the lower Tarim River (left) and spatial patterns of SINDVI from 2000 to 2007. (Reproduced from Sun et al. 2011 with permission from Elsevier.) SINDVI is NDVI integrated over the growing season. The Tarim River Basin is the largest continental river basin in Central Asia. Due to increasing water consumption and the construction of a reservoir in the 1970s, the groundwater table of the lower Tarim River (which is the study area considered here) fell substantially, leading to natural vegetation declining from 1972 to the late 1990s. In 2000, diversion of water from Bosten Lake to the lower Tarim River was implemented and eight intermittent water deliveries were put in place by the end of 2007. This figure allows assessing the effect of such interventions on the vegetation dynamics in the area. See also Plate 11.

in monitoring and managing mangrove forests for sustainable use in Malaysia (Hasmadi et al. 2011). In North American prairies, spatial and temporal variation in pond counts has been explained by environmental variables related to weather, phenology, and agricultural practices, with phenology-related variables obtained from satellite-derived NDVI data having the highest explanatory power (Herfindal et al. 2012). Other remote sensing products have been suggested to be better for the monitoring of wetlands: RADAR data have been suggested to be particularly useful for monitoring mangroves (Cornforth et al. 2013). However, few alternatives to the NDVI offer access to free information at the global scale and for up to 30 years, making the index perhaps an important component of current monitoring toolkits for wetlands.

10.2.4 NDVI and Reducing Emissions from Deforestation and forest Degradation (REDD)

10.2.4.1 What does REDD mean?

As acknowledged in Chapter 1 and in Section 10.1, there is a link between climate change and forest management, as forest loss and degradation contribute significantly to the release of greenhouse gas in the atmosphere. REDD is a proposed intergovernmental protocol seeking to reduce greenhouse gas emissions by giving forests a monetary value based on their capacity to store carbon. In its basic form, a REDD mechanism would lead to developed countries paying developing ones to reduce emissions caused by deforestation and forest degradation. 'Reducing emissions from deforestation in developing countries and approaches to stimulate action' has been discussed under the UNFCCC since it was placed on the agenda at the request of Papua New Guinea and Costa Rica at the UNFCCC COP 11 in Montreal in December 2005.

10.2.4.2 What does REDD involve?

The idea behind REDD is simple: under a REDD agreement, countries that are willing and able to reduce emissions from deforestation will be financially compensated for doing so. But the development of an international framework to support the implementation of such a scheme is far more complex, as this involves (i) agreeing on what is a forest and what is not a forest; (ii) agreeing on which countries will be eligible to receive compensation under the REDD agreement; (iii) defining a baseline against which deforestation and forest degradation will be estimated; (iv) converting forest loss and degradation into greenhouse gas emissions; (v) estimating the achieved reductions in emissions over an agreed time-interval (DeFries et al. 2007; Parker et al. 2009).

10.2.4.3 REDD and remote sensing

Since the idea of reducing forest emissions was first put forward at the eleventh COP to the UNFCCC, expert groups have been working to define how forest carbon monitoring schemes could be constructed (e.g. Achard et al. 2007; Gibbs and Herold 2007; Olander et al. 2008; Goetz et al. 2009). In 2009, an ad-hoc REDD working group, called the Global Observation of Forest and Land Cover Dynamics (GOFC-GOLD), compiled a working 'sourcebook' of details on how to set up forest monitoring programmes (GOFC-GOLD 2009). This group acknowledged that, in order to participate in any agreed REDD protocol, nations will need to have a reproducible, consistent, and transparent means of estimating forest carbon stocks, in line with internationally endorsed reporting guidelines; also highlighting that it will be important for countries to use consistent methods through time to enable comparison of forest carbon stocks between time-intervals (De Fries et al. 2007; GOFC-GOLD 2009). They concluded, therefore, that remote sensing-based methodologies were likely to offer the best options for supporting the implementation of a REDD protocol.

10.2.4.4 REDD and the NDVI

Changes in NDVI dynamics can be used to map forest loss and degradation (see, e.g., Roy et al. 1997 on European forest cover mapping; Potapov et al. 2008 on mapping boreal forest loss; Martinuzzi et al. 2008 on mapping tropical dry forests in Puerto Rico), but not always with very high accuracy (Roy et al. 1997; Martinuzzi et al. 2008). The NDVI's ability to map changes in forest cover is limited in several parts of the world, due to persistent cloud cover (DeFries et al. 2007). As we saw in Chapter 3, the relationship between the NDVI and vegetation biomass is asymptotic and this can limit the ability of the index to represent accurately changes in vegetation types such as forests (Ripple 1985). As DeFries et al. (2007) acknowledged, however, no single method is appropriate for all national circumstances. Although the NDVI is associated with constraints and limitations, the index can help to support forest stock assessments in certain parts of the world as long as the developed NDVI-based methods are reproducible, yield consistent results when applied at different times, are associated with ground-based measurements, and meet standards for assessment of mapping accuracy.

10.2.5 NDVI and the monitoring of ecosystem services

Ecosystem services can be broadly defined as the benefits that people obtain from ecosystems (see also Chapter 1). They can be divided into four groups: provisioning services (the products obtained from ecosystems, such as food, water, minerals, or pharmaceutical products); regulating services (the benefits obtained from regulation of ecosystem processes, such as carbon sequestration, waste decomposition air and water purification, pollination or pest and disease control); cultural services (the non-material benefits obtained from ecosystems, such as recreational experiences, scientific discovery or spiritual inspiration); and supporting services (the services necessary for the production of all other ecosystem services, such as nutrient cycling, seed dispersal, or primary production). The appearance of the term 'ecosystem services' can be related to a need to link ecosystems conceptually to human welfare, and the identification and management of ecosystem services are slowly becoming topics of increasing importance for ecological research (Norgaard 2009; see also Box 10.2).

It has been previously argued that remote sensing is the method of choice to address the basic issues associated with mapping ecosystem services and quantifying changes in the level of delivery (Feng et al. 2010). From the evidence presented in Chapters 5–8, it is clear that the NDVI is a component of the remote sensing toolkit that has great potential to inform ecosystem services assessment at local,

> **Box 10.2 Ecosystem services, natural capital, and the NDVI**
>
> In parallel with the concept of ecosystem services, the concept of natural capital is becoming increasingly widespread in the literature (e.g. Costanza 2003; Butcher 2006; Brand 2009; Campbell and Brown 2012). There are various definitions, but one easy way to introduce the natural capital idea is to see it as an extension of the economic notion of capital (manufactured means of production) to goods and services relating to the natural environment. The concept of natural capital distinguishes between stock and the annual flow of goods and services it produces; There is also a distinction between renewable and non-renewable natural capital (Costanza and Daly 1992). A subset of the natural capital, known as the critical natural capital, encompasses the part of nature that is life-supporting, 'essential' to human welfare, and irreplaceable (Ekins 2003). There are some evident links existing between the notions of ecosystem services and natural capital, and, as a consequence, the NDVI can be expected to play an important role in assessments of natural capital stocks, and especially of flows.

national, and potentially regional scales. Because of its correlation with the radiation intercepted by vegetation, the NDVI indeed supplies information on one fundamental service underlying terrestrial ecosystems, namely net primary production. Primary production is the energetic foundation of nearly all terrestrial animal communities and provides a basis for many other ecosystem services. High species richness is known to be associated with a high degree of ecosystem services (Balvanera et al. 2006; Benayas et al. 2009), and, as discussed in Chapter 7, the NDVI can supply data on potential species richness in many parts of the world. The NDVI might also help to monitor soil-based ecosystem services: Lobell et al. (2009) showed that vegetation indices such as the EVI (see Chapter 2) and the NDVI could be used to map soil salinity at regional scales, while Schnur et al. (2010) established a significant correlation between the NDVI and root-zone soil moisture in Texas, USA. As IPBES begins to compile evidence about the nature of the relationship between biodiversity and ecosystem services, and as progress is made worldwide in mapping and tracking changes in ecosystem services, one can safely bet that the NDVI will emerge as an important variable to help map and track supporting services (through its link with primary production), provisioning services (through its link with species richness, as well as through its link with various species' abundance and distribution), and regulating services (as primary production and climatic conditions are interdependent variables; see Chapter 4 for examples).

10.2.6 NDVI and the Convention on Migratory Species (CMS)

Migratory species can be a challenge for conservation, as their successful management relies not only on the current and future availability of healthy breeding grounds but also on habitat quality of stopover sites and winter grounds. NDVI conditions at the breeding grounds and winter grounds have been linked to migrants performance (Chapter 7): in springs with low NDVI values, for example, Danish barn swallows *Hirundo rustica* were reported to arrive at their breeding grounds earlier than in good NDVI years, and male adult survival was low in poor NDVI springs compared with good NDVI springs (Szep and Møller 2004). The NDVI was then successfully used by Singh and Milner-Gulland to inform future protected area needs for the saiga antelope *Saiga tatarica*, one of CMS's flagship species (Singh and Milner-Gulland 2011; Chapter 8). Additionally, the NDVI can be used to identify and monitor transit corridors for species such as African elephants in Tanzania (Pittiglio et al. 2012; Chapter 8). Thus the NDVI can help to support management decisions relating to the conservation of terrestrial migratory species throughout their range, by supplying relevant information on changes in the quality of their habitat.

10.3 Conclusions

There are numerous international conventions, as well as regional and national agreements, relevant to the preservation of biodiversity and ecosystem services. Numerous countries are parties to many of these relevant conventions and agreements: for some, meeting the technical, monitoring, and reporting expectations associated with the commitments made can be challenging, given the level of capacity required. There is a role for the international conservation science community to help countries meet these expectations.

The NDVI has the potential to offer access to information relevant to the aims and targets expressed by several multilateral agreements. In all cases, however, the relevance and accuracy of the NDVI data is likely to be context and location dependent, and in many situations reliance on other remote sensing products might be better advised. Nevertheless, the NDVI is frequently the only option to provide long-term, free, accessible data relevant to the monitoring of biodiversity and ecosystem services. In the world of conservation and remote sensing, one should never settle for less than one can get; one should also never dismiss opportunities when they exist.

CHAPTER 11

Future directions and challenges

Prediction is very difficult, especially when it's about the future. **Traditional Danish Proverb**

In the past two decades, a great deal of knowledge has been accumulated on what can be achieved with the NDVI, broadening our perception of its potential use in theoretical and applied sciences. This final chapter discusses ways forward, aiming to identify future opportunities and challenges for the NDVI (and, more broadly, for remote sensing information) to support ecological research and environmental management.

The first section discusses the current lack of recognition concerning the NDVI's role in informing environmental treaties. While elaborating on why this is so, I focus on identifying current and future approaches and mechanisms that could facilitate the introduction of the NDVI in the legal sphere. The second section overviews some of the opportunities for the NDVI to expand its current set of applications, elaborating on what the NDVI datasets of the future might be. Whereas the book has focused mostly on the NDVI for supporting terrestrial ecology and management, there are opportunities for the index to inform the ecology and management of the freshwater and marine environments—these are discussed in the third section. Alternatives to the NDVI do exist, and, in some situations, they should actively be considered: the nature, strengths, and weaknesses of these alternatives are briefly reviewed in the fourth section. The final section discusses the many logistical, technical, and financial challenges preventing an efficient integration of remote sensing information with other types of environmental and ecological data.

11.1 Availability of NDVI data for conservation work

11.1.1 NDVI as an essential biodiversity variable

Faced with accumulating evidence on the strength and speed of the current biodiversity crisis, policymakers have launched several international, regional, and national initiatives to halt, or at least decrease, the pace of biodiversity loss due to human activities (see Chapters 1 and 10). To inform these initiatives, multiple monitoring approaches have been undertaken, leading to more widely available information on the states and trends of single species or single communities. Yet much of the collected information is generally restricted, and gathered at small spatial scales, which has led to various stakeholders calling for the integration of these datasets into indicators of biodiversity components (Balmford et al. 2003a; Buckland et al. 2005; European Environment Agency 2007). An indicator, in this context, is defined as 'a measure based on verifiable data that conveys information about more than itself' (Biodiversity Indicators Partnership 2011).

It has been previously argued that increased integration of all available datasets should benefit all parties interested in biodiversity monitoring, as such a step has been shown to help gain a better comprehension of determinants of population and community trends for birds, which ultimately led to scientists being better able to predict the impact of several human activities on biodiversity (Henry et al.

2008). I would argue here that this line of thought also implies that traditional species-based indicators of biodiversity should be integrated with information on ecosystem processes, such as primary production dynamics.

It is quite remarkable that biodiversity is generally understood as species diversity and viability, and that few biodiversity indicators are related to ecosystems. Processes operating at the species level (e.g. demography, movement) have been perceived as being the most important source of information to track trends in biodiversity by many, and this is well exemplified by the current set of biodiversity indicators recognized by the Convention on Biological Diversity (Table 11.1). In the CBD global indicator framework, for example, three of the five indicators for monitoring the status and trends of the components of biological diversity are species-based; indicators of ecosystem integrity are separate from the indicators listed as providing information on the status and trends of the components of biodiversity (Biodiversity Indicators Partnership 2010).

Biodiversity has been previously defined by three components (namely composition, structure, and function) that should be evaluated at all levels of biological organization, from genes to populations, species, communities, or ecoregions (Noss 1990): composition refers to the identity and variety of entities in a collection; structure refers to the physical organization or pattern of a system (e.g. habitat complexity); function involves ecological and evolutionary processes (e.g. energy exchanges). It is therefore clear that any effort to assess biodiversity patterns and to monitor biodiversity change should consider ecosystem processes as well as species processes. Processes such as primary production not only support terrestrial biodiversity, but also yield information on the condition or 'health' of these ecosystems. Ecosystems are a biodiversity unit in their own right, much like species or genes: they are more than just a sum of organisms sharing the same environment, being defined by the complex set of relationships linking these organisms. One challenge in the coming years is to secure recognition of the monitoring of ecosystem processes as part of biodiversity monitoring, and to integrate this information with other available datasets.

In 2011, the Group on Earth Observations Biodiversity Observation Network initiated a programme to promote this holistic approach to biodiversity monitoring (GEO BON 2011). Recognizing that biodiversity change has many dimensions, including loss of genetic diversity, species extinctions, shifts in species ranges, changes in species abundance, and changes in ecosystem composition, structure, and function, the Group established the concept of Essential Biodiversity Variable (EBV). An EBV is defined as a quantity, based on observations for large parts of the Earth, which is required for the long-term management of biodiversity at national to global scales and especially for the detection of change (Pereira et al. 2013; Figure 11.1). This idea to define and identify EBVs is inspired by the development over the last decade of the Essential Climate Variables, which now guide the implementation by Parties to the UNFCCC of a comprehensive Global Climate Observing System.

There are potentially many hundreds of biodiversity variables, yet the aim of this Group is to identify the essential ones, using criteria such as temporal sensitivity, scalability, feasibility, and relevance to

Table 11.1 Convention on Biological Diversity global indicator framework.

Indicator on status and trends of biodiversity components	Applicable at national level?	Global indicator reliance on nationally reported data?
Extent of forests and forest types	Yes	Yes
Extent of marine habitats	Yes	Yes
Living Planet Index	Yes	No
Global Wild Bird Index	Yes	Yes
Coverage of protected areas	Yes	Yes
Protected area overlays with biodiversity	Yes	No
Management effectiveness of protected areas	Yes	Yes
IUCN Red List Index	Yes	No
Ex situ crop collections	Yes	Yes
Genetic diversity of terrestrial domesticated animals	Yes	No

Source: Biodiversity Indicators Partnership (2010).

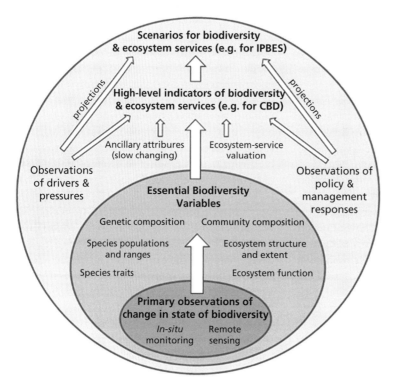

Figure 11.1 A framework for Essential Biodiversity Variables (EBVs). This framework allows positioning EBVs in the gradient of biodiversity-related information, ranging from primary observations of change in biodiversity (which are gathered using *in situ* monitoring and remote sensing) to high-level indicators of biodiversity and ecosystem services (such as the Living Planet or the Red List). This figure is courtesy of Henrique Pereira and the GEO-BON community. Pereira et al. (2013) should be consulted for more information on the EBV framework. (Reproduced with permission from the American Association for the Advancement of Science.)

important issues in biodiversity monitoring. Primary production is a key ecosystem process for the maintenance of biodiversity, and this process can be monitored at multiple spatial and temporal resolutions and scales using satellite-based information. It could therefore be argued that primary production is likely to be an EBV. However, there are various ways to monitor primary production using remote sensing information (see Chapter 2): the NDVI is only one of the many indices or parameters that could be used to monitor changes in this ecosystem process. Nevertheless the NDVI-based method probably offers the best added value. It is an index that has been correlated with more than just primary production, providing information that can be used directly to inform environmental and wildlife management and conservation. The NDVI is already part of the set of variables used by some regional assessments and monitoring programmes, such as the Assessment of African Protected Areas project (Hartley et al. 2007) and the recent Digital Observatory for Protected Areas (<http://dopa.jrc.ec.europa.eu>; see Section 10.2 and Box 11.1). The local, regional, and global integration of NDVI-based monitoring with available information on the states and trends of single species or single communities could be very powerful in gaining a better comprehension of determinants of population and community trends, increasing our ability to predict the impact of global environmental change on biodiversity.

11.1.2 Limitations on NDVI support for implementing conservation work

Chapters 5–10 argue that vegetation indices such as the NDVI have the potential to support the implementation of treaties, conventions, and other national agreements, yet the NDVI is not currently listed as a key indicator by any of the major international conventions. One major limitation is perhaps the technical and financial challenges associated with the access to, and manipulation of, remote sensing information. The methods, costs, and benefits of traditional biodiversity indicators are indeed relatively familiar to environmental decision-makers,

Box 11.1 The Digital Observatory for Protected Areas (DOPA)

Biodiversity is one of the nine key themes by which the activities of the intergovernmental Group on Earth Observations are organized. GEO BON is GEO's branch dealing with biodiversity issues: one of the network's aims is to integrate biodiversity data by bringing together the large variety of sensors, databases, and systems. DOPA is a platform conceived for such a purpose by the European Union's Joint Research Centre in collaboration with other international organizations including the Global Biodiversity Information Facility, the UNEP–World Conservation Monitoring Centre, Birdlife International, and the Royal Society for the Protection of Birds. It supports the assessment, monitoring, and forecasting of the state and pressure of protected areas at the global scale using a set of distributed databases combined with open, interoperable web services (Dubois et al. 2011). Within DOPA, remotely sensed data, derived from the Earth Observation measurement platforms such as SPOT Vegetation, the Spinning Enhanced Visible and Infrared Imager (SEVIRI)/Meteosat Second Generation Instruments (MSG) and TERRA-AQUA from MODIS, are processed to compute environmental trends and detect anomalies, for each protected area on Earth. Data on protected areas, species distributions, and socio-economic indicators are then combined with this remote sensing-based information to generate the necessary environmental indicators, maps and alerts. DOPA is still in its developing phase, and its usefulness depends on the biodiversity community's aptitude to work together and share information. The more information on biodiversity is shared on DOPA, the greater the ability for this platform to generate key information to support protected area management.

as opposed to the complexity associated with the various components of remote sensing, which may preclude its adoption (Strand et al. 2007). Conservation dollars to address the biodiversity crisis are limited and fragmented, and such limitations have a profound influence on the planning methods and conservation strategies of governmental and non-governmental organizations. As reported by Strand et al., member states of the Convention on Biological Diversity 'identified the costs associated with routine updating of satellite-based data to be a major challenge . . . and the capacity to use spatial data to identify, analyse and interpret decision makers' key questions was not always available.'

In order to gain widespread support for the use of remote sensing in management and conservation, remote sensing data and analyses need to become more accessible to the community of ecologists (see Section 11.5). A variety of products based on Earth observation data are freely available, such as most NDVI datasets at low-to-medium spatial resolutions. But the majority of products still need to be purchased, and certain trade-offs currently exist: typically the higher resolution the data, the higher the cost. The amount of data gathered by modern remote sensing systems is vast. For example, Landsat Thematic Mapper generates more than 104 MB of data per second, and much higher rates may be expected from the larger space platforms of the Earth Observation System (Barrett and Curtis 1999). Satellite-based data analysis has a non-negligible cost, depending upon access to the relevant data, access to computer and memory resources, access to software packages for data analysis and display, as well as access to the relevant expertise in large dataset processing and analysis. These costs can be considerable, especially so for developing countries where such monitoring would be the most beneficial. There are open-source solutions for the processing and analysis of remote sensing data, e.g. R (R Development Core Team 2013) and GRASS (GRASS Development Team 2013); but the frequent lack of clear documentation as well as the partial absence of a graphical user interface can render open-source solutions time-consuming. One hopes that the user-friendliness of open-source solutions will continue to increase. But for now, commercial software solutions such as ENVI (Exelis Visual Information Solutions 2013) and ArcGIS (ESRI 2013) can be expected to continue to dominate the education and training components of potential workforces.

As decision-makers ideally wish to use the same set of indicators in all countries and at all scales, another obstacle to the adoption of the NDVI for conservation agreements may be that its ability to monitor variation in primary production is not spatially and temporally constant (Chapter 3). However, the same argument could be used against some of the variables already classified as 'indicators' by the CBD. Moreover, the NDVI might offer advantages that other variables and indicators do not, such as access to more than 30 years of information or global coverage.

11.1.3 Not everything needs to be developed at the global scale

International treaties and conventions have shaped in many ways efforts to track biodiversity change and inform environmental management. Typically, the focus has been to identify biodiversity indicators that can be used everywhere, and to develop monitoring platforms that are associated with world coverage. DOPA (Box 11.1) is a good example. Factors such as average level of cloud contamination, average above-ground biomass density, study area size, or sought accuracy and resolutions are all factors that are spatially and temporally variable and that affect the quality of the information that can be gained from remote sensing data. Variables, information, and indicators associated with high levels of spatio-temporal inconsistency in their ability to perform are being ignored by international programmes for good reasons, but this has also resulted in them being ignored by more local initiatives. This can lead to a loss of opportunity, as variables, information, and indicators associated with spatio-temporal inconsistency at the global scale can also be associated with high level of reliability and accuracy in particular regions and locations. The point is that tailor-made solutions designed to perform best at regional, national, or even at local scales are part of the way forward: clearly, there needs to be a balance between the costs and benefits associated with the development of each platform, and in this respect it makes a lot of sense to focus on indicators that work everywhere. But there is also a niche for small-scale approaches and tailor-made initiatives, especially in ecologically important areas and conservation hotspots.

11.2 NDVI datasets in the future

11.2.1 NDVI at micro-scales?

LiDAR refers to a type of active sensor generally used by ecologists and environmental managers to retrieve forest parameters such as tree height, crown diameter, number of stems, stem diameter, basal area, vegetation structure or above-ground vegetation biomass (Wulder et al. 2012b; Chapter 1). Initiatives aiming to couple information derived from LiDAR with information derived from optical sensors, such as the NDVI, have started to appear (Tan and Narayanan 2004; Koetz et al. 2007; Morsdorf et al. 2009). In particular, projects aiming to utilize, in a single instrument, both the capacity of multispectral sensing to measure plant physiology (through, e.g., the NDVI) and the ability of LiDAR to measure vertical structure information are currently being developed (Woodhouse et al. 2011). Such technical advances are particularly exciting, as the successful development of such an instrument would mean that the NDVI could be measured through a canopy, allowing researchers to access a vertical profile, as well as the classical horizontal one, of the distribution of photosynthetic activity in a given pixel. Referred to as multispectral canopy LiDAR (MSCL), the current tunable laser operates at four wavelengths (531, 550, 660, and 780 nm) and has been successfully used on live trees in a laboratory setting to measure NDVI at various heights (Woodhouse et al. 2011).

11.2.2 Maintaining access to continuous, freely available NDVI datasets

Earth observation products such as the NDVI are central to our ability to monitor and predict the impact of global environmental change on biodiversity and the delivery of ecosystem services. The longest global-coverage NDVI time-series are provided by the joint USGS and NASA Landsat Program and the US NOAA AVHRR instrument. Ensuring the longevity and continuity of Earth observation programmes such as those is an imperative, yet not a given. Economic crisis, lack of political will, bureaucratic machinations, and technical problems can suddenly put a stop to data continuity. Difficulties of securing a successor to Landsat 7 have been well documented, and despite the incredible popularity and success of the Program, there was in 2012 no fully operational Landsat satellite (Wulder et al. 2012a). However, the situation should improve following the launch of Landsat 8 in May 2013. Having access to long-term, comparable Earth observation data is what enables ecologists and environmental managers to increase their understanding of the natural world, to make predictions about how this world might be transformed, and design efficient mitigation and adaptation strategies in the face of global environmental change. This richness of information cannot be found in data collected by new sensors gathering information in previously ignored bands. In this respect, accuracy needs to be balanced against continuity and inter-operability. The importance and difficulty of maintaining long-term datasets is not a new topic to ecologists: many have struggled for years to secure that extra bit of funding that would enable them to survey their population for another year; many have spent hours trying to convince possible funding agencies that it is worthwhile to continue investing in their project, even though their project has been continuing for ten years. This community has, however, clearly established how many ecological processes can only be revealed through the acquisition of long-term data, and is therefore very well placed to appreciate how the termination of data processing programs such as the one supporting the GIMMS dataset could have a devastating effect on ecological research. Fighting for the maintenance of long-term Earth observation data, such as the NDVI data, is a battle that conservationists and ecologists need to enter.

The cost of satellite imagery matters. Ecology is not a rich science, especially when entering the realm of conservation biology. Despite the vast amount of useful information that can be derived from satellite data, few projects are in a position to pay for the acquisition of remote sensing imagery. Several years ago, NASA decided to open its archives for free to the user community, and it was at about this time when satellite data started to become increasingly used in ecological studies (see, e.g., Woodcock et al. 2008; Wulder et al. 2012a; Figure 11.2). The

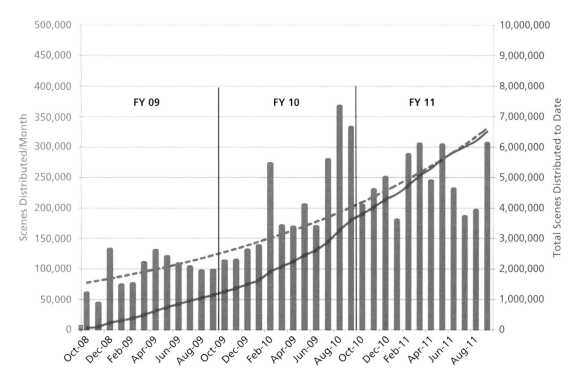

Figure 11.2 Monthly summary of scene downloads from the Earth Resources Observation and Science (EROS) Data Center, covering the period from October 2008 to September 2011, further delineated by US Government fiscal year (FY). (Reproduced from Wulder et al. 2012a with permission from Elsevier.) The open-access data policy <http://landsat.usgs.gov/documents/Landsat_Data_Policy.pdf> was adopted in 2008. Prior to October 2008, no calendar month ever recorded more than 3000 scenes sold in a given month.

majority of spatial agencies worldwide did not follow this bold move, and continued to charge users for data acquisition: NASA therefore experienced a loss of revenue, while being confronted with a steady increase in the need for technical support. Such increased pressure on NASA means that in future there may no longer be freely accessible Earth observation data. Data continuity, affordability, and access are essential to make satellite images a more established part of biodiversity monitoring, and thus support international efforts to reduce current rates of biodiversity loss (Leidner et al. 2012). As end-users, ecologists need to be aware of the pressures faced by space agencies, support open-access policies to be implemented by all major space agencies worldwide, and, whenever possible, voice their concerns over potential changes in data access.

11.3 NDVI in the freshwater and marine environments

This book has focused primarily on the NDVI for terrestrial ecologists, yet, as we saw in Chapter 10, opportunities do exist for the NDVI to support freshwater and marine ecology and conservation. The NDVI correlates with photosynthetic activity, and can therefore be used to identify the presence of photosynthetically active organisms on the water surface.

11.3.1 NDVI to detect and monitor aquatic vegetation

Aquatic vegetation is not as easily detectable as terrestrial vegetation in remotely sensed images, but in some situations NDVI data may be relevant for the aquatic realm.

Rodgers III et al. (2009) assessed the impact of hurricane Katrina on the coastal vegetation of the Weeks Bay Reserve, Alabama, from NDVI data. Hurricane Katrina hit the US Gulf coast on 29 August 2005 and the event was suspected to have a tremendous impact on the coastal vegetation in the area. Possible changes in estuary salinity and exposure to the hurricane-force winds were hypothesized to have negatively affected the vegetation of the Weeks Bay Reserve, an area designated as a National Estuarine Research Reserve since 1986. The aim was to assess the extent and magnitude of the vegetation damage within the reserve, using satellite-based information. They used Landsat satellite images acquired before landfall (24 March 2005), after landfall (16 September 2005), and eight months after landfall (28 April 2006), and analysed changes in NDVI patterns. Average NDVI values decreased by 49% after landfall, being 44% lower in April 2006 than they had been in March 2005. Among habitat types, estuarine emergent wetland experienced the largest average NDVI value decrease (–64%). Estuarine emergent wetland NDVI values continued to decrease by 27% from September 2005 to April 2006, whereas other habitats increased in NDVI. This continued reduction in greenness therefore confirmed expectations, further supporting the idea that increased salinity from the storm surge and regional drought conditions that occurred after landfall precipitated a decline in aquatic vegetation biomass in the estuarine emergent wetland (Rodgers III et al. 2009).

The NDVI has also been successfully used to map changes in wetland distribution and health (Klemas 2011), one recent example being the Sundarbans, the largest contiguous mangrove forest in the world situated in both India and Bangladesh (Rahman 2000). In 2007, a considerable area of the Sundarbans was affected by Cyclone Sidr (which struck on 15 November). Damage on the ground was reported, but degradation resulting from Sidr was also assessed for part of the ecosystem using remote sensing approaches. NDVI information extracted from MODIS imagery, a week before and after the cyclone, was used to assess the spread of damage, albeit at a very low spatial resolution (Bangladesh University of Engineering and Technology 2008).

In The Netherlands the NDVI was used to map salt-marsh plants (van der Wal and Herman 2012): yearly maps of the NDVI were derived from geo-corrected and atmospherically corrected airborne hyperspectral images with a 2–4 m spatial resolution. Using an NDVI-based threshold of 0.37, the information collected by the airborne sensors was used to map salt-marsh plant distribution, and the results were confirmed using ground photographs of the stations taken during sampling. In some situations, the NDVI can also be used to inform restoration efforts for wetlands: Tuxen et al. (2008) implemented an NDVI-based technique to document vegetation colonization in a restoring salt marsh.

11.3.2 Monitoring eutrophication

As human population size continues to increase, agricultural run-off, urban run-off, leaking septic systems, and sewage discharges are having an increasing impact on the rate at which nutrient and organic substances enter aquatic environments. Excess nutrients (e.g. nitrates, phosphates) stimulate aquatic plant growth, and this can interfere with the health and diversity of aquatic life. Vegetation blooms can negatively impact aquatic systems in several ways: they can cloud the water and block sunlight, which can lead to the death of underwater grasses and the destruction of important habitats such as fish-spawning and nursery areas; vegetation blooms can disrupt the stability of animal communities by favouring some species over others; the death and decomposition of this 'extra' vegetation can then create hypoxic conditions, which may lead to the death of many benthic organisms. This process of eutrophication has been cited many times as one of the most serious water quality problems worldwide. Satellites data and derived indices, such as the NDVI, may supply information on the spatio-temporal variation in the vegetation coverage of affected areas. In Venezuela, for instance, NDVI data derived from MODIS were used to monitor duckweed blooms and other floating vegetation in Lake Maracaibo (Kiage and Walker 2009). In Lake Victoria, East Africa, the index was used to explore the link between the occurrence of El Niño events and water hyacinth blooms in the Winam Gulf (Kiage and Obuoyo 2011). NDVI-based algorithms can also be used to detect harmful algal blooms, as acknowledged by Shen et al. (2012), who recently summarized the suitability of current satellite data sources and different algorithms for detecting such blooms. All in all, these studies highlight the potential for the NDVI to support eutrophication monitoring efforts in freshwater ecosystems.

11.3.3 Monitoring invasion

Invasive species can have a particularly strong impact on local biodiversity (Chapter 8), and this also holds in freshwater and marine ecosystems. *Caulerpa taxifolia*, for example, is a seaweed species thought to have been accidentally released into coastal waters of the Mediterranean in 1984. In less than a decade, this species had grown to cover thousands of acres, preventing native plants from growing. Referred to as 'killer algae,' *Caulerpa taxifolia* is one of two algae on the list of the world's 100 most harmful invasive species (IUCN Invasive Species Specialist Group 2012). As well as helping to monitor eutrophication, the NDVI has been useful in detecting and tracking invasions of photosynthetically active organisms in freshwater and marine ecosystems. In Lake Chivero, Zimbabwe, field data were combined with Landsat-based NDVI information to estimate the spatial extent and biomass of different aquatic weed species (Shekede et al. 2008; Figure 11.3). The weed coverage in Lake Chivero declined from 42% in 1976, 36% in 1989 to 22% in 2000. One of the aquatic weed species, *Typha capensis*, had more biomass than any other weed type in this ecosystem. The authors concluded that remote sensing and the NDVI are an invaluable asset for detecting invasions, assessing infestation levels, monitoring rate of spread, and determining the efficacy of weed mitigation measures.

11.3.4 NDVI associated with freshwater wildlife

Few authors have explored the link between terrestrial NDVI variability and freshwater wildlife; nevertheless some studies have posted encouraging results. In Finland, catchment productivity as indexed by the NDVI showed a relatively strong association with planktonic richness in 100 small lakes (Soininen and Luoto 2012). In Switzerland, spatial variation in abundance of salamander *Salamandra salamandra* larvae was positively and significantly correlated with photosynthetic biomass in the vineyards as indexed by the NDVI (Tanadini et al. 2012). Species richness of amphibians was positively correlated with both the annual NDVI and the NDVI during the course of the summer for the 245 localities considered across China (Qian et al. 2007). The

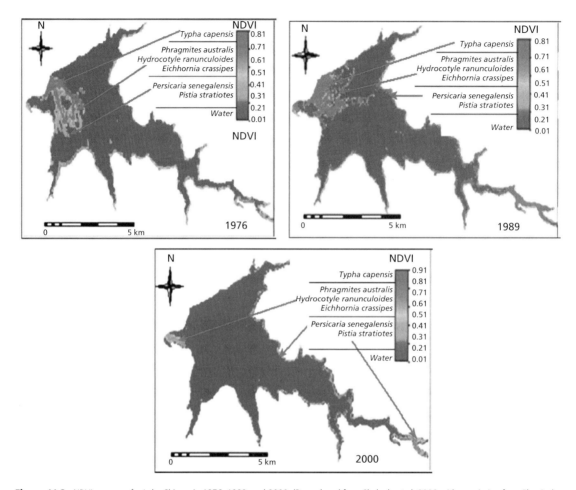

Figure 11.3 NDVI coverage for Lake Chivero in 1976, 1989, and 2000. (Reproduced from Shekede et al. 2008 with permission from Elsevier.) Information on the distribution of the main weed species is also shown. See also Plate 12.

NDVI was then shown to be an important variable structuring the distribution of Schistosomiasis risk estimation in Minas Gerais State, Brazil, a disease whose transmission is directly linked to the distribution of a freshwater snail in the genus *Biomphalaria* (Guimaraes et al. 2008). In Nebraska, larval densities of Platte River caddisfly *Ironoquia plattensis* were found to be higher in high-NDVI ungrazed areas, compared with grazed plots (Harner and Geluso 2012). This small compilation of case studies seems to suggest that catchment productivity influences freshwater wildlife distribution, meaning that further research efforts should consider the NDVI as a potentially important factor structuring observed spatial patterns in freshwater wildlife data.

11.4 Alternatives to the NDVI

11.4.1 NDVI versus EVI

The EVI is meant to take full advantage of the new state-of-the-art measurement capabilities of MODIS (Tucker and Yager 2011) and was originally designed to provide information on the spatial and temporal variation in the dynamics of the green biomass, while minimizing many of the contamination problems present in the NDVI (Huete et al. 2002) (see also Chapter 2). The EVI uses the blue band to remove residual atmospheric contamination due to smoke and sub-pixel thin clouds, and feedback adjustment to minimize canopy background variation and to enhance vegetation sensitivity from sparse to dense vegetation conditions. The MODIS NDVI and EVI are almost perfectly correlated for most vegetation conditions from sparse coverage to very dense plant canopies (Tucker and Yager 2011). However, it has been argued that the EVI is more responsive in high-biomass regions because it responds primarily to the near-infrared band, which does not saturate (Gamon et al. 1995; Huete et al. 2002). Nonetheless, several limitations to the use of the EVI have also been highlighted in the literature. First, the EVI has been developed on MODIS data, and these data are only available from 2000 onwards. Second, studies have shown that the relationship between the NDVI and the LAI (see Box 2.1 for a definition of the Leaf Area Index) is maintained throughout the year, whereas the relationship between the EVI and the LAI breaks down during the leaf senescence stage (Wang, Q. et al. 2005). Third, it has been shown that the relationship between the EVI and productivity is weak for forests with little seasonal variation (Sims et al. 2006), and Sesnie et al. (2012) recently challenged the usefulness of MODIS-derived EVI over MODIS-derived NDVI in tracking forage phenology for bighorn sheep *Ovis canadensis mexicana* in the Sonoran Desert. In this latter study, the authors highlighted the higher sensitivity of the MODIS-derived EVI—compared with the MODIS-derived NDVI—to Sun elevation, Sun elevation angle, and terrain effects.

11.4.2 NDVI versus MODIS-based NPP and GPP

With the launch of MODIS on board Terra in 1999, several new products have been made available to end-users, two of them being the gross primary productivity (GPP) and net primary productivity (NPP) datasets. Like the EVI, these were designed to better represent primary production and improve on the NDVI's noted weaknesses attributed to backscatter and saturation. The NPP and GPP datasets are derived from the same spectral bands as the NDVI, but they also incorporate information on land cover, climate, and vegetation characteristics. The GPP product has an eight-day temporal resolution and the NPP product has an annual resolution, both at 1 km. Some authors have argued that products such as NPP or GPP might have higher utility in ecological studies than the NDVI: Phillips et al. (2008) reported that the NDVI explained substantially less variation in bird species richness in North America than GPP or NPP in areas with more bare ground and in areas of dense forest. Such claims, however, need to be weighed against the reported challenges of validating GPP and NPP estimates across biomes, in different vegetation types, and in the tropics (e.g. Morisette et al. 2002; Turner et al. 2006; Tucker and Yager 2011). Moreover, like the EVI, these products are based on data collected by MODIS, meaning that no data are available pre-2000.

11.4.3 NDVI versus FPAR

The Fraction of Photosynthetically Active Radiation (FPAR; 0.4–0.7 μm) is a measure of the percentage of radiation available for photosynthesis that is absorbed by the vegetation, and can thus be used to characterize absorption capacity and greenness dynamics (Myneni et al. 2002). FPAR is expressed as a unitless fraction of the incoming radiation received by the land surface. Available FPAR datasets are generally derived from remote sensing-based information: FPAR data can be extrapolated for the entire world by relating reflectance data (or vegetation index data, such as the NDVI) to field measurements of FPAR using linear regression techniques. A typical FPAR dataset found in several ecological studies is the one extracted from the information collected by MODIS. As for the NDVI, several ecological studies have reported good correlations between wildlife distribution and FPAR variation: Ripple and Beschta (2012b) showed how herbivore densities correlated with FPAR variations in northern forest ecosystems; Herfindal et al. (2005) demonstrated a negative correlation between FPAR and lynx *Lynx lynx* home-range sizes, a correlation that was also reported to hold for other carnivore species (Nilsen et al. 2005); higher population densities of wild boar *Sus scrofa* were reported in areas with higher FPAR values in western Europe (Melis et al. 2006). Contrary to the EVI, FPAR products based on AVHRR data do exist, meaning that FPAR data pre-2000 is available. Recent analyses comparing FPAR derived from GIMMS NDVI and MODIS showed that both products are consistent—for example, in capturing well the seasonal variation in primary productivity for Eurasia (Peng et al. 2011).

11.4.4 Albedo

Surface albedo, defined as the fraction of solar energy reflected by the Earth back into space, is one of the key geophysical variables controlling the surface radiation budget, and is widely used in climate and hydrological models. In general, deserts as well as snow- and ice-covered areas have very high albedos, reflecting much solar energy back into space. Vegetation, however, has relatively low albedo values, with e.g. forest albedos being lower than those of grassland and croplands (Horning et al. 2010). Surface albedo can be derived from satellite data, and might be especially suitable for anyone dealing with ecosystems characterized by low vegetative cover: for instance, Otterman et al. (2002) used changes in albedo to detect overgrazed areas in deserts. Its use is also increasing in forested regions: Lyons et al. (2008) used changes in albedo to detect changes in fire disturbance regimes in forests (Figure 11.4). Though still rare in ecological research, albedo-based studies seem to be associated with real opportunities to inform environmental and wildlife management for several ecosystems.

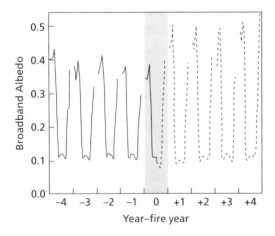

Figure 11.4 Mean broadband albedo before and after all fires that occurred in Alaska during the 2000–2004 period. (Reproduced from Lyons et al. 2008 with permission from John Wiley & Sons.) This figure combines MODIS observations of broadband albedo with information from the Alaska Large Fire Database, enabling the evaluation of changes in surface albedo dynamics after fire within interior Alaska.

11.4.5 Hyperspectral remote sensing products

As briefly introduced in Chapter 2, hyperspectral imagery relies on sensors typically measuring reflected light at wavelengths between 350 and 2500 nm using 150–300 contiguous bands of 5–10 nm bandwidths. Currently active hyperspectral sensors are mostly installed on board planes (with the exception of Hyperion), and include Airborne Visible/Infrared Imaging Spectrometer, Compact Airborne Spectrographic Imager, Airborne Imaging Spectroradiometer for Applications, and HyMap (He et al. 2011; see Table 11.2). These sensors generally collect information at high spatial resolution but low temporal resolution, and can enable the differentiation of plant species (Chapter 1; Ustin et al. 2004, 2009).

Chapters 6, 8, and 11 have briefly introduced how the NDVI can be used to detect, map, and predict the distribution of invasive species (terrestrial and aquatic). Although the NDVI can work in some situations, many authors have highlighted the real benefits associated with the use of hyperspectral sensors to monitor and predict the spread of invasive plant species (e.g. Underwood et al. 2003; Lass et al. 2005; Andrew and Ustin 2010; Ustin and Gamon 2010; Vitousek et al. 2011). An excellent example comes from Hawaii, where Asner et al. (2008) showed that the hyperspectral signatures of Hawaiian native trees were generally distinctive from those of introduced trees. These sensors enabled monitoring of the impact of the invasion at the ecosystem level, by collecting data on biochemical changes found at the foliar and canopy levels. Hyperspectral imagery might therefore represent a powerful alternative to the NDVI for invasive species monitoring. However, hyperspectral data can be expensive: although data collected by Hyperion at a 30 m resolution can now be obtained free of charge, the price of very high spatial resolution data can reach between $60,000 and $100,000 for a 20 × 20 km area at 2 to 3 m resolution (e.g. Lass et al. 2005). Data processing is moreover a complex task, such that most ecologists will require the services of experts specialized in hyperspectral imagery processing. Data storage capacity may also become an issue for many users, as data volume is generally huge (He et al. 2011).

Table 11.2 Hyperspectral sensor information in terms of the types of sensors, platforms, sensor characteristics, source of data and availability.

Sensor	Sensor characteristics	Source of data	Availability (reference site)
Airborne Imaging Spectroradiometer for Applications (AISA) hyperspectral imagery	492 bands; spectral range: 395–2503 nm; spatial resolution: 75 cm to 4 m	Specim, Spectral Imaging Ltd, Finland	<http://www.specim.fi/products/aisa-airborne-hyperspectral-systems/aisa-series.html>
Airborne Visible InfraRed Imaging Spectrometer (AVIRIS)	224 bands; spectral range: 400–2500 nm; spatial resolution: 3.5 m	NASA, USA	<http://aviris.jpl.nasa.gov/>
Compact Airborne Spectrographic Imager (CASI)	288 bands; spectral range: 430–870 nm; spatial resolution: 3 m	ITRES, Canada	<http://www.itres.com/>
EO-1 Hyperion hyperspectral sensor (Spaceborne)	220 bands; spectral range: 357–2576 nm; spatial resolution: 30 m	NASA, USA	<http://eo1.gsfc.nasa.gov/Technology/Hyperion.html>
HyMap (Hyperspectral Mapper, Airborne)	126 bands; spectral range: 450–2500 nm; spatial resolution: 3 m	HyVista, Integrated Spectronics Pty Ltd, Australia	<http://www.hyvista.com/>

Reproduced from He et al. (2011) with permission from John Wiley & Sons.

11.4.6 RADAR and LiDAR

Active sensors such as RADAR and LiDAR have now been clearly identified as the tool of choice for monitoring the above-ground biomass in forested areas (e.g. Goetz et al. 2009; GOFC-GOLD 2009). Funding permitted, the use of these instruments to track deforestation and forest degradation is highly recommended over the use of the NDVI. The adequacy of RADAR to monitor coastal retreat in forest mangroves has also been confirmed (Cornforth et al. 2013). Similarly, the potential of the network of weather radars in the USA to remotely sense animal phenologies was recently established, with Kelly et al. (2012) employing NEXRAD radar products for quantifying the phenology of the purple martin *Progne subis* at summer roost sites. Yet these active sensors too have their limits. RADAR can penetrate clouds, but has limited ability to monitor above-ground biomass in dense tropical forests (Kasischke et al. 1997). LiDAR is sensitive to clouds, which means that the successes shown in measuring the three-dimensional structure of dense forest only applies to cloud-free areas (Horning et al. 2010). At present, there are no functional, satellite-mounted LIDAR sensors. Access to RADAR and LiDAR data is generally not free and can be expensive, although there are exceptions (e.g. <http://lidar.cr.usgs.gov>).

11.4.7 The next generation of sensors

Several launches of regional and national satellites are scheduled for the coming years, which could greatly impact the set of environmental data available to ecologists. Most notably, NASA and the USGS have launched the most recent Landsat in May 2013, a mission known as the Landsat Data Continuity Mission or Landsat 8. At around the same time, the European Space Agency will start launching a series of satellite constellations (known as Sentinels). ESA's two Sentinel-2 missions, planned for launch in 2013 and 2014, will provide medium resolution (10 m to 60 m) imagery of global land surfaces and coastal waters every five days (Landsat currently has a revisit time of 16 days). Together with Landsat 8, these satellites will enable observation of any area on the Earth's surface with landscape-scale data almost daily—something that, if realized, will greatly improve opportunities for remote sensing information to support biodiversity research and conservation. Other notable events include the launch of VENµS (Vegetation and Environment monitoring on a New Micro-Satellite) in 2014, and the launch of EnMAP and HISUI onboard ALOS-3 (both to be launched in 2015). Highly specialized systems such as the SMAP (Soil Moisture Active Passive) system are also scheduled to be launched in 2014, which may supply valuable data on the Earth's spatio-temporal variability in soil moisture.

11.5 Constraints on the use of remote sensing information

11.5.1 Lessons from the NDVI and its success in ecology

The success of the NDVI in biodiversity research and conservation is likely to be shaped by many factors, but some in particular can be easily identified: as argued in Sections 11.1 and 11.2, ecology needs long-term, continuous, free, accessible information. Securing long-term data collection, and making remotely sensed data free and accessible to conservation and research projects, are key steps to facilitate progress by the research community to tackle some of the challenges our societies face. These are the steps that are likely to guarantee the success of any remote sensing product in ecological research and environmental management.

11.5.2 Remote sensing for all

The NDVI is far from being the only remote sensing product that has the potential to make a difference in ecological research and applications. There are many Earth Observation data products relevant to the biodiversity research and conservation applications communities, but often it is unclear which products are the best to use, or it is hard to understand the benefits and drawbacks of particular products. Relevant remote sensing datasets can also still be difficult to locate. User-friendly, intuitive, and centralized data portals need to be developed, as these could significantly increase product use. They could enable better communication and exchange of experiences about remote sensing products. It would of course be even more advantageous if these sites could be better linked or integrated into a single web service.

Training opportunities in remote sensing tailored for the biodiversity and conservation community are still extremely rare, and this lack of opportunities hampers the emergence of a new generation of scientists able to carry out integrated, multi-disciplinary approaches. Although open-access solutions are on the rise, there is also still a need to increase access to free tools, both in terms of open-source software development and of increased training opportunities focusing on the use of these free tools. This is especially important to reach a wide user community in applied conservation in developing countries, where access to expensive software and analysis tools can be limited. More broadly, capacity building within the ecological and conservation communities would address many of the barriers related to limited skills, lack of knowledge, and software/data access and usability. Additionally, developing new formal and informal networks between the biodiversity/conservation communities and remote sensing scientists will also dissolve many of the technical barriers that these communities have to face while using remotely sensed data, and can serve as a way for the biodiversity/conservation communities to share their experiences using Earth Observation data.

11.5.3 Bridging the gaps between communities

Despite the promotion in the literature of the NDVI, and more broadly, remote sensing science to inform conservation biology and environmental management, these two research communities have evolved independently for many years and have only recently started to co-ordinate their agendas. Such synchronization is key to improving the potential for remote sensing data to effectively support environmental management decisions. Integrative approaches are vital to allow a better ecological understanding of the mechanisms shaping current changes in biodiversity patterns. These are also the trigger behind the development of innovative approaches, new research directions in remote sensing science, and the development of new remote sensing products. Opportunities exist to facilitate the co-ordination of these research agendas. For instance, the Global Biodiversity Informatics Conference could shape the future of bioinformatics and further underline the steady growth of remote sensing from emerging technology (e.g. drones, or new sensors such as EnMap and Sentinel). The Group on Earth Observations Biodiversity Observation Network, and its current work on identifying Essential Biodiversity Variables, provide an excellent platform for the community to meet and develop collaboration, e.g. by addressing gaps in current Earth Observation products. Citizen science projects could also refine products and strengthen the link between different communities, as well as assist in building databases on wildlife distribution. This kind of approach may enhance the use of satellite imagery for biodiversity and conservation, for example, by providing more reference information or by enhancing online mapping contributions of roads and buildings (e.g. Bontemps et al. 2011). Open-access databases (such as GBIF or Movebank), liberal data policies (e.g. free Landsat archive and free access to data collected by MODIS), and open-source solutions are also becoming more popular and widespread, which makes data and knowledge exchange between and among communities easier.

11.6 General conclusions

There is no doubt that the conservation of biological diversity is a major public concern. The association between human wellbeing and biodiversity is becoming better understood, with the weight of scientific evidence suggesting that humans depend on many aspects of biodiversity, particularly the structure and functioning of ecosystems and the services that they provide (Cardinale et al. 2012). This book has attempted to highlight how the NDVI can inform environmental monitoring, wildlife management, and conservation. I hope that such a comprehensive review may convince many ecologists and practitioners that the NDVI could be of use to them. But this book is not intended to preach for the abolition of field data collection or for the supremacy of remote sensing over other forms of monitoring: remote sensing cannot solve the entire information collection issue associated with ecological monitoring. Whereas remote sensing-based techniques can address spatial and temporal domains inaccessible to traditional, on-the ground, approaches, remote sensing cannot match the accuracy, precision, and thematic richness of *in situ* measurement and monitoring (Gross et al. 2009). For example, satellite-based data yield information mostly on changes in canopy cover in terrestrial forested areas; yet biodiversity loss is known to happen in perfectly healthy forests, meaning that important ecological developments such as the long known empty forest syndrome (Redford 1992) cannot currently be monitored by remote sensing unless it triggers large-scale land-cover changes. This illustrates why NDVI information and remote sensing-based monitoring methods need to be integrated with field observations to maximize monitoring effectiveness.

Clearly, the NDVI is not the only useful remote sensing product for biological research and conservation (Pfeiffer et al. 2011; Nagendra et al. 2012), and this book does not aim to minimize the current and future contributions of other products in ecological research and environmental management. Rather, it aims to highlight the reasons why the NDVI has been so popular and to demonstrate how important it is, for successfully informing research and management, to have access to long-term, continuous information on changes in the environment at high temporal resolution and low-to-medium spatial resolution. Satellite imagery has traditionally been used for classifying, describing, and mapping vegetation structure and species habitats. Because of this, access to more detailed environmental information that would be collected at higher spatial resolution has been sought by many, and this need for finer scale and more accurate habitat maps has been expressed by numerous researchers and practitioners over the past decades, sometimes at the expense of appreciating the value of existing data. This book has tried to rectify the imbalance, arguing that we, the community of data producers and users, should also be concerned about the maintenance of time-series, such as the low-to-medium spatial resolution, global coverage, continuous, and long-term NDVI data.

Bibliography

Aanes, R., Saether, B.E., Smith, F.M., Cooper, E.J., Wookey, P.A., and Oritsland, N.A. (2002). The Arctic Oscillation predicts effects of climate change in two trophic levels in high arctic ecosystem. *Ecology Letters*, 5, 445–453.

Abramsky, Z. and Rosenzweig, M.L. (1984). Tilman's predicted productivity-diversity relationship shown by desert rodents. *Nature*, 309, 150–151.

Achard, F. and Blasco, F. (1990). Analysis of vegetation seasonal evolution and mapping of forest cover in West Africa with the use of NOAA AVHRR data. *Photogrammetric Engineering and Remote Sensing*, 56, 1359–1365.

Achard, F. and Estreguil, C. (1995). Forest classification of Southeast Asia using NOAA AVHRR data. *Remote Sensing of Environment*, 54, 198–208.

Achard, F., DeFries, R., Eva, H., Hansen, M., Mayaux, P., and Stibig, H.-J. (2007). Pan-tropical monitoring of deforestation. *Environmental Research Letters*, 2, 045022.

Adams, H.D., Guardiola-Clarmonte, M., Barron-Gafford, G.A., et al. (2009). Temperature sensitivity of drought-induced tree mortality portends increased regional die-off under global-change-type drought. *Proceedings of the National Academy of Sciences of the USA*, 106, 7063–7066.

Adams, J.M. and Woodward, F.I. (1989). Patterns in tree species richness as a test of the glacial extinction hypothesis. *Nature*, 339, 699–701.

African Union (2011). *African Convention on the Conservation of Nature and Natural Resources*. <http://www.africa-union.org/root/au/Documents/Treaties/Text/Convention_Nature%20&%20Natural_Resources.pdf>

Aguilar, C., Zinnert, J., Polo, M.J., and Young, D.R. (2012). NDVI as an indicator for changes in water availability to woody vegetation. *Ecological Indicators*, 23, 290–300.

Ahas, R. (1999). Long-term phyto-, ornitho- and ichthyo-phenological time-series analyses in Estonia. *International Journal of Biometeorology*, 42, 119–123.

Ahas, R., Aasa, A., Menzel, A., Fedotova, V.G., and Scheifinger, H. (2002). Changes in European spring phenology. *International Journal of Climatology*, 22, 1727–1738.

Ahmad, W. and Neil, D.T. (1994). An evaluation of Landsat Thematic Mapper (TM) digital data for discriminating coral reef zonation: Heron Reef (GBR). *International Journal of Remote Sensing*, 15, 2583–2597.

Alados, C.L., Puigdefabregas, J., and Martinez-Fernandez, J. (2011). Ecological and socio-economical thresholds of land and plant community degradation in semi-arid Mediterranean areas of southeastern Spain. *Journal of Arid Environments*, 75, 1368–1376.

Albon, S.D. and Langvatn, R. (1992). Plant phenology and the benefit of migration in a temperate ungulate. *Oikos*, 65, 502–513.

Albon, S.D., Mitchell, B., and Staines, B.W. (1983). Fertility and body mass in female red deer: a density-dependent relationship. *Journal of Animal Ecology*, 52, 969–980.

Alcaraz-Segura, D., Cabello, J., Paruelo, J.M., and Delibes, M. (2009). Use of descriptors of ecosystem functioning for monitoring a national park network: a remote sensing approach. *Environmental Management*, 43, 38–48.

Alcaraz-Segura, D., Chuvieco, E., Epstein, H.E., Kasischke, E.S., and Trishchenko, A. (2010a). Debating the greening vs. browning of the North American boreal forest: differences between satellite datasets. *Global Change Biology*, 16, 760–770.

Alcaraz-Segura, D., Liras, E., Tabik, S., Paruelo, J., and Cabello, J. (2010b). Evaluating the consistency of the 1982–1999NDVI trends in the Iberian Peninsula across four time-series derived from the AVHRR Sensor: LTDR, GIMMS, FASIR, and PAL-II. *Sensors*, 10, 1291–1314.

Allen, M.G., Bourke, R.M., Evans, B.R., et al. (2006). *Solomon Islands Smallholder Agriculture Study*, Volume 4, Provincial Reports. AusAID, Canberra. <http://www.ausaid.gov.au/publications/pdf/solomon_study_vol4.pdf>

Allen, C.D., Macalady, A.K., Chenchouni, H., et al. (2010). A global overview of drought and heat-induced tree mortality reveals emerging climate change risks for forests. *Forest Ecology and Management*, 259, 660–684.

Amaral, S., Bestetti Costa, C., and Daleles Renno, C. (2007). Normalized Difference Vegetation Index (NDVI) improving species distribution models: an example with the neotropical genus Coccocypselum (Rubiaceae). *Anais XIII Simpósio Brasileiro de Sensoriamento Remoto, Florianópolis, Brasil, 21–26 abril 2007*, INPE, pp. 2275–2282.

Ameca y Juárez, E.I., Mace, G.M., Cowlishaw, G., and Pettorelli, N. (2012). Natural population die-offs: causes and consequences for terrestrial mammals. *Trends in Ecology and Evolution*, 27, 272–277.

Amiro, B.D., Barr, A.G., Barr, J.G., et al. (2010). Ecosystem carbon dioxide fluxes after disturbance in forests of North America. *Journal of Geophysical Research*, 115, G00K02. <http://dx.doi.org/10.1029/2010JG001390>

Andersen, R., Gaillard, J.M., Linnell, J.D.C., and Duncan, P. (2000). Factors affecting maternal care in an income breeder, the European roe deer. *Journal of Animal Ecology*, 69, 672–682.

Andersen, R., Herfindal, I., and Saether, B.-E. (2004). When range expansion rate is faster in marginal habitats. *Oikos*, 107, 210–214.

Anderson, T.M., Hopcraft, J.G.C., Eby, S., Ritchie, M., Grace, J.B., and Olff, H. (2010). Landscape-scale analyses suggest both nutrient and antipredator advantages to Serengeti herbivore hotspots. *Ecology*, 91, 1519–1529.

Andren, H. (1994). Effects of habitat fragmentation on birds and mammals in landscapes with different proportions of suitable habitat—a review. *Oikos*, 71, 355–366.

Andreo, V., Lima, M., Provensal, C., Priotto, J., and Polop, J. (2009a). Population dynamics of two rodent species in agro-ecosystems of central Argentina: intra-specific competition, land-use, and climate effects. *Population Ecology*, 51, 297–306.

Andreo, V., Provensal, C., Scavuzzo, M., Lamfri, M., and Polop, J. (2009b). Environmental factors and population fluctuations of *Akodon azarae* (Muridae: Sigmontinae) in central Argentina. *Austral Ecology*, 34, 132–142.

Andrew, M.E. and Ustin, S.L. (2010). The effect of temporally variable dispersal and landscape structure on invasive species spread. *Ecological Applications*, 20, 593–608.

Anyamba, A. and Eastman, J.R. (1996). Interannual variability of NDVI over Africa and its relation to El Niño/Southern Oscillation. *International Journal of Remote Sensing*, 17, 2533–2548.

Anyamba, A. and Tucker, C.J. (2005). Analysis of Sahelian vegetation dynamics using NOAA-AVHRR NDVI data from 1981–2003. *Journal of Arid Environments*, 63, 596–614.

Anyamba, A., Tucker, C.J., and Eastman, J.R. (2001). NDVI anomaly patterns over Africa during the 1997/1998 ENSO warm event. *International Journal of Remote Sensing*, 22, 1847–1859.

Anyamba, A., Linthicum, K.J., Mahoney, R., and Tucker, C.J. (2002). Mapping potential risk of Rift Valley fever outbreaks in African savannas using vegetation index time series data. *Photogrammetric Engineering and Remote Sensing*, 68, 137–145.

Anyamba, A., Chretien, J.P., Small, J., Tucker, C.J., and Linthicum, K.J. (2006a). Developing climate anomalies suggest potential disease risks for 2006–2007. *International Journal of Health Geographics*, 5, 60.

Anyamba, A., Davies, G., Indeje, M., Ogallo, L.J., and Ward, M.N. (2006b). Predictability of the Normalized Difference Vegetation Index in Kenya and potential applications as an indicator of Rift Valley fever outbreaks in the Greater Horn of Africa. *Journal of Climate*, 19, 1673–1687.

Anyamba, A., Chretien, J.P., Small, J., et al. (2009). Prediction of a Rift Valley fever outbreak. *Proceedings of the National Academy of Sciences of the USA*, 106, 955–959.

Aplin, P. (2004). Remote sensing: land cover. *Progress in Physical Geography*, 28, 283–293.

Aplin, P. (2005). Remote sensing: ecology. *Progress in Physical Geography*, 29, 104–113.

Araujo, M.B. and Williams, P.H. (2000). Selecting areas for species persistence using occurrence data. *Biological Conservation*, 96, 331–345.

Arntzen, J.W. and Alexandrino, J. (2004). Ecological modelling of genetically differentiated forms of the Iberian endemic golden-striped salamander *Chioglossa lusitanica*. *Herpetological Journal*, 14, 137.

Ashburn, P. (1978). The vegetative index number and crop identification. In: *The LACIE Symposium, Proceedings of the Technical Session. Houston, Texas*, vol. II, pp. 843–856.

Asher, M., Oliveira, S.E., and Sachser, N. (2004). Social system and spatial organization of wild guinea pigs (*Cavia aperea*) in a natural low density population. *Journal of Mammalogy*, 85, 788–796.

Asner, G.P., Scurlock, J.M.O., and Hicke, J.A. (2003). Global synthesis of leaf area index observations: implications for ecological and remote sensing studies. *Global Ecology and Biogeography*, 12, 191–205.

Asner, G.P., Carlson, K.M., and Martin, R.E. (2005). Substrate age and precipitation effects on Hawaiian forest canopies from spaceborne imaging spectroscopy. *Remote Sensing of Environment*, 98, 457–467.

Asner, G.P., Jones, M.O., Martin, R.E., Knapp, D.E., and Hughes, R.F. (2008). Remote sensing of native and invasive species in Hawaiian forests. *Remote Sensing and Environment*, 112, 1912–1926.

Asrar, G., Fuchs, M., Kanemasu, E.T., and Hatfield J.L. (1984). Estimating absorbed photosynthetic radiation and leaf area index from spectral reflectance in wheat. *Agronomy Journal*, 76, 300–306.

Atkinson, P.M., Dash, J., and Jeganathan, C. (2011). Amazon vegetation greenness as measured by satellite sensors over the last decade. *Geophysical Research Letters*, 38, L19105. doi:10.1029/2011GL049118.

Ayoubi, S., Shahri, A.P., Karchegani, P.M., and Sahrawat, K.L. (2011). Application of Artificial Neural Network (ANN) to predict soil organic matter using remote sensing data in two ecosystems. In: *Biomass and Remote*

Sensing of Biomass, pp. 181–196. InTech Open Access <http://www.Intechopen.com>

Badhwar, G.D. (1981). The use of parameters to separate and identify spring small grains. Quarterly Technical Interchange Meeting, NASA-JSC, Houston, TX.

Bailey, S-A., Horner-Devine, M.C., Luck, G. et al. (2004). Primary productivity and species richness: relationships among functional guilds, residency groups and vagility classes at multiple spatial scales. *Ecography*, 27, 207–217.

Baillie, J.E.M., Collen, B., Amin, R., et al. (2008). Toward monitoring global biodiversity. *Conservation Letters*, 1, 18–26.

Baldi, G., Nosetto, M.D., Aragón, M.R., Aversa, F., Paruelo, J.M., and Jobbagy, E.G. (2008). Long-term satellite NDVI datasets: evaluating their ability to detect ecosystem functional changes in South America. *Sensors*, 8, 5397–5425.

Balmford, A., Gaston, K.J., Blyth, S., James, A., and Kapos, V. (2003a). Global variation in terrestrial conservation costs, conservation benefits, and unmet conservation needs. *Proceeding of the National Academy of Sciences of the USA*, 100, 1046–1050.

Balmford, A., Green, R.E., and Jenkins, M. (2003b). Measuring the changing state of nature. *Trends in Ecology and Evolution*, 18, 326–330.

Balmford, A., Crane, P., Dobson, A., Green, R.E., and Mace, G.M. (2005). The 2010 challenge: data availability, information needs and extraterrestrial insights. *Philosophical Transactions of the Royal Society of London*, 360, 221–228.

Balvanera, P., Pfisterer, A.B., Buchmann, N., et al. (2006). Quantifying the evidence for biodiversity effects on ecosystem functioning and services. *Ecology Letters*, 9, 1146–1156.

Bangladesh University of Engineering and Technology (2008). *Field Investigation on the Impact of Cyclone Sidr in the Coastal Region of Bangladesh*.

Bannari, A., Morin, D., and He, D.C. (1994). High spatial and spectral resolution remote sensing for the management of the urban environment. *First International Airborne Remote Sensing Conference and Exhibition*, Strasbourg, France, vol. III, pp. 247–260.

Bannari, A., Morin, D., Bonn, F., and Huete A.R. (1995). A review of vegetation indices. *Remote Sensing Reviews*, 13, 95–120.

Barbosa, P.M., Gregoire, J.M., and Pereira, J.M.C. (1999). An algorithm for extracting burned areas from time series of AVHRR GAC data applied at a continental scale. *Remote Sensing of Environment*, 69, 253–263.

Baret, F. and Guyot, G. (1991). Potentials and limits of vegetation indices for LAI and APAR assessment. *Remote Sensing of Environment*, 35, 161–173.

Baret, F., Guyot, G., and Major, D.J. (1989). TSAVI: A vegetation index which minimizes soil brightness effects on LAI and APAR estimation. In: *Proceedings of the 12th Canadian Symposium on Remote Sensing, Vancouver, Canada*, pp. 1355–1358.

Barker, D.H., Seaton, G.G.R., and Robinson, S.A. (1997). Internal and external photoprotection in developing leaves of the CAM plant *Cotyledon orbiculata*. *Plant Cell and Environment*, 20, 617–624.

Barnett, T.P., Adam, J.C., and Lettenmaier, D.P. (2005). Potential impacts of a warming climate on water availability in snow-dominated regions. *Nature*, 438, 303–309.

Barrett, E.C. and Curtis, L.F. (1999). *Introduction to Environmental Remote Sensing*, 4th edn. Stanley Thornes, Cheltenham.

Bart, J., Droege, S., Geissler, P., Peterjohn, B., and Ralph, C.J. (2004). Density estimation in wildlife surveys. *Wildlife Society Bulletin*, 32, 1242–1247.

Bartlett, D.S., Whiting, G.J., and Hartman, J.M. (1990). Use of vegetation indices to estimate intercepted solar radiation and net carbon dioxide exchange of a grass canopy. *Remote Sensing of Environment*, 30, 115–128.

Bartoń, K.A. and Zalewski, A. (2007). Winter severity limits red fox populations in Eurasia. *Global Ecology and Biogeography*, 16, 281–289.

Bartsch, A., Kumpula, T., Forbes, B.C., and Stammler, F. (2010). Detection of snow surface thawing and refreezing in the Eurasian Arctic with QuickSCAT: implications for reindeer herding. *Ecological Applications*, 20, 2346–2358.

Basille, M., Herfindal, I., Santin-Janin, H., et al. (2009). What shapes Eurasian lynx distribution in human dominated landscapes: selecting prey or avoiding people? *Ecography*, 32, 683–691.

Bavia, M.E., Malone, J.B., Hale, L., Dantas, A., Marroni, L., and Reis, R. (2001). Use of thermal and vegetation index data from earth observing satellites to evaluate the risk of schistosomiasis in Bahia, Brazil. *Acta Tropica*, 79, 79–85.

Bawa, K., Rose, J., Ganeshaiah, K.N., Barve, N., Kiran M.C., and Umashaanker, R. (2002). Assessing biodiversity from space: an example from the Western Ghats, India. *Conservation Ecology*, 6, 7.

Baylis, M. and Rawlings, P. (1998). Modelling the distribution and abundance of *Culicoides imicola* in Morocco and Iberia using climatic data and satellite imagery. In: Mellor, P.S., Baylis, M., Hamblin, C., Calisher, C.H., and Mertens, P.P.C. (eds), *African Horse Sickness*. Springer-Verlag, Vienna, pp. 137–153.

Baylis, M. and Rawlings, P. (1998). Modelling the distribution and abundance of *Culicoides imicola* in Morocco and Iberia using climatic data and satellite imagery. *Archives of Virology*, 14 (Supplement), 137–153.

Bayliss-Smith, T., Hviding, E., and Whitmore, T. (2003). Rainforest composition and histories of human disturbance in Solomon Islands. *Ambio*, 32, 346–352.

Bednorz, E. (2004). Snow cover in eastern Europe in relation to temperature, precipitation and circulation. *International Journal of Climatology*, 24, 591–601.

Begg, C.M., Begg, K.S., Du Toit, J.T., and Mills, M.G.L. (2005). Spatial organization of the honey badger Mellivora capensis in the southern Kalahari: home-range size and movement patterns. *Journal of Zoology*, 265, 23–35.

Begon, M., Harper, J.L., and Townsend, C.R. (1990). *Ecology: Individuals, Populations and Communities*, 2nd edn. Sinauer, Sunderland, MA.

Begon, M., Townsend, C., and Harper, J. (1996). *Ecology: Individuals, Populations and Communities*, 3rd edn. Blackwell Science, Oxford.

Behrenfeld, M.J., Randerson, J.T., McClain, C.R., et al. (2001). Biospheric primary production during an ENSO transition. *Science*, 291, 2594–2597.

Benavides, J.A., Huchard, E., Pettorelli, N., et al. (2012). From parasite encounter to infection: multiple-scale drivers of parasite richness in a wild social primate population. *American Journal of Physical Anthropology*, 147, 52–63.

Benayas J.M.R., Newton, A.C., Diaz, A., and Bullock, J.M. (2009). Enhancement of biodiversity and ecosystem services by ecological restoration. *Science*, 325, 1121–1124.

Bender, M.A., Knutson, T.R., Tuleya, R.E., et al. (2010). Modelled impact of anthropogenic warming on the frequency of intense atlantic hurricanes. *Science*, 327, 454–458.

Berger, J. (1992). Facilitation of reproductive synchrony by gestation adjustment in gregarious mammals: a new hypothesis. *Ecology*, 73, 323–329.

Bernhardsen, T. (2002). *Geographic Information Systems: An Introduction*, 3rd edn. John Wiley & Sons, New York.

Berryman, A. and Lima, M. (2006). Deciphering the effects of climate on animal populations: diagnostic analysis provides new interpretation of Soay sheep dynamics. *American Naturalist*, 168, 784–795.

Berteaux, D., Humphries, M.M., Krebs, C.J., et al. (2006). Constraints to projecting the effects of climate change on mammals. *Climate Research*, 32, 151–158.

Berube, C., Festa-Bianchet, M., and Jorgenson, J.T. (1999). Individual differences and reproductive senescence in bighorn ewes. *Ecology*, 80, 2555–2565.

Bhola, N., Ogutu, J.O., Said, M.Y., Piepho, H.-P., and Olff, H. (2012). The distribution of large herbivore hotspots in relation to environmental and anthropogenic correlates in the Mara region of Kenya. *Journal of Animal Ecology*, 81, 1268–1287.

Bierman, S.M., Fairbairn, J.P., Petty, S.J., Elston, D.A., Tidhar, D., and Lambin, X. (2006). Changes over time in the spatiotemporal dynamics of cyclic populations of field voles (*Microtus agrestis* L.). *American Naturalist*, 167, 583–590.

Bino, G., Levin, N., Darawshi, S., Van der Hal, N., Reich-Solomon, A., and Kark, S. (2008). Accurate prediction of bird species richness patterns in an urban environment using Landsat-derived NDVI and spectral unmixing. *International Journal of Remote Sensing*, 29, 3675–3700.

Biodiversity Indicators Partnership (2010). *Biodiversity Indicators and the 2010 Target: Experiences and Lessons Learnt from the 2010 Biodiversity Indicators Partnership*. Secretariat of the Convention on Biological Diversity, Montréal, Canada. Technical Series No. 53, 196 pp.

Biodiversity Indicators Partnership (2011). *Guidance for National Biodiversity Indicator Development and Use*. UNEP World Conservation Monitoring Centre, Cambridge, UK.

Birky, A.K. (2001). NDVI and a simple model of deciduous forest seasonal dynamics. *Ecological Modelling*, 143, 43–58.

Bischof, R., Loe, L.E., Meisingset, E.L., Zimmermann, B., Van Moorter, B., and Mysterud, A. (2012). A migratory northern ungulate in the pursuit of spring: jumping or surfing the green wave? *American Naturalist*, 180, 407–424.

Blackburn, T.M. and Gaston, K.J. (2001) Linking patterns in macroecology. *Journal of Animal Ecology*, 70, 338–352.

Blackburn, G.A. and Milton, E.J. (1995). Seasonal variations in the spectral reflectance of deciduous tree canopies. *International Journal of Remote Sensing*, 16, 709–720.

Blackburn, T.M., Pettorelli, N., Katzner, T., et al. (2010). Dying for conservation: eradicating invasive alien species in the face of opposition. *Animal Conservation*, 13, 227–228.

Blackburn, T.M., Pysek, P., Bacher, S., et al. (2011). A proposed unified framework for biological invasions. *Trends in Ecology and Evolution*, 26, 333–339.

Blackmer, T.M. and Schepers, J.S. (1995). Use of a chlorophyll meter to monitor nitrogen status and schedule fertigation for corn. *Journal of Production and Agriculture*, 8, 56–60.

Blake, S., Yackulic, C.B., Cabrera, F., et al. (2013). Vegetation dynamics drive segregation by body size in Galapagos tortoises migrating across altitudinal gradients. *Journal of Animal Ecology*, 82, 310–321.

Blanc, J.J., Barnes, R.F.W., Craig, G.C., et al. (2007). *African Elephant Status Report 2007: An Update from the African Elephant Database*. Occasional Paper Series of the IUCN Species Survival Commission, No. 33. IUCN/SSC African Elephant Specialist Group. IUCN, Gland, Switzerland.

Blumstein, D.T., Mennill, D.J., Clemins, P., et al. (2011). Acoustic monitoring in terrestrial environments using microphone arrays: applications, technological considerations and prospectus. *Journal of Applied Ecology*, 48, 758–767.

Boelman, N.T., Stieglitz, M., Rueth, H.M., et al. (2003). Response of NDVI, biomass, and ecosystem gas exchange to long-term warming and fertilization in wet sedge tundra. *Oecologia*, 135, 414–421.

Boelman, N.T., Gough, L., McLaren, J.R., and Greaves, H. (2011). Does NDVI reflect variation in the structural attributes associated with increasing shrub dominance in arctic tundra? *Environmental Research Letters*, 6, 035501 1. doi:0.1088/1748–9326/6/3/035501.

Boettiger, A.N., Wittemyer, G., Starfield, R., Volrath, F., Douglas-Hamilton, I., and Getz, W.M. (2011). Inferring ecological and behavioral drivers of African elephant movement using a linear filtering approach. *Ecology*, 92, 1648–1657.

Bojarska, K. and Selva, N. (2012). Spatial patterns in brown bear *Ursus arctos* diet: the role of geographical and environmental factors. *Mammal Review*, 42, 120–143.

Bokhorst, S., Bjerke, J.W., Tommervik, H., Callaghan, T.V., and Phoenix, G.K. (2009). Winter warming events damage sub-Arctic vegetation: consistent evidence from an experimental manipulation and a natural event. *Journal of Ecology*, 97, 1408–1415.

Bokhorst, S., Tommervik, H., Callaghan, T.V., Phoenix, G.K., and Bjerke, J.W. (2012). Vegetation recovery following extreme winter warming events in the sub-Arctic estimated using NDVI from remote sensing and handheld passive proximal sensors. *Environmental and Experimental Botany*, 81, 18–25.

Bontemps, S., Herold, M., Kooistra, L., *et al.* (2011). Revisiting land cover observations to address the needs of the climate modelling community. *Biogeosciences*, 8, 7713–7740.

Boone, J.D., Otteson, E.W., McGwire, K.C., Villard, P., Rowe, J.E., and St Jeor, S.C. (1998). Ecology and demographics of hantavirus infections in rodent populations in the walker River Basin of Nevada and California. *American Journal of Tropical Medicine and Hygiene*, 59, 445–451.

Boone, R.B., Thirgood, S.J., and Hopcraft, J.G.C. (2006). Serengeti wildebeest migratory patterns modelled from rainfall and new vegetation growth. *Ecology*, 87, 1987–1994.

Boutton, T.W. and Tieszen, L.L. (1983). Estimation of plant biomass by spectra reflectance in an East African grassland. *Journal of Range Management*, 36, 213–216.

Bouwman, L., Goldewijk, K.K., Van Der Hoek, K.W., *et al.* (2011). Exploring global changes in nitrogen and phosphorus cycles in agriculture induced by livestock production over the 1900–2050 period. *Proceedings of the National Academy of Sciences of the USA* 2011 May 16 (Epub ahead of print). doi:10.1073/pnas.1012878108.

Boyce, M. (1979). Seasonality and patterns of natural-selection for life-histories. *American Naturalist*, 114, 569–583.

Bradley, B.A. and Mustard, J.F. (2006). Characterizing the landscape dynamics of an invasive plant and risk invasion using remote sensing. *Ecological Applications*, 16, 1132–1147.

Brand, F. (2009). Critical natural capital revisited: ecological resilience and sustainable development. *Ecological Economics*, 68, 605–612.

Brinkman, H.-J. and Hendrix, J.C. (2011). *Food insecurity and violent conflict: Causes, consequences, and addressing the challenges*. Occasional Paper No. 24. World Food Programme, Rome, Italy.

Britch, S.C., Linthicum, K.J., Anyamba, A., *et al.* (2008) Satellite vegetation index data as a tool to forecast population dynamics of medically important mosquitoes at military installations in the continental United States. *Military Medicine*, 173, 677–683.

Bro-Jørgensen, J., Brown, M., and Pettorelli, N. (2008). Using NDVI to explain ranging patterns in a lek-breeding antelope: the importance of scale. *Oecologia*, 158, 177–182.

Brooks, T.M., Mittermeier, R.A., da Fonseca, G.A.B., *et al.* (2006). Global biodiversity conservation priorities. *Science*, 313, 58–61.

Brown, J.H. (1988). Species diversity. In: Myers, A.A. and Giller, P.S. (eds), *Analytical Biogeography: An Integrated Approach to the Study of Animal and Plant Distributions*. Chapman & Hall, London, pp. 57–89.

Brown, M.E. (2008). *Famine Early Warning Systems and Remote Sensing Data*. Springer-Verlag, Heidelberg.

Brown, M.E., Pinzon. J.E., Didan, K., Morisette, J.T., and Tucker, C.J. (2006). Evaluation of the consistency of long-term NDVI time series derived from AVHRR, SPOT-Vegetation, SeaWiFS, MODIS, and Landsat ETM+ sensors. *IEEE Transactions Geoscience and Remote Sensing*, 44, 1787–1793.

Brown, M.E., de Beurs, K., and Vrieling, A. (2010). The response of African land surface phenology to large scale climate oscillations. *Remote Sensing of Environment*, 114, 2286–2296.

Brönnimann, S., Xoplaki, E., Casty, C., Pauling, A., and Luterbacher, J. (2007). ENSO influence on Europe during the last centuries. *Climate Dynamics*, 28, 181–197.

Buckland, S.T., Magurran, A.E., Green, R.E., and Fewster, R.M. (2005). Monitoring change in biodiversity through composite indices. *Philosophical Transactions of the Royal Society of London Series B*, 360, 243–254.

Buermann, W., Wang, Y.J., Dong, J.R., *et al.* (2002). Analysis of a multiyear global vegetation leaf area index data set. *Journal of Geophysical Research Atmospheres*, 107, 4646.

Buermann, W., Anderson, B., Tucker, C.J., *et al.* (2003). Interannual covariability in Northern Hemisphere air temperatures and greenness associated with El Niño–Southern Oscillation and the Arctic Oscillation. *Journal of Geophysical Research*, 108, 4396.

Bullock, S.H., Mooney, H.A., and Medina, E. (1995). *Seasonally Dry Tropical Forests*. Cambridge University Press, Cambridge.

Bunnefeld, N., Linnell, J.D.C., Odden, J., van Duijn, M.A.J., and Andersen, R. (2006). Risk taking by Eurasian lynx in a human-dominated landscape: effects of sex and reproductive status. *Journal of Zoology (London)*, 270, 31–39.

Burton, A.C., Sam, M.K., Balangtaa, C., and Brashares, J.S. (2012). Hierarchical multi-species modeling of carnivore responses to hunting, habitat and prey in a West African protected area. *PLoS ONE*, 7(5), e38007. doi:10.1371/journal.pone.0038007

Buschman, C. and Nagel, E. (1993). In vivo spectroscopy and internal optics of leaves as a basis for remote sensing of vegetation. *International Journal of Remote Sensing*, 14, 711–722.

Butchart, S.H.M., Walpole, M., Collen, B., *et al.* (2010). Global biodiversity: indicators of recent declines. *Science*, 328, 1164–1165

Butcher, J. (2006). Natural capital and the advocacy of ecotourism as sustainable development. *Journal of Sustainable Tourism*, 14, 529–544.

Cairns, J., McCormick, P.V., and Niederlehner, B.R. (1993). A proposed framework for developing indicators of ecosystem health. *Hydrobiologia*, 236, 1–44.

Calder, W.A. (1984). *Size, Function and Life History*. Harvard University Press, Cambridge, MA.

Camberlin, P., Janicot, S., and Poccard, I. (2001). Seasonality and atmospheric dynamics of the teleconnection between African rainfall and tropical sea-surface temperature: Atlantic vs. ENSO. *International Journal of Climatology*, 21, 973–1005.

Campbell, J.B. (2007). Introduction to remote sensing, 4th Edition. Guilford Press.

Campbell, E.T. and Brown, M.T. (2012). Environmental accounting of natural capital and ecosystem services for the US National Forest System. *Environment, Development and Sustainability*, 14, 691–724.

Canadian Council of Forest Ministers (2002). *National Forestry Database Program*. Canadian Forest Service, Natural Resources Canada. <http://nfdp.ccfm.org>

Carbone, C. and Gittleman, J.L. (2002). A common rule for the scaling of carnivore density. *Science*, 295, 2273–2276.

Cardinale, B.J., Duffy, J.E., Gonzalez, A., *et al.* (2012). Biodiversity loss and its impact on humanity. *Nature*, 486, 59–67.

Carlson, T.N. and Ripley, D.A. (1997). On the relation between NDVI, fractional vegetation cover, and leaf area index. *Remote Sensing of Environment*, 62, 241–252.

Carpenter, S.R., DeFries, R., Dietz, T., *et al.* (2006). Millennium ecosystem assessment: research needs. *Science*, 314, 257–258.

Carter, G.A. (1991). Primary and secondary effects of water content on the spectral reflectance of leaves. *American Journal of Botany*, 78, 916–924.

Carter, P. and Gardner, W.E. (1977). An image-processing system applied to earth-resource imagery. In: Barrett, E.C. and Curtis, L.F. (eds), *Environmental Remote Sensing: Practices and Problems*. Arnold, London, pp. 143–162.

Cayuela, L., Benayas, J.M., Justel, A., and Salas-Rey, J. (2006). Modelling tree diversity in a highly fragmented tropical montane landscape. *Global Ecology and Biogeography*, 15, 602–613.

Ceballos, G., Garcia, A., and Ehrlich, P.R. (2010). The sixth extinction crisis: loss of animal populations and species. *Journal of Cosmology*, 8, 1821–1831.

Ceccato, P. (2004). Operational early warning system using SPOT-VEGETATION and TERRA-MODIS to predict desert locust outbreaks. In: *Proceedings of the 2nd International vegetation user conference, Antwerp, Belgium, 24–26 March 2004*.

Ceccato, P., Flasse, S., Tarantola, S., Jacquemoud, S., and Gregoire, J. (2001). Detecting vegetation leaf water content using reflectance in the optical domain. *Remote Sensing of Environment*, 77, 22–33.

Ceccato, P., Flasse, S., and Gregoire, J. (2002). Designing a spectral index to estimate vegetation water content from remote sensing data: Part 2. Validation and applications. *Remote Sensing of Environment*, 82, 198–207.

Chafer, C.J., Noonan, M., and Mcnaught, E. (2004). The post-fire measurement of fire severity and intensity in the Christmas 2001 Sydney wildfires. *International Journal of Wildland Fire*, 13, 227–240.

Chamaille-Jammes, S., Valeix, M., and Fritz, H. (2007). Managing heterogeneity in elephant distribution: interactions between elephant population density and surface-water availability. *Journal of Applied Ecology*, 44, 625–633.

Chamard, P., Courel, M.F., Ducousso, M., *et al.* (1991). Utilisation des bandes spectrales du vert et du rouge pour une meilleure évaluation des formations végétales actives. *Teledetection et Cartographie*, Ed. AUPELF-UREF, pp. 203–209.

Chan, K.-S., Mysterud, A., Oristland, N.A., Severinsen, T., and Stenseth, N.C. (2005). Continuous and discrete extreme climatic events affecting the dynamics of a high-arctic reindeer population. *Oecologia*, 145, 556–563.

Chape, S., Harrison, J., Spalding, M., and Lysenko, I. (2005). Measuring the extent and effectiveness of protected areas as an indicator for meeting global biodiversity targets. *Philosophical Transactions of the Royal Society, Series B*, 360, 443–455.

Chape, S., Spalding, M., and Jenkins, M. (2008). *The World's Protected Areas: Status, Values and Prospects in the Twenty-first Century*. University of California Press, Berkeley.

Chen, D., Cane, M.A., Kaplan, A., Zebiak, S.E., and Huang, D. (2004). Predictability of El Niño over the past 148 years. *Nature*, 428, 733–736.

Chen, D., Huang, J., and Jackson, T.J. (2005). Vegetation water content estimation for corn and soybeans using spectral indices derives from MODIS near- and short-wave infrared bands. *Remote Sensing of Environment*, 98, 225–236.

Chen, J., Jonsson, P., Tamura, M., Gu, Z., Matsushita, B., and Eklundh, L. (2004). A simple method for reconstructing a high quality NDVI time-series data set based on the Savitzky–Golay filter. *Remote Sensing of Environment*, 91, 332–344.

Chen, J.M. and Cihlar, J. (1996). Retrieving leaf area index of boreal conifer forests using Landsat TM images. *Remote Sensing of Environment*, 55, 153–162.

Cheng, Y.-B., Zarco-Tejada, P.J., Riano, D., Rueda, C.A., and Ustin, S.L. (2006). Estimating vegetation water content with hyperspectral data for different canopy scenarios: relationships between AVIRIS and MODIS indexes. *Remote Sensing of Environment*, 105, 354–366.

Cheng, Y.-B., Ustin, S.L., Riano, D., and Vanderbilt, V.C. (2008). Water content estimation from hyperspectral images and MODIS indexes in Southeastern Arizona. *Remote Sensing of Environment*, 112, 363–374.

Chmielewski, F.M. and Rotzer, T. (2002). Annual and spatial variability of the beginning of growing season in Europe in relation to air temperature changes. *Climate Research*, 19, 257–264.

Cho, J., Yeh, P.J.F., Lee, Y.-W., *et al.* (2010). A study on the relationship between Atlantic sea surface temperature and Amazonian greenness. *Ecological Informatics*, 5, 367–378.

Chomitz, K. (2007). *At Loggerheads? Agricultural Expansion, Poverty Reduction, and Environment in the Tropical Forests*. World Bank Policy Research Report 308, Washington, DC.

Chuvieco, E., Cocero, D., Riano, D., *et al.* (2004). Combining NDVI and surface temperature for the estimation of live fuels moisture content in forest fire danger rating. *Remote Sensing of Environment*, 92, 322–331.

Cihlar, J., St-Laurent, L., and Dyer, J.A. (1991). Relation between the normalised difference vegetation index and ecological variables. *Remote Sensing of Environment*, 35, 279–798.

Clark, D.B., Xue, Y.K., Harding, R.J., and Valdes, P.J. (2001). Modeling the impact of land surface degradation on the climate of tropical North Africa. *Journal of Climate*, 14, 1809–1822.

Clarke, K.C. (2007). *Remote Sensing of the Environment. An Earth Resource Perspective*, 2nd edn. Prentice-Hall, Upper Saddle River, NJ.

Clavero, M. and García-Berthou, E. (2005). Invasive species are a leading cause of animal extinctions. *Trends in Ecology and Evolution*, 20, 110.

Clay, D.E., Kim, K.-I., Chang, S.A., and Dalsted, K. (2006). Characterizing water and nitrogen stress in corn using remote sensing. *Agronomy Journal*, 98, 579–587.

Clerici, N., Weissteiner, C.J., and Gerard, F. (2012). Exploring the Use of MODIS NDVI-based phenology indicators for classifying forest general habitat categories. *Remote Sensing*, 4, 1781–1803.

Clerici, M., Skøien, J., Combal, B., *et al.* (in press). The eStation, an Earth Observation processing server for supporting ecological monitoring. *Ecological Informatics*.

Clevers, J.P.W. (1986). The application of a vegetation index in correcting the infrared reflectance for soil background. *International Archives of Photogrammetry and Remote Sensing*, 26, 221–226.

Clobert, J., Danchin, E., Dhondt, A.A., and Nichols, J.D. (2001). *Dispersal*. Oxford University Press, New York.

Clutton-Brock, T.H. (1991). *The Evolution of Parental Care*. Princeton University Press, Princeton, NJ.

Coad, L., Burgess, N.D., Bombard, B., and Besancon, C. (2009). *Progress towards the Convention on Biological Diversity's 2010 and 2012 targets for protected area coverage*. Technical report for the IUCN international workshop 'Looking at the future of the CBD Programme of work on protected areas,' Jeju Island, Republic of Korea, 14–17 September 2009. UNEP World Conservation Monitoring Centre, Cambridge, UK.

Cohen, W.B. and Goward, S.N. (2004). Landsat's role in ecological applications of remote sensing. *BioScience*, 54, 535–545.

Cohen, W.B., Maiersperger, T.K., Gower, S.T., and Turner, D.P. (2003). An improved strategy for regression of biophysical variables and Landsat ETM+ data. *Remote Sensing of Environment*, 84, 561–571.

Colditz, R.R., Schmidt, M., Conrad, C., Hansen, M.C., and Dech, S. (2011). Land cover classification with coarse spatial resolution data to derive continuous and discrete maps for complex regions. *Remote Sensing of Environment*, 115, 3264–3275.

Coley, P.D. and Kusar, T.A. (1996). Anti-herbivore defenses of young tropical leaves: physiological constraints and ecological tradeoffs. In: Mulkey, S.S., Chazdon, R.L., and Smith, A.P. (eds), *Tropical Forest Plant Ecophysiology*. Chapman & Hall, New York, pp. 305–335.

Collen, B., Pettorelli, N., Baillie, J.E.M., and Durant, S.M. (2013). *Biodiversity Monitoring and Conservation: Bridging the Gaps Between Global Commitment and Local Action*. Wiley–Blackwell, Cambridge, UK.

Colvin, B.A., Fall, M.W., Fitzgerald, L.A., and Loope, L.L. (2005). *Review of Brown Tree snake Problems and Control Programs: Report of Observations and Recommendations*. US Department of Interior, Office of Insular Affairs, Brown Treesnake Control Committee, Washington, DC.

Colwell, R.K. (1974). Predictability, constancy, and contingency of periodic phenomena. *Ecology*, 55, 1148–1153.

Colwell, R.N. (1983). *Manual of Remote Sensing*. American Society of Photogrammetry, Falls Church, VA.

Conservation International (2007). *Makira, Solomon Islands*. <https://library.conservation.org/Published%20Documents/2009/Makira%20Solomon%20Islands-2007.pdf>

Convention on Biological Diversity (1992). *Convention on Biological Diversity*. Article 2, CBD <http://www.cbd.int/convention/articles/?a=cbd-02>

Convention on Biological Diversity (2012). Press release: At United Nations Biodiversity Conference, countries agree to double resources for biodiversity protection by 2015. <http://www.cbd.int/doc/press/2012/pr-2012-10-20-cop-11-en.pdf>

Cook, B., Bolstad, P., Martin, J., *et al.* (2008). Using light-use and production efficiency models to predict photosynthesis and net carbon exchange during forest canopy disturbance. *Ecosystems*, 11, 26–44.

Coops, N.C., Johnson, M., Wulder, M.A., and White, J.C. (2006). Assessment of Quickbird high spatial resolution imagery to detect red attack damage due to mountain pine beetle infestation. *Remote Sensing of Environment*, 103, 67–80.

Cornélis, D., Benhamou, S., Janeau, G., Morellet, N., Ouedraogo, M., and de Visscher, M.-N. (2011). Spatiotemporal dynamics of forage and water resources shape space use of West African savanna buffaloes. *Journal of Mammalogy*, 92, 1287–1297.

Cornforth, W.A., Fatoyinbo, T.E., Freemantle, T.P., and Pettorelli, N. (2013). ALOS PALSAR to inform the conservation of mangroves: Sundarbans as a case study. *Remote Sensing*, 5, 224–237.

Coronel Arellano, H., Lopez González, C.A., and Moreno Arzate, C.N. (2009). Can landscape variables predict white-tailed deer abundance? Northwestern Mexico as a case study. *Tropical Conservation Science*, 2, 229–236.

Costanza, R. (2003). Social goals and the valuation of natural capital. *Environmental Monitoring and Assessment*, 86, 19–28.

Costanza, R. and Daly, H. (1992). Natural capital and sustainable development. *Conservation Biology*, 6, 37–46.

Costanza, R., d'Arge, R., de Groot, R., *et al.* (1997). The value of the world's ecosystem services and natural capital. *Nature*, 387, 253–260.

Costanza, J.K., Moody, A., and Peet, R.K. (2011). Multi-scale environmental heterogeneity as a predictor of plant species richness. *Landscape Ecology*, 26, 851–864.

Couturier, S., Côté, S.D., Otto, R.D., Weladji, R.B., and Huot, J. (2009). Variation in calf body mass in migratory caribou: the role of habitat, climate, and movements. *Journal of Mammalogy*, 90, 442–452.

Cox, P.M., Betts, R.A., Collins, M., Harris, C., Huntingford, C., and Jones, C.D. (2004). Amazonian forest dieback under climate-carbon cycle projections for the 21st century. *Theoretical and Applied Climatology*, 78, 137–156.

Crawley, M.J. (1983). Herbivory: the dynamics of animal–plant interactions. Blackwell, Oxford, UK.

Creel, S., Christianson, D., Liley, S., and Winnie Jr, J.A. (2007). Predation risk affects reproductive physiology and demography of elk. *Science*, 315, 960.

Crippen, R.E. (1990). Calculating the vegetation index faster. *Remote Sensing of Environment*, 34, 71–73.

Cumming, G.S. (2002). Comparing climate and vegetation as limiting factors for species ranges of African ticks. *Ecology*, 83, 255–268.

Cumming, G.S., Cumming, D.H.M., and Redman, C.L. (2006). Scale mismatches in social-ecological systems: causes, consequences, and solutions. *Ecology and Society*, 11, 14.

Cunha, M., Marcal, A.R.S., and Silva, L. (2010). Very early prediction of wine yield based on satellite data from vegetation. *International Journal of Remote Sensing*, 31, 3125–3142.

Curran, P.J. (1980). Multi-spectral remote sensing of vegetation amount. *Progress in Physical Geography*, 4, 315–321.

Curran, R.J. (1989). Satellite-borne lidar observations of the Earth: requirements and anticipated capabilities. *IEEE Proceedings*, 77, 478–490.

Curran, P. and Hay, A.M. (1986). The importance of measurement error for certain procedures in remote sensing at optical wavelengths. *Photogrammetric Engineering Remote Sensing*, 52, 229–241.

Curran, P.J., Dungan, J.L., and Gholz, H.L. (1990). Exploring the relationship between reflectance red edge and chlorophyll content in slash pine. *Tree Physiology*, 7, 33–48.

Currie, D.J. (1991). Energy and large-scale patterns of animal and plant species richness. *American Naturalist*, 137, 27–49.

Cushing, D.H. (1990). Plankton production and year-class strength in fish populations—an update of the match mismatch hypothesis. *Advances in Marine Biology*, 26, 249–293.

Da Conceicao Prates-Clark, C., Saatchi, S.S., and Agosti, D. (2008). Predicting geographical distribution models of high-value timber trees in the Amazon Basin using remotely sensed data. *Ecological Modelling*, 211, 309–323.

Dai, A., Trenberth K.E., and Qian, T. (2004). A global dataset of palmer drought severity index for 1870–2002: relationship with soil moisture and effects of surface warming. *Journal of Hydrometeorology*, 5, 1117–1130.

Daily, G.C. (1997). *Nature's Services: Societal Dependence on Natural Ecosystems*. Island Press, Washington, DC.

Dale, V.H. and Beyeler, S.C. (2001). Challenges in the development and use of ecological indicators. *Ecological Indicators*, 1, 3–10.

Dall'Olmo, G. and Karnieli, A. (2002). Monitoring phenological cycles of desert ecosystems using NDVI and LST data derived from NOAA-AVHRR imagery. *International Journal of Remote Sensing*, 23, 4055–40741.

Damuth, J. (1981). Home ranges, home range overlap, and species energy use among herbivorous mammals. *Biological Journal of the Linnean Society*, 15, 185–193.

Danielsen, F., Filardi, C.E., Jonsson, K.A., et al. (2010). Endemic avifaunal biodiversity and tropical forest loss in Makira, a mountainous Pacific island. *Singapore Journal of Tropical Geography*, 31, 100–114.

Danson, F.M. and Curran, P.J. (1993). Factors affecting the remotely sensed response of coniferous forest plantations. *Remote Sensing of Environment*, 43, 55–65.

Danson, F.M., Steven, M.D., Malthus, T.J., and Clark, J.A. (1992). High-spectral resolution data for determining leaf water content. *International Journal of Remote Sensing*, 13, 461–470.

D'Arrigo, R.D., Malstrom, C.M., Jacoby, G.C., Los, S.O., and Bunker, D.E. (2000). Correlation between maximum latewood density of annual tree rings and NDVI based estimates of forest productivity. *International Journal of Remote Sensing*, 21, 2329–2336.

Datt, B. (1998). Remote sensing of chlorophyll *a*, chlorophyll *b*, chlorophyll *a* + *b*, and total carotenoid content in eucalyptus leaves. *Remote Sensing of Environment*, 66, 111–121.

Davenport, M.L. and Nicholson, S.E. (1993). On the relation between rainfall and the Normalized Difference Vegetation Index for diverse vegetation types in East Africa. *International Journal of Remote Sensing*, 14, 2369–2389.

Davidson, A., Wang, S., and Wilmshurst, J. (2006). Remote sensing of grassland-shrubland vegetation water content in the shortwave domain. *International Journal of Applied Earth Observation and Geoinformation*, 8, 225–236.

Davies, R.G., Orme, C.D.L., Storch, D., et al. (2007) Topography, energy and the global distribution of bird species richness. *Proceedings of the Royal Society of London, Series B*, 274, 1189–1197.

Davis M.A. (2009) *Invasion Biology*. Oxford University Press, Oxford.

de Beurs, K.M., Wright, C.K., and Henebry, G.M. (2009). Dual scale trend analysis for evaluating climatic and anthropogenic effects on the vegetated land surface in Russia and Kazakhstan. *Environmental Research Letters*, 4, 045012. doi:10.1088/1748-9326.

de Boer, W.F., Vis, M.J.P., de Knegt, H.J., et al. (2010). Spatial distribution of lion kills determined by the water dependency of prey species. *Journal of Mammalogy*, 91, 1280–1286.

DeFries, R., Hansen, M., and Townshend, J.R.G. (1995). Global discrimination of land cover types from metrics derived from AVHRR Pathfinder data. *Remote Sensing of Environment*, 54, 209–222.

DeFries, R.S., Townshend, J.R.G., and Hansen, M.C. (1999). Continuous fields of vegetation characteristics at the global scale at 1-km resolution. *Journal of Geophysical Research—Atmospheres*, 104, 16911–16923

DeFries, R., Hansen, A., Newton, A.C., and Hansen, M.C. (2005). Increasing isolation of protected areas in tropical forests over the past twenty years. *Ecological Applications*, 15, 19–26.

DeFries, R., Achard, F., Brown, S., et al. (2007). Earth observations for estimating greenhouse gas emissions from deforestation in developing countries. *Environmental Science and Policy*, 10, 385–394.

DeFries, R.S., Rudel, T., Uriarte, M., and Hansen, M. (2010). Deforestation driven by urban population growth and agricultural trade in the twenty-first century. *Nature Geoscience*, 3, 178–181.

Deering, D.W. (1978). Rangeland reflectance characteristics measured by aircraft and spacecraft sensors. PhD dissertation, Texas A & M University, College Station, TX.

Deering, D.W., Rouse, J.W., Haas, R.H., and Schell, J.A. (1975). Measuring "Forage Production" of Grazing Units From Landsat MSS Data. *Proceedings of the 10th International Symposium on Remote Sensing of Environment*, II, pp. 1169–1178.

Defries, R.S., and Townshend, J.R.G. (1994). NDVI-derived land cover classifications at a global scale. *International Journal of Remote Sensing*, 15, 3567–3586.

De La Maza, M., Lima, M., Meserve, P.L., Gutierrez, J.R., and Jaksic, M. (2009). Primary production dynamics and climate variability: ecological consequences in semiarid Chile. *Global Change Biology*, 15, 1116–1126.

Demmig-Adams, B., and Adams III, W.W. (1996). The role of xanthophyll cycle carotenoids in the protection of photosynthesis. *Trends in Plant Science*, 1, 21–27.

de Sherbinin, A., Kline, K., and Raustiala, K. (2002). Remote sensing data: valuable support for environmental treaties. *Environment*, 44, 22–31.

De Souza Gomes Guarino, E., Barbosa, A.M., and Waechter, J.L. (2012) Occurrence and abundance models of threatened plant species: applications to mitigate the impact of hydroelectric power dams. *Ecological Modelling*, 230, 22–33.

Devictor, V., Julliard, R., Couvet, D., and Jiguet, F. (2008). Birds are tracking climate warming, but not fast enough. *Proceedings of the Royal Society of London, Series B*, 275, 2743–2748.

De Wulf, R.R., Goossens, R.E., MacKinnon, J.R., and Cai, W.C. (1988). Remote sensing for wildlife management: Giant panda habitat mapping from LANDSAT MSS images. *Geocarto International*, 3, 41–50.

Diallo, O., Diouf, A., Hanan, N.P., Ndiaye, A., and Prevost, Y. (1991). AVHRR monitoring of savanna primary production in Senegal, West Africa: 1987–1988. *International Journal of Remote Sensing*, 12, 1259–1297.

Diamond, J. (1988). Factors controlling species diversity: Overview and synthesis. *Annals of the Missouri botanical Garden*, 75, 117–129.

Diaz-Delgado, R., Lloret, F., and Pons X. (2003). Influence of fire severity on plant regeneration by means of remote sensing imagery. *International Journal of Remote Sensing*, 24, 1751–1763.

Ding, T.S., Yuan, H.W., Geng, S., Koh, C.N., and Lee, P.F. (2006). Macroscale bird species richness patterns of the East Asian mainland and islands: energy, area and isolation. *Journal of Biogeography*, 33, 683–693.

Dingle, H. and Drake, V.A. (2007). What is migration? *Bioscience*, 57, 113–121.

Diouf, A. and Lambin, E.F. (2001). Monitoring land-cover changes in semi-arid regions: Remote sensing data and field observations in the Ferlo, Senegal. *Journal of Arid Environments*, 48, 129–148.

Dogan, H.M., Celep, F., and Karaer, F. (2009). Evaluation of the NDVI in plant community composition mapping a case study of Tersakan Valley, Amasya County, Turkey. *International Journal of Remote Sensing*, 30, 3769–3798.

Doiron, M., Legagneux, P., Gauthier, G., and Lévesque, E. (2013). Broad-scale satellite Normalized Difference Vegetation Index data predict plant biomass and peak date of nitrogen concentration in Arctic tundra vegetation. *Applied Vegetation Science*, 16, 343–351.

Doraiswamy, P.C. and Cook, P.W. (1995). Spring wheat yield assessment using NOAA AVHRR data. *Canadian Journal of Remote Sensing*, 21, 43–51.

Doraiswamy, P.C., Moulin, S., Cook, P.W., and Stern, A. (2003). Crop Yield Assessment from Remote Sensing. *Photogrammetric Engineering and Remote Sensing*, 69, 665–674.

Dubayah, R.O. and Drake, J.B. (2000). Lidar remote sensing for forestry. *Journal of Forestry*, 98, 44–46.

Dubois, G., Clerici, M., Pekel, J.F., *et al*. (2011). On the contribution of remote sensing to DOPA, a Digital Observatory for Protected Areas. In: *Proceedings of the 34th International Symposium on Remote Sensing of Environment*, 10 April 2011, Sydney, Australia.

Dudley, N. (2008). *Guidelines for Applying Protected Area Management Categories*. IUCN, Gland, Switzerland, 86 pp.

Dudley, N., Belokurov, A., Borodin, O., *et al*. (2004). *Are Protected Areas Working? An Analysis of Protected Areas*. WWF International, Gland, Switzerland.

Dudley, N., Stolton, S., Belokurov, A., *et al*. (2010). *Natural Solutions: Protected Areas Helping People Cope with Climate Change*. IUCN-WCPA, TNC, UNDP, WCS, World Bank and WWF, Gland/Washington, DC/New York.

Duffy, J.P. and Pettorelli, N. (2012). Exploring the relationship between NDVI and African elephant population density in protected areas. *African Journal of Ecology*, 50, 455–463.

Dugan, P. (1993). *Wetlands in Danger: A World Conservation Atlas*. Oxford University Press, New York.

Dugdale, G., Hardy, S., and Milford, J.R. (1991). Daily catchment rainfall estimated from Meteosat. *Hydrological Processes*, 5, 261–270.

Duncan, J., Stow, D., Franklin, J., and Hope, A. (1993). Assessing the relationship between spectral vegetation indices and shrub cover in the Jordana Basin, New Mexico. *International Journal of Remote Sensing*, 14, 3395–3416.

Duncan, R.P., Bomford, M., Forsyth, D.M., and Conibear, L. (2001). High predictability in introduction outcomes and the geographical range size of introduced Australian birds: a role for climate. *Journal of Animal Ecology*, 70, 621–632.

Duncan, R.P., Blackburn, T.M., and Sol, D. (2003). The ecology of bird introductions. *Annual Review of Ecology, Evolution, and Systematics*, 34, 71–98.

Dungan, J.L., Perry, J.N., Dale, M.R.T., *et al*. (2002). A balanced view of scale in spatial statistical analysis. *Ecography*, 25, 626–640.

Durant, J.M., Hjermann, D.Ø., Anker-Nilssen, T., *et al*. (2005). Timing and abundance as key mechanisms affecting trophic interactions in variable environments. *Ecology Letters*, 8, 952–958

Durant, S.M., Dickman, A.J., Maddox, T., *et al*. (2010). Past, present and future of cheetahs in Tanzania: their behavioural ecology and conservation. In: MacDonald, D.W. and Loveridge, A.J. (eds), *Biology and Conservation of Wild Felids*. Oxford University Press, Oxford, pp. 373–382.

Duro, D., Coops, N.C., Wulder, M.A., and Han, T. (2007). Development of a large area biodiversity monitoring driven by remote sensing. *Progress in Physical Geography*, 31, 235–260.

Easterling, D.R., Meehl, G.A., Parmesan, C., Changnon, S.A., Karl, T.R., and Mearns, L.O. (2000). Climate extremes: observations, modelling, and impacts. *Science*, 289, 2068–2074.

Ebi, K.L., Lewis, N.D., and Corvalan, C. (2006). Climate variability and change and their potential health effects in small island states: information for adaptation planning in the health sector. *Environmental Health Perspectives*, 114, 1957–1963.

Ehrlich, P.R. and Ehrlich, A.H. (1981). *Extinction: The Causes and Consequences of the Disappearance of Species*. Random House, New York.

Ekins, P. (2003). Identifying critical natural capital. *Ecological Economics*, 44, 277–292.

Eklundh, L., Johansson, T., and Solberg, S. (2009). Mapping insect defoliation in Scots pine with MODIS time-series data. *Remote Sensing of Environment*, 113, 1566–1573.

Elith, J. and Leathwick, J.R. (2009). Species distribution models: ecological explanation and prediction across space and time. *Annual Review of Ecology, Evolution, and Systematics*, 40, 677–697.

Elvidge, C.D. and Lyon, R.J.P. (1985). Influence of rock-soil spectral variation on the assessment of green biomass. *Remote Sensing of Environment*, 17, 265–279.

English, A.K., Chauvenet, A.L.M., Safi, K., and Pettorelli, N. (2012). Reassessing the Determinants of Breeding Synchrony in Ungulates. *PLoS ONE*, 7(7), e41444. doi:10.1371/journal.pone.0041444.

ENVI (2011). *Envy Users Guide*. <http://geol.hu/data/online_help/Vegetation_Indices.html>

Erasmi, S., Propastin, P., Kappas, M., and Panferov, O. (2009). Spatial patterns of NDVI variation over indonesia and their relationship to ENSO warm events during the period 1982–2006. *Journal of Climate*, 22, 6612–6623.

Ervin, J. (2003a). Protected area assessments in perspective. *Bioscience*, 53, 819–822.

Ervin, J. (2003b). Rapid assessment of protected area management effectiveness in four countries. *BioScience*, 53, 833–841.

Escadafal, R. and Huete, A.R. (1991). Etude des propriétés spectrales des sols arides appliquée à l'amélioration des indices de végétation obtenus par télédétection. *Compte Rendu de l'Academie des Sciences de Paris* 312, 1385–1391.

ESRI (2013). *ArcGIS*. <http://www.esri.com/software/arcgis>

Estes, J.A., Crooks, K., and Holt, R. (2001). Predators, ecological role of. In: Levin S. (ed.), *Encyclopedia of Biodiversity*, vol. 4. Academic Press, San Diego, pp. 280-1–280-22.

Estrada-Peña, A. (1999). Geostatistics and remote sensing using NOAA-AVHRR satellite imagery as predictive tools in tick distribution and habitat suitability estimations for *Boophilus microplus* (Acari: Ixodidae) in South America. *Veterinary Parasitology*, 1, 73–82.

Estrada-Pena, A. and Thuiller, W. (2008). An assessment of the effect of data partitioning on the performance of modelling algorithms for habitat suitability for ticks. *Medical and Veterinary Entomology*, 22, 248–257.

Estrada-Peña, A., Nava, S., Horak, I.G., and Guglielmone, A.A. (2010). Using ground-derived data to assess the environmental niche of the spinose ear tick, *Otobius megnini*. *Entomologia Experimentalis et Applicata*, 137, 132–142.

European Environment Agency (2007). *Halting the loss of biodiversity by 2010: proposal for a first set of indicators to monitor progress in Europe*. Office for Official Publications of the European Communities <http://reports.eea.europa.eu>

Evangelista, P.H., Stohlgren, T.J., Morisette, J.T., and Kumar, S. (2009). Mapping invasive tamarisk: a comparison of single-scene and time-series analyses of remotely sensed data. *Remote Sensing*, 1, 519–533.

Evans, J.R. (1983). Nitrogen and photosynthesis in the flag leaf of wheat (*Triticum aestivum* L.). *Plant Physiology*, 72, 297–302.

Evans, K.L., James, N.A., and Gaston, K.J. (2006). Abundance, species richness and energy availability in the North American avifauna. *Global Ecology and Biogeography*, 15, 372–385.

Evans, K.L., Newson, S.E., Storch, D., Greenwood, J.J.D., and Gaston, K.J. (2008). Spatial scale, abundance and the species–energy relationship in British birds. *Journal of Animal Ecology*, 77, 395–405.

Exelis Visual Information Solutions (2013) *ENVI*. <http://www.exelisvis.com/ProductsServices/ENVI.aspx>

Ezebilo, E.E. and Mattsson, L. (2010). Socio-economic benefits of protected areas as perceived by local people around Cross River National Park, Nigeria. *Forest Policy and Economics*, 12, 189–193.

Fahrig, L. (2003). Effects of habitat fragmentation on biodiversity. *Annual Review of Ecology and Systematics*, 34, 487–515.

Fairbanks, D.H.K. and McGwire, K.C. (2004). Patterns of floristic richness in vegetation communities of California: regional scale analysis with multi-temporal NDVI. *Global Ecology and Biogeography*, 13, 221–235.

Fassnacht, K.S., Gower, S.T., MacKenzie, M.D., Nordheim, E.V., and Lillesand, T.M. (1997). Estimating the leaf area index of north central Wisconsin forests using the Landsat Thematic Mapper. *Remote Sensing of Environment*, 61, 229–245.

Fastring, D.R. and Griffith, J.A. (2009). Malaria incidence in Nairobi, Kenya and dekadal trends in NDVI and climatic variables. *Geocarto International*, 24, 207–221.

Fazey, I., Latham, I., Hagasua, J., and Wagatora, D. (2007). *Kahua Research Preliminary Report: Livelihoods and Change in Kahua, Solomon Islands*. Aberystwyth University, Aberystwyth.

Fazey, I., Pettorelli, N., Kenter, J.O., Wagatora, D., and Schuett, D. (2011). Maladaptive trajectories of change in Makira, Solomon Islands. *Global Environmental Change*, 21, 1275–1289.

Feddema, J.J., Oleson, K.W., Bonan, G.B., et al. (2005). The importance of land-cover change in simulating future climates. *Science*, 310, 1674–1678.

Feeley, K.J., Gillespie, T.W., and Terborgh, J.W. (2005). The utility of spectral indices from Landsat ETM+ for measuring the structure and composition of tropical dry forests. *Biotropica*, 37, 508–519.

Feng, X., Fu, B., Yang, X., and Lu, Y. (2010). Remote sensing of ecosystem services: an opportunity for spatially explicit assessment. *Chinese Geography Sciences*, 20, 522–535.

Fensholt, R., Nielsen, T.T., and Stisen, S. (2006a). Evaluation of AVHRR PAL and GIMMS 10-day composite NDVI time series products using SPOT-4 vegetation data for the African continent. *International Journal of Remote Sensing*, 27, 2719–2733.

Fensholt, R., Sandholt, I., and Stisen, S. (2006b). Evaluating MODIS, MERIS, and VEGETATION vegetation indices using in situ measurements in a semiarid environment. *IEEE Transactions on Geoscience and Remote Sensing*, 44, 1774–1786.

Fensholt, R., Rasmussen, K., Nielsen, T.T., and Mbow, C. (2009). Evaluation of earth observation based long term vegetation trends—ntercomparing NDVI time series trend analysis consistency of Sahel from AVHRR GIMMS, Terra MODIS and SPOT VGT data. *Remote Sensing of Environment*, 113, 1886–1898.

Ferguson, A.W., Currit, N.A., and Weckerly, F.W. (2009). Isometric scaling in home-range size of male and female bobcats. *Canadian Journal of Zoology*, 87, 1052–1060.

Festa-Bianchet, M. and Jorgenson, J.T. (1998). Selfish mothers: reproductive expenditure and resource availability in bighorn ewes. *Behavioural Ecology*, 9, 144–150.

Fieberg, J. and Börger, L. (2012). Could you please phrase "home range" as a question? *Journal of Mammalogy*, 93, 890–902.

Field, R., Hawkins, B.A., Cornell, H.V., *et al*. (2009). Spatial species-richness gradients across scales: a meta-analysis. *Journal of Biogeography*, 36, 132–147.

Fiske, S.J. (1992). Sociocultural aspects of establishing marine protected areas. *Ocean and Coastal Management*, 18, 25–46.

Foden, W., Mace, G., Vié, J-C., *et al.* (2008). Species susceptibility to climate change impacts. In: Vié, J-C., Hilton-Taylor, C., and Stuart, S.N. (eds), *The 2008 Review of The IUCN Red List of Threatened Species*. IUCN, Gland, Switzerland.

Foley, J.A., Levis, S., Costa, M.H., Cramer, W., and Pollard, D. (2000). Incorporating dynamic vegetation cover within global climate models. *Ecological Applications*, 10, 1620–1632.

Foley, J.A., Asner, G.P., Costa, M.H., *et al.* (2007). Amazonia revealed: forest degradation and loss of ecosystem goods and services in the Amazon Basin. *Frontiers in Ecology and the Environment*, 5, 25–32.

Food and Agriculture Organization (2009). *The State of Food and Agriculture 2009: Livestock in the Balance*. FAO, Rome.

Foody, G.M. (2004). Spatial nonstationary and scale-dependency in the relationship between species richness and environmental determinants for the sub-Saharan endemic fauna. *Global Ecology and Biogeography*, 13, 315–320.

Foody, G.M. and Cutler, M.E. (2006). Mapping the species richness and composition of tropical forests from remotely sensed data with neural networks. *Ecological Modelling*, 195, 37–42.

Foran, B.D. (1987) Detection of yearly cover change with Landsat MSS on pastoral landscapes in Central Australia. *Remote Sensing of Environment*, 23, 333–350.

Foran, B.D. and Pickup, G. (1984) Relationships of aircraft radiometric measurements to bare ground on semi-desert landscapes in central Australia. *Australian Rangelands Journal*, 6, 59–68.

Forister, M.L. and Shapiro, A.M. (2003). Climatic trends and advancing spring flight of butterflies in lowland California. *Global Change Biology*, 9, 1130–1135.

Forster, B (1984). Derivation of atmospheric correction procedures for LANDSAT MSS with particular reference to urban data. *International Journal of Remote Sensing*, 5, 799–818.

Freemantle, T.P., Wacher, T., Newby, J., and Pettorelli, N. (in press). Earth observation: overlooked potential to support species reintroduction programmes. *African Journal of Ecology*, 51, 482–492.

Fretwell, P.T. and Trathan, P.N. (2009). Penguins from space: faecal stains reveal the location of emperor penguin colonies. *Global Ecology and Biogeography*, 18, 543–552.

Fretwell, P.T., LaRue, M.A., Morin, P., *et al.* (2012). An emperor penguin population estimate: the first global, synoptic survey of a species from space. *PLoS ONE* 7(4), e33751. doi:10.1371/journal.pone.0033751.

Friedl, M.A., Michaelsen, J., Davis, F.W., Walker, H., and Schimel, D.S. (1994). Estimating grassland biomass and leaf area index using ground and satellite data. *International Journal of Remote Sensing*, 15, 1401–1420.

Frolking, S., Palace, M.W., Clark, D.B., Chambers, J.Q., Shugart, H.H., and Hurtt, G.C. (2009). Forest disturbance and recovery: a general review in the context of spaceborne remote sensing of impacts on aboveground biomass and canopy structure. *Journal of Geophysical Research*, 114, G00E02.<http://dx.doi.org/10.1029/2008JG000911>

Fuentes, M.V., Malone, J.B., and Mas-Coma, S. (2001). Validation of a mapping and prediction model for human fasciolosis transmission in Andean very high altitude endemic areas using remote sensing data. *Acta Tropica*, 79, 87–95.

Fuller, D.O. and Murphy, K. (2006). The ENSO-fire dynamic in insular Southeast Asia. *Climate Change*, 74, 435–455.

Funk, C.C. and Brown, M.E. (2006). Intra-seasonal NDVI change projections in semi-arid Africa. *Remote Sensing of Environment*, 101, 249–256.

GEO BON (2011). *Adequacy of Biodiversity Observation Systems to Support the CBD 2020 targets*. Report for the Convention on Biological Diversity, 106 pp.

GOFC-GOLD (2009). *Reducing Greenhouse Gas Emissions from Deforestation and Degradation in Developing Countries: A Sourcebook of Methods and Procedures for Monitoring, Measuring and Reporting*. Report version COP14–2.

Gaillard, J.-M., Delorme, D., Van Laere, G., Duncan, P., and Lebreton, J.D. (1997). Early survival in roe deer: causes and consequences of cohort variation in two contrasted populations. *Oecologia*, 112, 502–513.

Gaillard, J.-M., Festa-Bianchet, M., Yoccoz, N.G., Loison, A., and Toigo, C. (2000). Temporal variation in fitness components and population dynamics of large herbivores. *Annual Review of Ecology and Systematics*, 31, 367–393.

Gallai, N., Salles, J.M., Settele, J., and Vaissière B.E. (2009). Economic valuation of the vulnerability of world agriculture confronted with pollinator decline. *Ecological Economics*, 68, 810–821.

Gamon, J.A., Field, C.B., Goulden, M.L., et al. (1995). Relationships between NDVI, canopy structure and photosynthesis in three californian vegetation types. *Ecological Applications*, 5, 28–41.

Gao, B.-C. (1996). NDWI—A normalized difference water index for remote sensing of vegetation liquid water from space. *Remote Sensing of Environment*, 58, 257–266.

Garbulsky, M.F. and Paruelo, J.M. (2004). Remote sensing of protected areas to derive baseline vegetation functioning characteristics. *Journal of Vegetation Science*, 15, 711–720.

Garel, M., Gaillard, J.-M., Jullien, J.-M., Dubray, D., Maillard, D., and Loison, A. (2011). Population abundance and early spring conditions determine variation in body mass of juvenile chamois. *Journal of Mammalogy*, 92, 1112–1117.

Garonna, I., Fazey, I., Brown, M., and Pettorelli, N. (2009). Rapid primary productivity change in one of the last coastal rainforests: the case of Kahua, Solomon Islands. *Environmental Conservation*, 36, 253–260.

Garrity, S.R., AllenC.D., Brumby S.P., Gangodagamage C., McDowell N.G., and Cai D.M. (2013). Quantifying tree mortality in a mixed species woodland using multitemporal high spatial resolution satellite imagery. *Remote Sensing of Environment*, 129, 54–65.

Gaston, K.J. and Spicer, J.I. (2004). *Biodiversity: An Introduction*, 2nd edn. Blackwell, Oxford.

Gaston, K.J., Charman, K., Jackson, S.F., et al. (2006). The ecological effectiveness of protected areas: the United Kingdom. *Biological Conservation*, 132, 78–87.

Gaston, K.J., Jackson, S.F., Cantu-Salazar, L., and Cruz-Pinon, G. (2008). The ecological performance of protected areas. *Annual Review of Ecology, Evolution and Systematics*, 39, 93–113.

Gavashelishvili, A. and Lukarevskiy, V. (2008). Modelling the habitat requirements of leopard Panthera pardus in west and central Asia. *Journal of Applied Ecology*, 45, 579–588.

Gavier-Pizarro, G.I., Radeloff, V.C., Stewart, S.I., Huebner, C.D., and Keuler, N.S. (2010). Housing is positively associated with invasive exotic plant species richness in New England, USA. *Ecological Applications*, 20, 1913–1925.

Gebert, C. and Verheyden-Tixier, H. (2001). Variations of diet composition of red deer (*Cervus elaphus* L.) in Europe. *Mammal Review*, 31, 189–201.

Genovesi, P. (2005). Eradications of invasive alien species in Europe: a review. *Biological Invasions*, 7, 127–133.

Getz, L.L., Oli, M.K., Hofmann, J.E., McGuire, B., and Ozgul, A. (2005). Factors influencing movement distances of two species of sympatric voles. *Journal of Mammalogy*, 86, 647–654.

Gibbs, H.K. and Herold, M. (2007). Tropical deforestation and greenhouse gas emissions. *Environmental Research Letters*, 2, 045023.

Gilabert, M.A., González-Piqueras, J., García-Haro F.J., and Meliá, J. (2002). A generalized soil-adjusted vegetation index. *Remote Sensing of Environment*, 82, 303–310.

Gillespie, T.W. (2005). Predicting woody-plant species richness in tropical dry forests: a case study from south Florida, USA. *Ecological Applications*, 15, 27–37.

Gillespie, T.W., Foody, G.M., Rocchini, D., Giorgi, A.P., and Saatchi, S. (2008). Measuring and modelling biodiversity from space. *Progress in Physical Geography*, 32, 203–221.

Gillespie, T.W., Saatchi, S., Pau, S., Bohlman, S., Giorgi, A.P., and Lewis, S. (2009). Towards quantifying tropical tree species richness in tropical forests. *International Journal of Remote Sensing*, 30, 1629–1634.

Gilliam, F.S. (2007). The ecological significance of the herbaceous layer in temperate forest ecosystems. *Bioscience*, 57, 845–858.

Giri, C., Ochieng, E., Tieszen, L.L., et al. (2011). Status and distribution of mangrove forests of the world using earth observation satellite data. *Global Ecology and Biogeography*, 20, 154–159.

Gitelson, A.A. (2004). Wide dynamic range vegetation index for remote quantification of biophysical characteristics of vegetation. *Journal of Plant Physiology*, 161, 165–173.

Gitelson, A. and Merzlyak, M.N. (1994). Spectral reflectance changes associated with autumn senescence of *Aesculus hippocastanum* L., and *Acer platanoides* L. leaves. Spectral features and relation to chlorophyll estimation. *Journal of Plant Physiology*, 143, 286–292.

Gitelson, A. and Merzlyak, M.N. (1997). Remote estimation of chlorophyll content in higher plant leaves. *International Journal of Remote Sensing*, 18, 2691–2697.

Gitelson, A.A., Kaufman Y.J., and Merzlyak, M.N. (1996). Use of a green channel in remote sensing of global vegetation from EOS-MODIS. *Remote Sensing of Environment*, 58, 289–298.

Gitelson, A.A., Viňa, A., Rundquist, D.C., Ciganda, V., and Arkebauer, T.J. (2005). Remote estimation of canopy chlorophyll content in crops. *Geophysical Research Letters*, 32, L08403.

Glantz, M.H. (1996). *Currents of Change: El Niño's Impact on Climate and Society*. Cambridge University Press, Cambridge, 194 pp.

Glenn, E.P., Huete, A.R., Nagler, P.L., and Nelson, S.G. (2008). Relationship between remotely-sensed vegetation indices, canopy attributes and plant physiological processes: what vegetation indices can and cannot tell us about the landscape. *Sensors*, 8, 2136–2160.

Goetz, S.J., Baccini, A., Laporte, N.T., *et al*. (2009). Mapping and monitoring carbon stocks with satellite observations: a comparison of methods. *Carbon Balance and Management*, 4, 1–7.

Gong, D. and Shi, P. (2003). Northern hemispheric NDVI variations associated with large-scale climate indices in spring. *International Journal of Remote Sensing*, 24, 2559–2566.

Good, P., Lowe, J.A., Collins, M., and Moufouma-Okia, W. (2008). An objective tropical Atlantic sea surface temperature gradient index for studies of south Amazon dry-season climate variability and change. *Philosophical Transactions of the Royal Society, Series B*, 363, 1761–1766.

Goodwin, N.R., Coops, N.C., Wulder, M.A., Gillanders, S., Schroeder, T.A., and Nelson, T. (2008). Estimation of insect infestation dynamics using a temporal sequence of Landsat data. *Remote Sensing of Environment*, 112, 3680–3689.

Gordon, I.J., Pettorelli, N., Katzner, T., *et al*. (2010). International Year of Biodiversity—missed targets and the need for better monitoring, real action and global policy. *Animal Conservation*, 13, 113–114.

Gordon, I.J., Acevedo-Whitehouse, K., Altwegg, R., *et al*. (2012). What the 'food security' agenda means for animal conservation in terrestrial ecosystems. *Animal Conservation*, 15, 115–116.

Gotelli, N.J. and Colwell, R.K. (2001). Quantifying biodiversity: procedures and pitfalls in the measurement and comparison of species richness. *Ecology Letters*, 4, 379–191.

Gould, W. (2000). Remote sensing of vegetation, plant species richness, and regional biodiversity hotspots. *Ecological Applications*, 10, 1861–1870.

Gouveia, C., Trigo, R.M., DaCamara, C.C., Libonati, R., and Pereira, J.M.C. (2008). The North Atlantic Oscillation and European vegetation dynamics. *International Journal of Climatology*, 28, 1835–1847.

Goward, S.N., Tucker, C.J., and Dye, D.G. (1985). North American vegetation patterns observed with the NOAA-7 advanced very high resolution radiometer. *Vegetation*, 64, 3–14.

Goward, S.N., Markham, B., Dye, D.G., Dulaney, W., and Yang, J. (1991). Normalized difference vegetation index measurements from the Advanced Very High Resolution Radiometer. *Remote Sensing of Environment*, 35, 257–277.

Grande, J.M., Serrano, D., Tavecchia, G., *et al*. (2009). Survival in a long-lived territorial migrant: effects of life-history traits and ecological conditions in wintering and breeding areas. *Oikos*, 118, 580–590.

Grantz, K., Rajagopalan, B., Clark, M., and Zagona, E. (2007). Seasonal Shifts in the North American Monsoon. *Journal of Climate*, 20, 1923–1935.

GRASS Development Team (2013). *Geographic Resources Analysis Support System*. <http://grass.osgeo.org/>

Green, A., Lokani, P., Atu, W., Ramohia, P., Thomas, P., and Almany, J. (2006). *Solomon Islands Marine Assessment*. Technical report of survey conducted May 13–June 17, 2004. The Nature Conservancy, Brisbane.

Griffith, B., Douglas, D., Walsh, N.E., *et al*. (2002). The Porcupine caribou herd. In: Douglas, D.C. *et al*. (eds), *Arctic Refuge Coastal Plain Terrestrial Wildlife Research Summaries*. US Geological Survey, Biological Resources Division, Biological Science Report USGS/BRD/BSR–2002–0001, pp. 8–37.

Grimm, N.B., Foster, D., Groffman, P., *et al*. (2008). The changing landscape: ecosystem responses to urbanization and pollution across climatic and societal gradients. *Frontiers in Ecology and the Environment*, 6, 264–272.

Groisman, P.Y., Knight, R.W., Easterling, D.R., Karl, T.R., Hegerl, G.C., and Razuvaev, V.N. (2005). Trends in intense precipitation in the climate record. *Journal of Climate*, 18, 1326–1350.

Gross, J.E., Goetz, S.J., and Cihlar, J. (2009). Application of remote sensing to parks and protected area monitoring: introduction to the special issue. *Remote Sensing of Environment*, 113, 1343–1345.

Guimaraes, R.J.P.S., Freitas, C.C., Dutra, L.V., *et al*. (2008). Schistosomiasis risk estimation in Minas Gerais State, Brazil, using environmental data and GIS techniques. *Acta Tropica*, 108, 234–241.

Guisan, A. and Thuiller, W. (2005). Predicting species distribution: offering more than simple habitat models. *Ecology Letters*, 8, 993–1009.

Gunawan, D., Gravenhorst, G., and Jacobs, D. (2003). Rainfall variability studies in South Sulawesi using regional climate model REMO. *Journal of Meteorology*, 4, 65–70.

Gutman, G. (1991). Vegetation indices from AVHRR—an update and future prospects. *Remote Sensing of Environment*, 35, 121–136.

Gutman, G. (1999). On the use of long-term global data of land reflectances and vegetation indices derived from the AVHRR. *Journal of Geophysical Research*, 104, 6241–6255.

Gutman, G. and Ignatov, A. (1996). The relative merit of cloud/clear identification in the NOAA/NASA Pathfinder AVHRR Land 10-day composites. *International Journal of Remote Sensing*, 17, 3295–3304.

Gutman, G., Tarpley, D., Ignatov, A., and Olson, S. (1995). The enhanced NOAA global land dataset from Advanced Very High Resolution Radiometer. *Bulletin of the American Meteorological Society*, 76, 1141–1156.

H-Acevedo, D. and Currie, D.J. (2003). Does climate determine broad-scale patterns of species richness? A test of the causal link by natural experiment. *Global Ecology and Biogeography*, 12, 461–473.

Haarsma, R.J., Selten, F.M., Weber, S.L., and Kliphuis, M. (2005). Sahel rainfall variability and response to greenhouse warming. *Geophysical Research Letters*, 32, L17702, doi:10.1029/2005GL023232.

Hallett, T.B., Coulson, T., Pilkington, J.G., Clutton-Brock, T.H., Pemberton, J.M., and Grenfell, B.T. (2004). Why large-scale climate indices seem to predict ecological processes better than local weather. *Nature*, 430, 71–75.

Hamel, S., Garel, M., Festa-Bianchet, M., Gaillard, J.-M., and Côté, S.D. (2009). Spring Normalized Difference Vegetation Index (NDVI) predicts annual variation in timing of peak faecal crude protein in mountain ungulates. *Journal of Applied Ecology*, 46, 582–589.

Hammill, K.A. and Bradstock, R.A. (2006). Remote sensing of fire severity in the Blue Mountains: influence of vegetation type and inferring fire intensity. *International Journal of Wildland Fire*, 15, 213–226.

Han, S., Hendrickson, L.L., and Ni, B. (2002). Comparison of satellite and aerial imagery for detecting leaf chlorophyll content in corn. *Transactions of the ASAE*, 45, 1229–1236.

Hannah, L., Midgley, G., Andelman, S., et al. (2007). Protected area needs in a changing climate. *Frontiers in Ecology and the Environment*, 5, 131–138.

Hansen, B.B., Aanes, R., Herfindal, I., Sæther, B.E., and Henriksen, S. (2009a). Winter habitat-space use in a large arctic herbivore facing contrasting forage abundance. *Polar Biology*, 32, 971–984.

Hansen, B.B., Herfindal, I., Aanes, R., Sæther, B.E., and Henriksen, S. (2009b). Functional response in habitat selection and the tradeoffs between foraging niche components in a large herbivore. *Oikos*, 118, 859–872.

Hansen, M.C., Defries, R.S., Townshend, J.R.G., and Sohlberg, R. (2000). Global land cover classification at 1 km spatial resolution using a classification tree approach. *International Journal of Remote Sensing*, 21, 1331–1364.

Harestad, A.S. and Bunnell, F.L. (1979). Home range and body weight—a re-evaluation. *Ecology*, 60, 389–402.

Harner, M.J. and Geluso, K. (2012). Effects of cattle grazing on Platte River caddisflies (*Ironoquia plattensis*) in central Nebraska. *Freshwater Science*, 31, 389–394.

Harrison, S., Davies, K.F., Safford, H.D., and Viers, J.H. (2006). Beta diversity and the scale-dependence of the productivity-diversity relationship: a test in the Californian serpentine flora. *Journal of Ecology*, 94, 110–117.

Hartley, A.J., Nelson, A., Mayaux, P., and Gregoire, J.M. (2007). *The Assessment of African Protected Areas, Scientific and Technical Report*. Office for Official Publications of the European Communities, Luxembourg.

Hasmadi, M., Pakhriazad, H.Z., and Norlida, K. (2011). Remote sensing for mapping RAMSAR heritage site at Sungai Pulai Mangrove Forest Reserve, Johor, Malaysia. *Sains Malaysiana*, 40, 83–88.

Hawkins, B.A. (2004). Summer vegetation, deglaciation and the anomalous bird diversity gradient in eastern North America. *Global Ecology and Biogeography*, 13, 321–325.

Hawkins, B.A., Porter, E.E., and Diniz-Filho, J.A.F. (2003). Productivity and history as predictors of the latitudinal diversity gradient of terrestrial birds. *Ecology*, 84, 1608–1623.

Hay, C.M., Kuretz, C.A., Odenweller, J.B., Scheffner E.J., and Wood, B. (1979). *Development of AI Procedures for Dealing with the Effects of Episodal Events on Crop Temporal Spectral Response*. AgRISTARS SR-Bg-00434, Contract NAS 9–14565.

Hayes, L.D., Chesh, A.S., and Ebensperger, L.A. (2007). Ecological predictors of range areas and use of burrow systems in the diurnal rodent, Octodon degus. *Ethology*, 113, 155–165.

He, J. and Shao, X. (2006). Relationships between tree-ring width index and NDVI of grassland in Delingha. *Chinese Science Bulletin*, 51, 1106–1114.

He, K.S., Zhang, J., and Zhang, Q. (2009). Linking variability in species composition and MODIS NDVI based on beta diversity measurements. *Acta Oecologica*, 35, 14–21.

He, K.S., Rocchini, D., Neteler, M., and Nagendra, H. (2011). Benefits of hyperspectral remote sensing for tracking plant invasions. *Diversity and Distributions*, 17, 381–392.

Hebblewhite, M., Merrill, E., and McDermid, G. (2008). A multi-scale test of the forage maturation hypothesis in a partially migratory ungulate population. *Ecological Monographs*, 78, 141–166.

Hendon, H.H. (2003). Indonesian rainfall variability: Impacts of ENSO and local air—sea interaction. *Journal of Climate*, 16, 1775–1790.

Henry, P.-Y., Lengyel, S., Nowicki, P., et al. (2008). Integrating ongoing biodiversity monitoring: potential benefits and methods. *Biodiversity and Conservation*, 17, 3357–3382.

Hepinstall, J.A. and Sader, S.A. (1997) Using Bayesian statistics, Thematic Mapper satellite imagery, and breeding bird survey data to model bird species probability of occurrence in Maine. *Photogrammetry Engineering and Remote Sensing*, 63, 1231–1237.

Herfindal, I., Linnell, J.D.C., Odden, J., Nilsen, E.B., and Andersen, R. (2005). Prey density and environmental productivity explain variation in Eurasian lynx home range size at two spatial scales. *Journal of Zoology (London)*, 265, 63–71.

Herfindal, I., Drever, M.C., Høgda, K.-A., et al. (2012). Landscape heterogeneity and the effect of environmental conditions on prairie wetlands. *Landscape Ecology*, 27, 1435–1450.

Herfindal, I., Solberg, E.J., Saether, B.E., Hogda, K.A., and Andersen, R. (2006a). Environmental phenology and geographical gradients in moose body mass. *Oecologia*, 150, 213–224.

Herfindal, I., Sæther, B.E., Solberg, E.J., Andersen, R., and Hogda, K.A. (2006b). Population characteristics predict responses in moose body mass to temporal variation in the environment. *Journal of Animal Ecology*, 75, 1110–1118.

Herrmann, S.M., Anyamba A., and Tucker, C.J. (2005). Recent trends in vegetation dynamics in the African Sahel and their relationship to climate. *Global Environmental Change*, 15, 394–404.

Herrmann, S.M. and Tappan, G.G. (2013). Vegetation impoverishment despite greening: a case study from central Senegal. *Journal of Arid Environments*, 90, 55–66.

Hewison, A.J.M. (1996). Variation in the fecundity of roe deer in Britain: effects of age and body weight. *Acta Theriologica*, 41, 187–198.

Hicke, J.A., Asner, G.P., Randerson, J.T., et al. (2002). Trends in North American net primary productivity derived from satellite observations, 1982–1998. *Global Biogeochemical Cycles*, 16, 1018–1032.

Hickling, R., Roy, D.B., Hill, J.K., and Thomas, C.D. (2005). A northward shift of range margins in British Odonata. *Global Change Biology*, 11, 502–506.

Hielkema, J.U. (1990). Operational satellite environmental monitoring for food security by FAO, The ARTEMIS System, FAO Remote Sensing Centre, Rome, Italy.

Hielkema, J.U., Prince, S.D., and Astle, W.L. (1986). Rainfall and vegetation monitoring in the savannah zone of the Democratic Republic of Sudan using NOAA AVHRR. *International Journal of Remote Sensing*, 7, 1499–1513.

Hirzel, A.H., Hausser, J., Chessel, D., and Perrin, N. (2002). Ecological-niche factor analysis: How to compute habitat-suitability maps without absence data? *Ecology*, 83, 2027–2036.

Hjort, J. (1914). Fluctuations in the great fisheries of Northern Europe viewed in the light of biological research. *Volume 20 of Rapports et procès-verbaux des réunions*, 1–228.

Hobbs, R.J. (1992). The role of corridors in conservation: solution or bandwagon? *Trends in Ecology and Evolution*, 7, 389–392.

Hobbs, R.J. and Humphries, S.E. (1995). An integrated approach to the ecology and management of plant invasions. *Conservation Biology*, 9, 761–770.

Hockings, M., Stolton, S., and Dudley, N. (2004). Management effectiveness: assessing management of protected areas. *Journal of Environmental Policy and Planning*, 6, 157–174.

Hockings, M., Stolton, S., Leverington, F., Dudley, N., and Corrau, J. (2006). *Evaluating Effectiveness: A Framework for Assessing Management Effectiveness of Protected Areas*, 2nd edn. IUCN, WCPA and University of Queensland, Gland, Switzerland, and Brisbane Australia.

Hoegh-Guldberg, O. (2007). Coral reefs under rapid climate change and ocean acidification. *Science*, 318, 1737–1742.

Hoehn, P., Tscharntke, T., Tylianakis, J.M., and Ingolf, S.-D. (2008). Functional group diversity of bee pollinators increases crop yield. *Proceedings of the Royal Society of London, Series B*, 275, 2283–2291.

Hoerling, M., Eischeid, J., Perlwitz, J., Quan X., Zhang T., Pegion P. (2011). On the increased frequency of mediterranean drought. *Journal of Climate*, 25, 2146–2161.

Hogg, J.T., Hass, C.C., and Jenni, D.A. (1992). Sex-biased maternal expenditure in Rocky Mountain bighorn sheep. *Behavioural Ecology and Sociobiology*, 31, 243–251.

Holben, B.N. (1986) Characteristics of maximum-value composite images from temporal AVHRR data. *International Journal of Remote Sensing*, 7, 1417–1434.

Holben, B.N. and Fraser, R.S. (1984). Red and near-infrared sensor response to off-nadir viewing. *International Journal of Remote Sensing*, 5, 145–160.

Holben, B.N. and Justice, C.O. (1981) An examination of spectral band ratioing to reduce the topographic effect on remotely sensed data. *International Journal of Remote Sensing*, 2, 115–133.

Holland, G.J. and Webster, P.J. (2007). Heightened tropical cyclone activity in the North Atlantic: natural variability or climate trend? *Philosophical Transactions of the Royal Society, Series A*, 365, 2695–2716.

Hooper, D.U., Chapin, F.S., Ewel, J.J., et al. (2005). Effects of biodiversity on ecosystem functioning: a consensus of current knowledge. *Ecological Monographs* 75, 3–35.

Horler, D.N.H., Dockray, M., and Barber, J. (1983). The red edge of plant leaf reflectance. *International Journal of Remote Sensing*, 4, 273–288.

Horning, N., Robinson, J.A., Sterling, E.J., Turner, W., and Spector, S. (2010). *Remote Sensing for ecology and Conservation*. Oxford University Press, Oxford, 467 pp.

Howard, W.E. (1960). Innate and environmental dispersal of individual vertebrates. *American Midland Naturalist*, 63, 152–161.

Hoyos, L.E., Gavier-Pizarro, G.I., Kuemmerle, T., Bucher, E.H, Radeloff, V.C., and Tecco, P.A. (2010). Invasion of glossy privet (*Ligustrum lucidum*) and native forest loss in the Sierras Chicas of Cordoba, Argentina. *Biological Invasions*, 12, 3261–3275.

Hu, J., and Jiang, Z. (2011). Climate change hastens the conservation urgency of an endangered ungulate. *PLoS ONE*, 6(8), e22873. doi:10.1371/journal.pone.0022873.

Hua, W., Fan, G., Zhou, D., et al. (2008). Preliminary analysis on the relationships between Tibetan Plateau NDVI change and its surface heat source and precipitation of China. *Science in China, Part D*, 51, 677–685.

Huber, S. and Fensholt, R. (2011). Analysis of teleconnections between AVHRR-based sea surface temperature and vegetation productivity in the semi-arid Sahel. *Remote Sensing of Environment*, 115, 3276–3285.

Huete, A.R. (1988) A soil-adjusted vegetation index (SAVI). *Remote Sensing of Environment*, 25, 295–309.

Huete, A.R. and Tucker, C.J. (1991). Investigation of soil influences in AVHRR red and near-infrared vegetation index imagery. *International Journal of Remote Sensing*, 12, 1223–1242.

Huete, A.R., Jackson, R.D., and Post, D.F. (1985). Spectral response of a plant canopy with different soil backgrounds. *Remote Sensing of Environment*, 17, 37–53.

Huete, A.R., Liu, H., Batchily, K., and van Leeuwen, W. (1997). A Comparison of Vegetation Indices Over a Global Set of TM Images for EOS-MODIS. *Remote Sensing of Environment*, 59: 440–451.

Huete, A.R, Didan, K., Miura, T., Rodriguez, E.P., Gao, X., and Ferreira, L.G. (2002). Overview of the radiometric and biophysical performance of the MODIS vegetation indices. *Remote Sensing of Environment*, 83, 195–213.

Huete, A.R., Didan, K., Shimabukuro, Y.E., et al. (2006). Amazon rainforests green-up with sunlight in dry season. *Geophysical Research Letters*, 33, L06405. doi:10.1029/2005GL025583.

Hunter, M.L. (1996). *Fundamentals of Conservation Biology*. Blackwell, Oxford.

Hurlbert, A.H. (2004). Species–energy relationships and habitat complexity in bird communities. *Ecology Letters*, 7, 714–720.

Hurlbert, A.H. and Haskell, J.P. (2003). The effect of energy and seasonality on avian species richness and community composition. *American Naturalist*, 161, 83–97.

Hurrell, J.W. (1995). Decadal trends in the North Atlantic Oscillation regional temperatures and precipitation. *Science*, 269, 676–679.

Hurrell, J.W., Kushnir, Y., Ottersen, G., and Visbeck, M. (2003). *An Overview of the North Atlantic Oscillation. The North Atlantic Oscillation: Climatic Significance and Environmental Impact. Geophysical Monograph* 134. American Geophysical Union, 10.1029/134GM01.

Höglund, J. and Alatalo, R.V. (1995). *Leks*. Princeton University Press, Princeton.

Hüttich, C., Herold, M., Schmullius, C., Egorov, V., and Bartalev, S.A. (2007). Indicators of Northern Eurasia's land-cover change trends from SPOT-VEGETATION time-series analysis 1998–2005. *International Journal of Remote Sensing*, 28, 4199–4206.

Ichii, K., Kawabata, A., and Yamaguchi, Y. (2002). Global correlation analysis for NDVI and climatic variables and NDVI trends: 1982–1990. *International Journal of Remote Sensing*, 23, 3873–3878.

Indeje, M., Ward, M.N., Ogallo, L.J., Davies, G., Dilley, M., and Anyamba, A. (2006). Predictability of the Normalized Difference Vegetation Index in Kenya and Potential Applications as an Indicator of Rift Valley Fever Outbreaks in the Greater Horn of Africa. *Journal of Climate*, 19, 1673–1687.

IPCC (2001). *Climate Change 2001. Synthesis report*. Cambridge University Press, Cambridge.

IPCC (2007). *Climate Change 2007: Synthesis report*. Fourth Assessment Report of the Intergovernmental Panel on Climate Change. IPCC, Gland, Switzerland.

IPCC (2012). Summary for Policymakers. In: C.B. Field, V. Barros, T.F. Stocker, et al. (eds), *Managing the Risks of Extreme Events and Disasters to Advance Climate Change Adaptation. A Special Report of Working Groups I and II of the Intergovernmental Panel on Climate Change*. Cambridge University Press, Cambridge, UK, and New York, NY, pp. 1–19.

IUCN (1987). *IUCN Position Statement on the Translocation of Living Organisms: Introductions, Reintroductions, and Re-stocking*. Prepared by the Species Survival Commission in collaboration with the Commission on Ecology and the Commission on Environmental Policy, Law and Administration. IUCN, http://www.iucnsscrsg.org/.

IUCN (1994). *Guidelines for Protected Area Management Categories*. IUCN, Gland, Switzerland/World Conservation Union, Cambridge, UK.

IUCN (1998). *Guidelines for Reintroductions*. IUCN/SSC Reintroduction Specialist Group, Gland, Switzerland.

IUCN (2012). *IUCN Guidelines for Reintroductions and Other Conservation Translocations*. IUCN/SSC Reintroduction Specialist Group. IUCN, Gland, Switzerland.

IUCN Invasive SpeciesSpecialist Group (2012). <http://www.issg.org/database/species/search.asp?st=100ss>

Ishiguro, E., Ogawa, Y., Miyazato, M., and Chen, Y.J. (1994). Discrimination of the frost damaged tea fields using LANDSAT-5/TM data. *Bulletin of the Faculty of Agriculture—Kagoshima University*, 44, 35–41.

Ito, T.Y., Miura, N., Lhagvasuren, B., et al. (2006). Satellite tracking of Mongolian gazelles (Procapra gutturosa) and habitat shifts in their seasonal ranges. *Journal of Zoology (London)*, 269, 291–298.

Pinter Jr, P.J., Jackson, R.D., Ezra, C.E., and Gausman, H.W. (1985). Sun angle and canopy architecture effects on the reflectance of six wheat cultivars. *International Journal of Remote Sensing*, 6, 1813–1825.

Jackson, R.D. and Huete, A.R. (1991). Interpreting vegetation indices. *Preventive Veterinary Medicine*, 11, 185–200.

Jackson, R.D., Slater, P.N., and Pinter, P.J. (1983). Discrimination of growth and water stress in wheat by various vegetation indices through clear and turbid atmospheres. *Remote Sensing of Environment*, 13, 187–208.

James, M.E. and Kalluri, S.N.V. (1994). The Pathfinder AVHRR land data set: an improved coarse resolution data set for terrestrial monitoring. *International Journal of Remote Sensing*, 15, 3347–3363.

Janicot, S., Trzaska, S., and Poccard, I. (2001). Summer SAHEL-ENSO teleconnection and decadal time scale SST variations. *Climate Dynamics*, 18, 303–320.

Jensen, J.R. (1996). *Introductory Digital Image Processing. A Remote Sensing Perspective*, 2nd edn. Prentice-Hall, Upper Saddle River, NJ.

Jensen, J.R. (2007). *Remote Sensing of the Environment: An Earth Resource Perspective*, 2nd edn. Prentice-Hall, Upper Saddle River, NJ.

Jeong, S.-J., Ho, C.-H., Brown, M.E., Kug, J.-S., and Piao, S. (2011). Browning in desert boundaries in Asia in recent decades. *Journal of Geophysical Research*, 116, D02103. doi:10.1029/2010JD014633.

Jepsen, J.U., Hagen, S.B., Hogda, K.A., *et al.* (2009). Monitoring the spatiotemporal dynamics of geometrid moth outbreaks in birch forest using MODIS-NDVI data. *Remote Sensing of Environment*, 113, 1939–1947.

Jia, L., Yu, Z., Li, F., *et al.* (2012). Nitrogen status estimation of winter wheat by using an IKONOS satellite image in the North China Plain. *IFIP Advances in Information and Communication Technology*, 369, 174–184.

Jian, J., Jiang, H., Zhou, G., *et al.* (2011). Mapping the vegetation changes in giant panda habitat using Landsat remotely sensed data. *International Journal of Remote Sensing*, 32, 1339–1356.

Jiang, Z., Huete, A. R., Didan, K., and Miura, T. (2008). Development of a two-band enhanced vegetation index without a blue band. *Remote Sensing of Environment*, 112, 3833–3845.

Jimenez-Valverde, A., and Lobo, J.M. (2006). Distribution determinants of endangered Iberian spider *Macrothele calpeiana* (Araneae, Hexathelidae). *Environmental Entomology*, 35, 1491–1499.

Jin, M. (2004). Analysis of land skin temperature using AVHRR observations. *Bulletin of the American Meteorological Society*, 85, 587.

Johns, D. (2010). International year of biodiversity—from talk to action. *Conservation Biology*, 24, 238–240.

Jones, H.G. and Vaughan, R.A. (2010). *Remote Sensing of Vegetation: Principles, Techniques and Applications*. Oxford University Press, Oxford.

Jones, H.P., Tershy, B.R., Zavaleta, E.S., *et al.* (2008). Severity of the effects of invasive rats on seabirds: a global review. *Conservation Biology*, 22, 16–26.

Jordan, C.F. (1969). Derivation of leaf area index from quality of light on the forest floor. *Ecology*, 50, 663–666.

Jorgensen, A.F. and Nohr, H. (1996). The use of satellite images for mapping of landscape and biological diversity in the Sahel. *International Journal of Remote Sensing*, 17, 91–109.

Julien, Y., Sobrino, J.A. and Verhoef, W. (2006). Changes in land surface temperatures and NDVI values over Europe between 1982 and 1999. *Remote Sensing of Environment*, 103, 43–55.

Justice, C.O., Holben, B.N., and Gwynne, M.D. (1986). Monitoring East African vegetation using AVHRR data. *International Journal of Remote Sensing*, 7, 1453–1474.

Justice, C.O., Townshend, J.R.G., Holben, B.N., and Tucker, C.J. (1985). Analysis of the phenology of global vegetation using meteorological satellite data. *International Journal of Remote Sensing*, 6, 1271–1318.

Jönsson, P. and Eklundh, L. (2002). Seasonality extraction by function fitting to time-series of satellite sensor data. *IEEE Transactions on Geoscience and Remote Sensing*, 40, 1824–1832.

Kalluri, S., Gilruth, P., Rogers, D., and Szczur, M. (2007). Surveillance of arthropod vector-borne infectious diseases using remote sensing techniques: a review. *PLoS Pathogene*, 3(10), e116. doi:10.1371/journal.ppat.0030116.

Kanemasu, E.T. (1974). Seasonal canopy reflectance patterns of wheat, sorghum, and soybean. *Remote Sensing of Environment*, 3, 43–47.

Karl, T.R. and Knight, R.W. (1998). Secular trends of precipitation amount, frequency, and intensity in the United States. *Bulletin of the American Meteorology Society*, 79, 231–241.

Karnieli, A., Kaufman, Y.J., Remer, L., and Wald, A. (2001). AFRI-aerosol free vegetation index. *Remote Sensing of Environment*, 77, 10–21.

Karr, J.R. and Chu, E.W. (1995). Ecological integrity: reclaiming lost connections. In WestraL. and Lemons J. (eds), *Perspectives on Ecological Integrity*. Kluwer Academic, Dordrecht, pp. 34–48.

Karr, J.R. and Dudley, D.R. (1981). Ecological perspective on water quality goals. *Environmental Management*, 5, 55–68.

Kasischke, E.S. and French, N.H.F. (1997). Constraints on using AVHRR composite index imagery to study patterns of vegetation cover in boreal forests. *International Journal of Remote Sensing*, 18, 2403–2426.

Kasischke, E.S., Melack, J.M., and Dobson, M.C. (1997). The use of imaging radars for ecological applications—a review. *Remote Sensing of Environment*, 59, 141–156.

Kaufman, Y.J. and Holben, B.N. (1993). Calibration of the AVHRR visible and near-IR bands by atmospheric scattering, ocean glint and desert reflection. *International Journal of Remote Sensing*, 14, 21–52.

Kaufman, Y.J. and Tanré, D. (1992). Atmospherically resistant vegetation index (ARVI) for EOS-MODS. *IEEE Transactions on Geoscience and Remote Sensing*, 30, 261–270.

Kaufmann, R.K., D'Arrigo, R.D., Laskowski, C., Myneni, R.B., Zhou, L., and Davi, N.K. (2004). The effect of growing season and summer greenness on northern forests. *Geophysical Research Letters*, 31, L09205, doi:10.1029/2004GL019608

Kausrud, K., Viljugrein, H., Frigessi, A., *et al*. (2007). Climatically-driven synchrony of gerbil populations allows large-scale plague outbreaks. *Proceedings of the Royal Society of London serie B*, 274, 1963–1969.

Kausrud, K., Mysterud, A., Steen, H., *et al*. (2008). Linking climate change to lemming cycles. *Nature*, 456, 93–97.

Kauth, R.J. and Thomas, G.S. (1976). The tasselled cap—a graphic description of the spectral-temporal development of agricultural crops as seen by Landsat. In: *Proceedings of the Symposium on Machine Processing of Remotely Sensed Data, Purdue University, West Lafayette, Indiana*, pp. 41–51.

Kearney, M. (2006). Habitat, environment and niche: what are we modelling? *Oikos*, 115, 186–191.

Keddy, P.A. (2007). *Plants and Vegetation*. Cambridge University Press, Cambridge.

Keddy, P.A. (2010). *Wetland Ecology: Principles and Conservation*, 2nd edn. Cambridge University Press, Cambridge.

Kelly, J.F., Shipley, J.R., Chilson, P.B., Howard, K.W., Frick, W.F., and Kunz, T.H. (2012). Quantifying animal phenology in the aerosphere at a continental scale using NEXRAD weather radars. *Ecosphere*, 3, art16 <http://dx.doi.org/10.1890/ES11-00257.1>.

Kennedy, P. (1989). Monitoring the vegetation of Tunisian grazing lands using NDVI. *Ambio*, 18, 119–123.

Kerr, J.T. and Currie, D.J. (1995). Effects of human activity on global extinction risk. *Conservation Biology*, 9, 1528–1538.

Kerr, J.T. and Ostrovsky, M. (2003). From space to species: ecological applications for remote sensing. *Trends in Ecology and Evolution*, 18, 299–305.

Key, C.H. and Benson, N.C. (2004). Remote sensing measure of severity: the normalized burn ratio. *FIREMON Landscape Assessment (LA) V4 Sampling and Analysis Methods*. pp. LA1–16.

Kharuk, V.I., Alshansky, A.M., and Yegorv, V.V. (1992). Spectral characteristics of vegetation cover: factors of variability. *International Journal of Remote Sensing*, 13, 3236–3272.

Kiage, L.M. and Obuoyo, J. (2011). The potential link between El Niño and water hyacinth blooms in Winam Gulf of Lake Victoria, East Africa: evidence from satellite imagery. *Water Resources Management*, 25, 3931–3945.

Kiage, L.M. and Walker, N.D. (2009). Using NDVI from MODIS to monitor duckweed bloom in lake Maracaibo, Venezuela. *Water Resources Management*, 23, 1125–1135.

Kimes, D.S., Holben, B.N., Tucker, C.J., and Newcomb, W.W. (1984). Optimal directional view angles for remote sensing missions. *International Journal of Remote Sensing*, 5, 887–908.

Kinnear, J.E., Sumner, N.R., and Onus, M.L. (2002). The red fox in Australia—an exotic predator turned biocontrol agent. *Biological Conservation*, 108, 335–359.

Kinyanjui, M.J. (2011). NDVI-based vegetation monitoring in Mau forest complex, Kenya. *African Journal of Ecology*, 49, 165–174.

Klaassen, R.H.G., Strandberg, R., Hake, M., Olofsson, P., Tottrup, A.P., and Alerstam, T. (2010). Loop migration in adult marsh harriers *Circus aeruginosus*, as revealed by satellite telemetry. *Journal of Avian Biology*, 41, 200–207.

Klemas, V. (2011). Remote sensing techniques for studying coastal ecosystems: an overview. *Journal of Coastal Research*, 27, 2–14.

Knight, J.R., Folland, C.K., and Scaife, A.A. (2006). Climate impacts of the Atlantic multidecadal oscillation. *Geophysical Research Letters*, 33, L17706. doi:10.1029/2006GL026242.

Knipling, E.B. (1970). Physical and physiological basis for the reflectance of visible and near infrared radiation from vegetation. *Remote Sensing of Environment*, 1, 155–159.

Kodani, E., Awaya, Y., Tanaka, K., and Matsumura, N. (2002). Seasonal patterns of canopy structure, biochemistry and spectral reflectance in a broad-leaved deciduous *Fagus crenata* canopy. *Forest Ecology and Management*, 167, 233–249.

Koetz, B., Sun, G., Morsdorf, F., *et al*. (2007). Fusion of imaging spectrometer and lidar data over combined radiative transfer models for forest canopy characterization. *Remote Sensing of Environment*, 106, 449–459.

Kogan, F.N. (1997). Global drought watch from space. *Bulletin of the American Meteorological Society*, 78, 621–636.

Kogan, F.N. (1998). Global drought and flood-watch from NOAA polar-orbiting satellites. *Advances in Space Research*, 21, 477–480.

Kogan, F.N. (2000). Contribution of remote sensing to drought early warning. In: Wilhite, D.A., Sivakumar, M.V.K., and Wood, D.A. (eds), *Early Warning Systems for Drought Preparedness and Drought Management. Proceedings of an Expert Group Meeting, 5–7 September, Lisbon, Portugal*. World Meterological Organization, Geneva, pp. 86–100.

Kogan, F.N. and Zhu, X. (2001). Evolution of long-term errors in NDVI time series: 1985–1999. *Advances in Space Researches*, 28, 149–153.

Koh, C.N., Lee, P.F., and Lin, R.S. (2006). Bird species richness patterns of northern Taiwan: primary productivity, human population density, and habitat heterogeneity. *Diversity and Distributions*, 12, 546–554.

Krishnaswamy, J., Bawa, K.S., Ganeshaiah, K.N., and Kiran, M.C. (2009). Quantifying and mapping biodiversity and ecosystem services: Utility of a multi-season

NDVI based Mahalanobis distance surrogate. *Remote Sensing of the Environment*, 113, 857–867.

La Sorte, F.A., Lee, T.M., Wilman, H., and Jetz, W. (2009). Disparities between observed and predicted impacts of climate change on winter bird assemblages. *Proceedings of the Royal Society of London, Series B*, 276, 3167–3174.

Lamoreux, J.F., Morrison, J.C., Ricketts, T.H., *et al.* (2006). Global tests of biodiversity concordance and the importance of endemism. *Nature*, 440, 212–214.

Lande, R. (1988). Genetics and demography in biological conservation. *Science*, 241, 1455–1460.

Langvatn, R., Mysterud, A., Stenseth, N.C., and Yoccoz, N.G. (2004). Timing and synchrony of ovulation in red deer constrained by short northern summers. *American Naturalist*, 163, 763–772.

Larigauderie, A. and Mooney, H.A. (2010). The Intergovernmental science-policy Platform on Biodiversity and Ecosystem Services: moving a step closer to an IPCC-like mechanism for biodiversity. *Current Opinion in Environmental Sustainability*, 2, 9–14.

Larsen, S., Andersen, T., and Hessen, D.O. (2011). Climate change predicted to cause severe increase of organic carbon in lakes. *Global Change Biology*, 17, 1186–1192.

Lass, W.L., Prather, T.S., Glenn, N.F., Weber, K.T., Mundt, J.T., and Pettingill, J. (2005). A review of remote sensing of invasive weeds and example of early detection of spotted knapweed (*Centaurea maculosa*) and babysbreath (*Gypsophila paniculata*) with a hyperspectral sensor. *Weed Science*, 53, 242–251.

Lassau, S.A. and Hochuli, D.F. (2008). Testing predictions of beetle community patterns derived empirically using remote sensing. *Diversity and Distributions*, 14, 138–147.

Laurance, W.F. (2007). A new initiative to use carbon trading for tropical forest conservation. *Biotropica*, 39, 20–24.

Law, B.E. and Waring, R.H. (1994). Remote sensing of leaf area index and radiation intercepted by understory vegetation. *Ecological Applications*, 4, 272–279.

Lee, P.F., Ding, T.S., Hsu, F.H., and Geng, S. (2004). Breeding bird species richness in Taiwan: distribution on gradients of elevation, primary productivity and urbanization. *Journal of Biogeography*, 31, 307–314.

Leidner, A.K., Turner, W., Pettorelli, N., Leimgruber, P., and Wegmann, M. (2012). Satellite remote sensing for biodiversity research and conservation applications: a Committee on Earth Observation Satellites (CEOS) workshop. <http://remote-sensing-biodiversity.org/images/workshops/ceos/CEOS_SBA_Biodiversity_WorkshopReport_Oct2012_DLR_Munich.pdf>

Lenoir, J., Gegout, J.C., Marquet, P.A., de Ruffray, P., and Brisse, H. (2008). A significant upward shift in plant species optimum elevation during the 20th century. *Science*, 320, 1768–1771.

Lepczyk, C.A., Flather, C.H., Fadeloff, V.C., Pidgeon, A.M., Hammer, R.B., and J. Liu (2008). Human impacts on regional avian diversity and abundance. *Conservation Biology*, 22, 405–416.

Levanoni, O., Levin, N., Peer, G., Turbe, A., and Kark, S. (2011). Can we predict butterfly diversity along an elevation gradient from space? *Ecography*, 34, 372–383.

Levin, S.A. (1992). The problem of pattern and scale in ecology. *Ecology*, 73, 1943–1967.

Levin, N., Shmida, A., Levanoni, O., Tamari, H., and Kark, S. (2007). Predicting mountain plant richness and rarity from space using satellite-derived vegetation indices. *Diversity and Distributions*, 13, 692–703.

Li, D. (2006). A study on the impact pattern of the North Atlantic Oscillation (NAO) on global land NDVI changes. Abstract from the *36th COSPAR Scientific Assembly, 16–23 July 2006*, Beijing, *China*.

Li, H. and Reynolds, J.F. (1993). A new contagion index to quantify spatial patterns of landscapes. *Landscape Ecology*, 8, 155–162.

Li, J., Da, L., Wang, Y., and Song, Y. (2006) Vegetation classification of East China with multi-temporal NOAA-AVHRR data. *Frontiers of Biology in China*, 3, 303–309.

Li, W.H., Dickinson, R.E., Fu, R., Niu, G.-Y., Yang, Z.-L., and Canadell, J.G. (2007). Future precipitation changes and their implications for tropical peatlands. *Geophysical Research Letters*, 34, L01403. doi:10.1029/2006GL028364.

Lichtenthaler, H.K., Gitelson, A., and Lang, M. (1996). Non-destructive determination of chlorophyll content of leaves of a green and an aurea mutant of tobacco by reflectance measurements. *Journal of Plant Physiology*, 148, 483–493.

Lillesand, T., Kiefer, R.W., and Chipman, J. (2008). *Remote Sensing and Image Interpretation*, 6th edn. John Wiley & Sons, New York.

Lindstedt, S.L., Miller, B.J., and Buskirk, S.W. (1986). Home range, time, and body size in mammals. *Ecology*, 67, 413–418.

Lindström, J. (1999). Early development and fitness in birds and mammals. *Trends in Ecology and Evolution*, 14, 343–348.

Lioubimtseva, E. and Henebry, G.M. (2009). Climate and environmental change in arid Central Asia: impacts, vulnerability, and adaptations. *Journal of Arid Environments*, 73, 963–977.

Liu, Y., Wang, X., Guo, M., Tani, H., Matsuoka, N., and Matsumura, S. (2011). Spatial and temporal relationships among NDVI, climate factors, and land cover changes in northeast Asia from 1982 to 2009. *GIScience and Remote Sensing*, 48, 371–393.

Lobell, D.B., Lesch, S.M., Corwin, D.L., *et al.* (2009). Regional-scale Assessment of soil salinity in the Red River Valley using multi-year MODIS EVI and NDVI. *Journal of Environmental Quality*, 39, 35–41.

Loe, L.E., Bonenfant, C., Mysterud, A., *et al.* (2005). Climate predictability and breeding phenology in red deer:

timing and synchrony of rutting and calving in Norway and France. *Journal of Animal Ecology*, 74, 579–588.

Loison, A. and Langvatn, R. (1998). Short- and long-term effects of winter and spring weather on growth and survival of red deer in Norway. *Oecologia*, 116, 489–500.

Lomolino, M.V. (2000). Ecology's most general, yet protean pattern: the species-area relationship. *Journal of Biogeography*, 27, 17–26.

Longley, P.A., Goodchild, M.F., Maguire, D.J., and Rhind, D.W. (2005). *Geographic Information Systems and Science*, 2nd edn. John Wiley & Sons, New York.

Lopatin, E., Kolstrom, T., and Spiecker, H. (2006). Determination of forest growth trends in Komi Republic: combination of tree-ring analysis and remote sensing data. *Boreal Environment Research*, 11, 341–353.

Los, S.O. (in press). Analysis of trends in fused AVHRR and MODIS NDVI data for 1982–2006: indication for a CO_2 fertilization effect in global vegetation. *Global Biogeochemical Cycles*.

Los, S.O., Collatz, G.J., Sellers, P.J., *et al.* (2000). A global 9-year biophysical land-surface data set from NOAA AVHRR data. *Journal of Hydrometeorology*, 1, 183–199.

Los, S.O., Collatz G.J., Bounoua L., Sellers P.J., and Tucker, C.J. (2001). Global interannual variations in sea surface temperature and land surface vegetation, air temperature, and precipitation. *Journal of Climate*, 14, 1535–1549.

Lourenco, P.M., Sousa, C.A., Seixas, J., Lopes, P., Novo, M.T., and Almeida, P.G. (2011). Anopheles atroparvus density modeling using MODIS NDVI in a former malarious area in Portugal. *Journal of Vector Ecology*, 36, 279–291.

Loveland, P. and Webb, J. (2003). Is there a critical level of organic matter in the agricultural soils of temperate regions: a review. *Soil and Tillage Research*, 70, 1–18.

Lu, D. and Weng, Q. (2007). A survey of image classification methods and techniques for improving classification performance. *International Journal of Remote Sensing*, 28, 823–870.

Lyons, E.A., Jin, Y., and Randerson, J.T. (2008). Changes in surface albedo after fire in boreal forest ecosystems of interior Alaska assessed using MODIS satellite observations. *Journal of Geophysical Research*, 113, G02012. doi:10.1029/2007JG000606.

Ma, J., Han, X., Hasibagan, A., *et al.* (2005). Monitoring East Asian migratory locust plagues using remote sensing data and field investigations. *International Journal of Remote Sensing*, 26, 629–634.

MacArthur, R.H. (1965). Patterns of species diversity. *Biological Reviews of the Cambridge Philosophical Society*, 40, 510–533.

MacArthur, R.H. and MacArthur, J.W. (1961). On bird species diversity. *Ecology*, 42, 594–598.

MacDonald, R.B. (1979). *The LACIE Symposium, Proceedings of Technical Sessions, 23–26 October 1978, NASA, Lyndon B. Johnson Space Center, Houston, Texas*, 1125 pp.

Mace, G.M. and Baillie, J.E.M. (2007). The 2010 biodiversity indicators: challenges for science and policy. *Conservation Biology*, 21, 1406–1413.

Malhi, Y., Roberts, J.T., Betts, R.A., Killeen, T.J., Li, W., and Nobre, C.A. (2008). Climate change, deforestation, and the fate of the Amazon. *Science*, 319, 169–172.

Malo, A.R. and Nicholson, S.E. (1990). A study of rainfall and vegetation dynamics in the African Sahel using normalized difference vegetation index. *Journal of Arid Environments*, 19, 1–24.

Malone, J.B., Abdel-Rahman, M.S., El Bahy, M.M., Huh, O.K., Shafik, M., and Bavia, M. (1997). Geographic information systems and the distribution of *Schistosoma mansoni* in the Nile delta. *Parasitology Today*, 13, 112–119.

Malone, J.B., Gommes, R., Hansen, J., *et al.* (1998). A geographic information system on the potential distribution and abundance of *Fasciola hepatica* and *F. gigantica* in east Africa based on Food and Agriculture Organization databases. *Veterinary Parasitology*, 78, 87–101.

Malone, J.B., Yilma, J.M., McCarroll, J.C., Erko, B., Mukaratirwa, S., and Zhou, X. (2001). Satellite climatology and the environmental risk of *Schistosoma mansoni* in Ethiopia and east Africa. *Acta Tropica*, 79, 59–72.

Malstrom, C.M., Thompson, M.V., Juday, G.P., Los, S.O., Randerson, J.T., and Field, C.B. (1997). Interannual variation in global-scale net primary production: testing model estimates. *Global Biogeochemical Cycles*, 11, 367–392.

Manly, B., McDonald, L., and Thomas, D. (1993). *Resource Selection by Animals: Statistical Design and Analysis for Field Studies*. London: Chapman & Hall.

Mann, M.E. and Emanuel, K.A. (2006). Atlantic hurricane trends linked to climate change. EOS, 87, 233–244.

Mao, J., Shi, X., Thornton, P.E., Piao, S. and Wang, X. (2012). Causes of spring vegetation growth trends in the northern mid–high latitudes from 1982 to 2004. *Environmental Research Letters*, 7, 014010. doi:10.1088/1748-9326/7/1/014010.

Marable, M.K., Belant, J.L., Godwin, D., and Wang, G. (2012). Effects of resource dispersion and site familiarity on movements of translocated wild turkeys on fragmented landscapes. *Behavioural Processes*, 91, 119–124.

Markon, C.J., Fleming, M.D., and Binnian, E.F. (1995). Characteristics of vegetation phenology over the Alaskan landscape using AVHRR time-series data. *Polar Records*, 31, 179–190.

Markon, C.J. and Peterson, K.M. (2002). The utility of estimating net primary productivity over Alaska using baseline AVHRR data. *International Journal of Remote Sensing*, 23, 4571–4596.

Marshal, J.P., Bleich, V.C., Krausman, P.R., Reed, M.L., and Andrew, N.G. (2006). Factors affecting habitat use and distribution of desert mule deer in an arid environment. *Wildlife Society Bulletin*, 34, 609–619.

Marshal, J.P., Rajah, A., Parrini, F., Henley, M., Henley, S.R., and Erasmus, B.F.N. (2011). Scale-dependant selection of greenness by African elephants in the Kruger-private reserve transboundary region, South Africa. *European Journal of Wildlife Research*, 57, 537–548.

Martinuzzi, S., Gould, W.A., Ramos Gonzalez, O.M., *et al.* (2008). Mapping tropical dry forest habitats integrating Landsat NDVI, Ikonos imagery, and topographic information in the Caribbean Island of Mona. *Revista de Biologia Tropical*, 56, 625–639.

Martinez-Jauregui, M., San Miguel-Ayanz, A., Mysterud, A., *et al.* (2009). Are local weather, NDVI and NAO consistent determinants of red deer weight across 3 contrasting European countries? *Global Change Biology*, 15, 1727–1738.

Maselli, F., Romanelli, S., Bottai, L., and Zipoli, G. (2003). Use of NOAA-AVHRR NDVI images for the estimation of dynamic fire risk in Mediterranean areas. *Remote Sensing of Environment*, 86, 187–197.

Mattson, W.J. (1980). Herbivory in relation to plant nitrogen content. *Annual Review of Ecology and Systematic*, 11, 119–161.

Maynard Smith, J. (1974). The theory of games and the evolution of animal conflicts. *Journal of Theoretical Biology*, 47, 209–221.

McCarthy, M.A., Thompson, C.J., and Williams, N.S.G. (2006). Logic for designing nature reserves for multiple species. *American Naturalist*, 167, 717–727.

McNab, B.K. (1963). Bioenergetics and the determination of home range size. *American Naturalist*, 97, 113–140.

McNairn, H. and Protz, R. (1993). Mapping corn residue cover on agricultural fields in Oxford County, Ontario, using Thematic Mapper. *Canadian Journal of Remote Sensing*, 19, 152–159.

McNaughton, S.J. (1990). Mineral nutrition and seasonal movements of African migratory ungulates. *Nature*, 345, 613–615.

McPhaden, M.J., Zebiak, S.E., and Glantz, M.H. (2006). ENSO as an integrating concept in Earth science. *Science*, 314, 1740–1745.

Meehl, G.A., Washington W.M., Collins W.D., *et al.* (2005). How much more global warming and sea level rise? *Science*, 307, 1769–1772.

Meffe, G.K. and Carroll, C.R. (1994). The design of conservation reserves. In: Meffe, G.K. and Carroll, C.R. (eds), *Principles of Conservation Biology*. Sinauer Associates, Sunderland, MA, pp. 265–306.

Meffe, G.K. and Groom, M.J. (2006). *Principles of Conservation Biology*, 3rd edn. Sinauer Associates, Sunderland, MA.

Melis, C., Basille, M., Herfindal, I., *et al.* (2010). Roe deer population growth and lynx predation along a gradient of environmental productivity and climate in Norway. *Ecoscience*, 17, 166–174.

Melis, C., Herfindal, I., Kauhala, K., Andersen, R., and Hogda, K.A. (2010). Predicting animal performance through climatic and plant phenology variables: the case of an omnivore hibernating species in Finland. *Mammalian Biology*, 75, 151–159.

Melis, C., Szafrańska, P.A., Jędrzejewska, B., and Bartoń, K. (2006). Biogeographical variation in the population density of wild boar (*Sus scrofa*) in western Eurasia. *Journal of Biogeography*, 33, 803–811.

Mendez, M., Gwynn, J.D., and Manetas, Y. (1999). Enhanced UV-B radiation under field conditions increases anthocyanin and reduces the risk of photoinhibition but does not affect growth in the carnivorous plant *Pinguicula vulgaris*. *New Phytologist*, 144, 275–282.

Menzel, A., Jakobi, G., Ahas, R., Scheifinger, H., and Estrella, N. (2003). Variations of the climatological growing season (1951–2000) in Germany compared with other countries. *International Journal of Climatology*, 23, 793–812.

Menzel, A., Sparks, T.H., Estrella, N., *et al.* (2006). European phenological response to climate change matches the warming pattern. *Global Change Biology*, 12, 1969–1976.

Mikesic, D.G. and Drickamer, L.C. (1992). Factors affecting home-range size in house mice (*Mus musculus domesticus*) living in outdoor enclosures. *American Midland Naturalist*, 127, 31–40.

Milchunas, D.G. and Lauenroth, W.K. (1995). Inertia in plant community structure: state changes after cessation of nutrient enrichment stress. *Ecological Applications*, 5, 1195–2005.

Millennium Ecosystem Assessment Board (2005). *Ecosystems and Human Well-being: Biodiversity Synthesis*. World Resources Institute, Washington, DC.

Miller, J.R., White, H.P., Chen, J.M., *et al.* (1997). Seasonal change in understory reflectance of boreal forests and influence on canopy vegetation indices. *Journal of Geophysical Research*, 102, 29475–29482.

Milner-Gulland, E.J., Kholodova, M.V., Bekenov, A.B., *et al.* (2001). Dramatic declines in saiga antelope populations. *Oryx*, 35, 340–345.

Misra, P.N., Wheeler, S.G., and Oliver, R.E. (1977). *Kauth–Thomas brightness and greenness axes*. Contract NASA 9-14350, RES 23–46.

Mitchard, E.T.A., Meir, P., Ryan, C.M., *et al.* (2013). A novel application of satellite radar data: measuring carbon sequestration and detecting degradation in a community forestry project in Mozambique. *Plant Ecology & Diversity*, 6, 159–170.

Miura, T., Huete, A.R., and van Leeuwen, W.J.D. (1998). Vegetation detection through smoke-filled AVHRIS images: an assessment using MODIS band passes. *Journal of Geophysical Research*, 103, 32001–32011.

Monteith, J.L. (1981). Climatic variation and the growth of crops. *Quarterly Journal of the Royal Meteorological Society*, 107, 749–774.

Mooney, H.A. and Mace, G. (2009). Biodiversity policy challenges. *Science*, 325, 1474.

Morellet, N., Gaillard, J.-M., Hewison, A.J.M., *et al.* (2007). Indicators of ecological change: new tools for managing populations of large herbivores. *Journal of Applied Ecology*, 44, 634–643.

Morin, P.J. (2011). *Community Ecology*, 2nd edn. Wiley, Chichester, 407 pp.

Morisette, J.T., Privette, J., and Justice, C. (2002). A framework for the validation of MODIS Land products. *Remote Sensing of Environment*, 83, 77–96.

Morisette, J.T., Jarnevich, C.S., Ullah, A., *et al.* (2006). A tamarisk habitat suitability map for the continental United States. *Frontiers in Ecology and the Environment*, 4, 11–17.

Morrison, M.L., Marcot, B.G., and Mannan, R.W. (1992). *Wildlife-habitat relationships: concepts and applications*. University of Wisconsin Press, Madison, WI, 343 pp.

Morsdorf, F., Nichol, C., Malthus, T., and Woodhouse, I.H. (2009). Assessing forest structural and physiological information content of multi-spectral LiDAR waveforms by radiative transfer modelling. *Remote Sensing of Environment*, 113, 2152–2163.

Mueller, T., Olson, K.A., Fuller, T.K., Schaller, G.B., Murray, M.G., and Leimgruber, P. (2008). In search of forage: predicting dynamic habitats of Mongolian gazelles using satellite-based estimates of vegetation productivity. *Journal of Applied Ecology*, 45, 649–658.

Muldavin, E.H., Neville, P., and Harper, G. (2001). Indices of grassland biodiversity in the Chihuahuan Desert ecoregion derived from remote sensing. *Conservation Biology*, 15, 844–855.

Muller, J. and Brandl, R. (2009). Assessing biodiversity by remote sensing in mountainous terrain: the potential of LiDAR to predict forest beetle assemblages. *Journal of Applied Ecology*, 46, 897–905.

Munday, P.L., Jones, G., Pratchett, M., and Williams, A. (2008). Climate change and the future for coral reef fishes. *Fish Fisheries*, 9, 261–285.

Musiega, D.E. and Kazadi, S.-N. (2004). Simulating the East African wildebeest migration patterns using GIS and remote sensing. *African Journal of Ecology*, 42, 355–362.

Myneni, R.B., Hall, F.G., Sellers, P.J., and Marshak, A.L. (1995). The interpretation of spectral vegetation indexes. *IEEE Transactions Geoscience and Remote Sensing*, 33, 481–486.

Myneni, R.B., Los, S.L., and Tucker, C.J. (1996). Satellite based identification of linked vegetation index and sea surface temperature anomaly areas from 1982 to 1990 for Africa, Australia and South America. *Geophysical Research Letters*, 23, 729–732.

Myneni, R.B., Nemani, R.R., and Running, S.W. (1997). Estimation of global leaf area index and absorbed par using radiative transfer models. *IEEE Transactions Geoscience and Remote Sensing*, 35, 1380–1393.

Myneni, R.B., Hoffmann, S., Knyazikhin, Y., *et al.* (2002). Global products of vegetation leaf area and fraction absorbed PAR from year one of MODIS data. *Remote Sensing of Environment*, 83, 214–231.

Mysterud, A. (2010). Still walking on the wild side? Management actions as steps towards semi-domestication of hunted ungulates. *Journal of Applied Ecology*, 47, 920–925.

Mysterud, A., Stenseth, N.C., Yoccoz, N.G., Langvatn, R., and Steinheim, G. (2001). Nonlinear effects of large-scale climatic variability on wild and domestic herbivores. *Nature*, 410, 1096–1099.

Mysterud, A., Tryjanowski, P., Panek, M., Pettorelli, N., and Stenseth, N.C. (2007). Inter-specific synchrony of 2 contrasting ungulates: wild boar and roe deer. *Oecologia*, 151, 232–239.

Mysterud, A., Yoccoz, N.G., Langvatn, R., Pettorelli, N., and Stenseth, N.C. (2008). Hierarchical path analysis of deer responses to direct and indirect effects of climate in northern forest. *Philosophical Transactions of the Royal Society of London, Series B*, 363, 2359–2368.

Naeem, S., Chapin, C.F.S. III, Costanza, R., *et al.* (1999). Biodiversity and ecosystem functioning: maintaining natural life support processes. *Ecological Issues*, 4, 1–14.

Nagendra, H. (2001). Using remote sensing to assess biodiversity. *International Journal of Remote Sensing*, 22, 2377–2400.

Nagendra, H., Lucas, R., Honrado, J.P., *et al.* (2012) Remote sensing for conservation monitoring: assessing protected areas, habitat extent, habitat condition, species diversity and threats. *Ecological Indicators*, 2012 Oct 29 (Epub ahead of print).

Neigh, C.S.R., Tucker, C.J., and Townshend, J.R.G. (2008). North American vegetation dynamics observed with multiresolution satellite data. *Remote Sensing of Environment*, 112, 1749–1772.

Nemani, R.R., and Running, S.W. (1997). Land cover characterization using multitemporal red, NIR and thermal IR data from NOAA/AVHRR. *Ecological Applications*, 7, 79–90.

Nemani, R., Keeling, C.D., Hashimoto, H., *et al.* (2003). Climate-driven increases in global terrestrial net primary production from 1982 to 1999. *Science*, 300, 1560–1563.

Nemani, R., Hashimoto, H., Votava, P., *et al.* (2009). Monitoring and forecasting ecosystem dynamics using the Terrestrial Observation and Prediction System (TOPS). *Remote Sensing of Environment*, 113, 1497–1509.

Newby, J.E. (1980). The birds of the Ouadi Rime Ouadi Achim Faunal Reserve: a contribution to the study of the Chadian avifauna. *Malimbus*, 2, 29–50

Newmark, W.D. (2008). Isolation of African protected areas. *Frontiers in Ecology and the Environment*, 6, 321–328.

Nicholson, S.E. (2005). On the question of the 'recovery' of the rains in the West African Sahel. *Journal of Arid Environments*, 63, 615–641.

Nicholson, S.E. and Farrar, T.J. (1994). The influence of soil type on the relationship between NDVI, rainfall and soil moisture in semi-arid Botswana. *Remote Sensing of Environment*, 50, 107–120.

Nicholson, S.E., Tucker, C.J., and Ba, M.B. (1998). Desertification, drought and surface vegetation: An example from the West African Sahel. *Bulletin of the American Meteorological Society*, 79, 815–829.

Nielsen, A., Yoccoz, N.G., Steinheim, G., et al. (2012). Are responses of herbivores to environmental variability spatially consistent in alpine ecosystems? *Global Change Biology*, 18, 3050–3062.

Nilsen, E.B., Herfindal, I., and Linnell, J.D.C. (2005). Can intra-specific variation in carnivore home-range size be explained using remote-sensing estimates of environmental productivity? *Écoscience*, 12, 68–75.

Nilsson, J-Å. (1998). Fitness consequences of timing of reproduction. In: Adams, N.J. and Slotow, R.H. (eds), *Proceedings of the 22nd International Ornithological Congress*. BirdLife South Africa, Durban, pp. 234–247.

Norgaard, R.B. (2009). Ecosystem services: From eye-opening metaphor to complexity blinder. *Ecological Economics*, 69, 1219–1227.

Norris, D., Rocha-Mendes, F., Marques, R., de Almeida Nobre, R., and Galetti, M. (2011). Density and spatial distribution of buffy-tufted-ear marmosets (Callithrix aurita) in a Continuous Atlantic Forest. *International Journal of Primatology*, 32, 811–829.

Noss, R.F. (1990). Indicators for monitoring biodiversity—a hierarchical approach. *Conservation Biology*, 4, 355–364.

Oba, G., Post, E., and Stenseth, N.C. (2001). Sub-saharan desertification and productivity are linked to hemispheric climate variability. *Global Change Biology*, 7, 241–246.

O'Connell, A.F., Nichols, J.D., and Karanth, K.U. (2011). *Camera Traps in Animal Ecology*. Springer, 280 pp.

Oesterheld, M., Dibella, C.M., and Kerdiles, H. (1998). Relation between NOAA-AVHRR satellite data and stocking rate of rangelands. *Ecological Applications*, 8, 207–212.

Oindo, B.O. and Skidmore, A.K. (2002). Interannual variability of NDVI and species richness in Kenya. *International Journal of Remote Sensing*, 23, 285–298.

Olander, L.P., Gibbs, H.K., Steininger, M., Swenson, J.J., and Murray, B.C. (2008). Reference scenarios for deforestation and forest degradation in support of REDD: a review of data and methods. *Environmental Research Letters*, 3, 025011.

Olofsson, J., Tommervik, H., and Callaghan, T.V. (2012). Vole and lemming activity observed from space. *Nature Climate Change*, 2, 880–883.

Olson, K.A., Mueller, T., Bolortsetseg, S., Leimgruber, P., Fagan, W.F., and Fuller, T.K. (2009). Short Communication A mega-herd of more than 200000 Mongolian gazelles *Procapra gutturosa*: a consequence of habitat quality. *Oryx*, 43, 149–153.

Olsson, L., and Eklundh, L. (1994). Fourier transformations for the analysis of temporal sequences of satellite imagery. *International Journal of Remote Sensing*, 15, 3735–3741.

O'Neill, R.V., Krummel, J.R., Gardner, R.H., et al. (1988). Indices of landscape pattern. *Landscape Ecology*, 1, 153–162.

Otterman, J. and Tucker, C.J. (1982). Satellite monitoring of surface temperatures in semidesert: Differences between anthropogenically impacted terrain and protected natural vegetation. *Advances in Space Research*, 2, 51–55.

Otterman, J., Karnieli, A., Brakke, T., et al. (2002). Desert scrub optical density and spectral-albedo ratios of impacted-to-protected areas by model inversion. *International Journal of Remote Sensing*, 23, 3959–3970.

Owen, J.G. (1988). On productivity as a predictor of rodent and carnivore diversity. *Ecology*, 69, 1161–1165.

Owen, H.J. and Norton, D.A. (1995). The diet of introduced brushtail possums Trichosurus vulpecula in a low-diversity New Zealand Nothofagus forest and possible implications for conservation management. *Biological Conservation*, 71, 339–345.

Palandro, D.A., Andrefouet, S., Hu, C., et al. (2008). Quantification of two decades of shallow-water coralreef habitat decline in the Florida Keys National Marine Sanctuary using Landsat data (1984–2002). *Remote Sensing of Environment*, 112, 3388–3399.

Parent, M.B., and Verbyla, D. (2010). The browning of Alaska's boreal forest. *Remote Sensing*, 2, 2729–2747.

Parker, C., Mitchell, A., Trivedi, M., and Mardas, N. (2009). *The Little REDD Book*. Global Canopy Foundation, Oxford.

Parks Canada Agency (2000). *Unimpaired for Future Generations? Protecting Ecological Integrity with Canada's National Parks*, vol. I: *A Call to Action*; vol. II: *Setting a New Direction for Canada's National Parks*. Report of the Panel on Ecological Integrity of Canada's National Parks, Ottawa, Ontario.

Parks Canada Agency (2005). *Monitoring and Reporting Ecological Integrity in Canada's National Parks, Guiding Principles*, vol. 1. Parks Canada Agency, Hull, Québec.

Parks, S.A. and Harcourt, A.H. (2002). Reserve size, local human density, and mammalian extinctions in U.S. protected areas. *Conservation Biology*, 16, 800–808.

Parmesan, C. (2006). Ecological and evolutionary responses to recent climate change. *Annual Review of Ecology and Evolutionary Systematics*, 37, 637–669.

Parra, J.L., Graham, C.C., and Freile, J.F.(2004). Evaluating alternative data sets for ecological niche models of birds in the Andes. *Ecography*, 27, 350–360.

Parrish, J., Braun, D., and Unnasch, R. (2003). Are we conserving what we say we are? Measuring ecological integrity within protected areas. *BioScience*, 53, 851–860.

Paruelo, J.M., Epstein, H.E., Lauenroth, W.K., and Burke, I.C. (1997). ANPP estimates from NDVI for the central grassland region of the United States. *Ecology*, 78, 953–958.

Paruelo, J.M., Jobbagy, E.G., and Sala, O.E. (2001). Current distribution of ecosystem functional types in temperate South America. *Ecosystems*, 4, 683–698.

Parviainen, M., Luoto, M., and Heikkinen, R.K. (2009). The role of local and landscape level measures of greenness in modelling boreal plant species richness. *Ecological Modelling*, 220, 2690–2701.

Pau, S., Gillespie, T.W., and Wolkovich, E.M. (2012). Dissecting NDVI–species richness relationships in Hawaiian dry forests. *Journal of Biogeography*, 39, 1678–1686.

Pausas, J.G. and Ribeiro, E. (2013). The global fire–productivity relationship. *Global Ecology and Biogeography*, 22, 728–736.

Pavelka, M.S.M. and Behie, A.M. (2005). The effect of hurricane Iris on the food supply of black howlers (*Alouatta pigra*) in southern Belize. *Biotropica*, 37, 102–108.

Payero, J.O., Neale, C.M.U., and Wright, J.L. (2004). Comparison of eleven vegetation indices for estimating plant height of alfalfa and grass. *Applied Engineering in Agriculture*, 20, 385–393.

Pearson, R.L. and Miller, L.D. (1972). Remote mapping of standing crop biomass for estimation of the productivity of the short-grass prairie. In: *Proceedings of the 8th International Symposium on Remote Sensing of Environment, Environmental Research Institute of Michigan, Ann Arbor*, pp. 1357–1381.

Pearson, R.L., Tucker, C.J., and Miller, L.C. (1976). Spectral mapping of short grass prairie biomass. *Photogrammetric Engineering and Remote Sensing*, 42, 317–323.

Pedelty, J., Devadiga, S., Masuoka, E., *et al.* (2007). Generating a long-term land data record from the AVHRR and MODIS Instruments, Geoscience and Remote Sensing Symposium. IGARSS 2007. *IEEE International*, 2007; pp. 1021–1025.

Pedersen, A.O., Jepsen, J.U., Yoccoz, N.G., and Fuglei, E. (2007). Ecological correlates of the distribution of territorial Svalbard rock ptarmigan. *Canadian Journal of Zoology*, 85, 122–132.

Pelkey, N.W., Stoner, C.J., and Caro, T.M. (2000). Vegetation in Tanzania: assessing long term trends and effects of protection using satellite imagery. *Biological Conservation*, 94, 297–309.

Pelkey, N.W., Stoner, C.J., and Caro, T.M. (2003). Assessing habitat protection regimes in Tanzania using AVHRR NDVI composites: comparisons at different spatial and temporal scales. *International Journal of Remote Sensing*, 24, 2533–2558.

Peng, D., Liu, L., Zhang, B., and Shen, Q. (2011). Comparisons of FPAR derived from GIMMS AVHRR NDVI and MODIS product. Geoscience and Remote Sensing Symposium (IGARSS), 2011 IEEE International Conference, 24–29 July 2011 pp. 1842–1845.

Pereira, M.M. and Cooper, H.D. (2006). Towards the global monitoring of biodiversity change. *Trends in Ecology and Evolution*, 21, 123–129.

Pereira, H.M., Ferrier, S., Walters, M., *et al.* (2013). Essential bodiversity variables. *Science*, 339, 277.

Perry, C.R. and Lautenschlager, L.F. (1984). Functional equivalence of spectral vegetation indices. *Remote Sensing of Environment*, 14, 169–182.

Peters, R.H. (1983). *The Ecological Implications of Body Size*. Cambridge University Press, Cambridge.

Peters, A.J. and Eve, M.D. (1995). Satellite monitoring of desert plant community response to moisture availability. *Environmental Monitoring and Assessment*, 37, 273–287.

Peterson, U. (1992). Seasonal reflectance factor dynamics in boreal forest clear-cut communities. *International Journal of Remote Sensing*, 13, 753–772.

Peterson, E.B. (2005). Estimating cover of an invasive grass (*Bromus tectorum*) using tobit regression and phenology derived from two dates of Landsat ETM plus data. *International Journal of Remote Sensing*, 26, 2491–2507.

Pettorelli, N., Mysterud, A., Yoccoz, N.G., Langvatn, R., and Stenseth, N.C. (2005a). Importance of climatological downscaling and plant phenology for red deer in heterogeneous landscapes. *Proceedings of the Royal Society of London, Series B*, 272, 2357–2364.

Pettorelli, N., Vik, J., Mysterud, A., Gaillard, J.-M., Tucker, C., and Stenseth, N. (2005b). Using the satellite-derived NDVI to assess ecological responses to environmental change. *Trends in Ecology and Evolution*, 20, 503–510.

Pettorelli, N., Weladji, R., Holand, Ø., Mysterud, A., Breie, H., and Stenseth, N.C. (2005c). The relative role of winter and spring conditions: linking climate and landscape-scale plant phenology to alpine reindeer body mass. *Biology Letters*, 1, 24–26.

Pettorelli, N., Gaillard, J.M., Mysterud, A., *et al.* (2006). Using a proxy of plant productivity (NDVI) to track animal performance: the case of roe deer. *Oikos*, 112, 565–572.

Pettorelli, N., Pelletier, F., Von Hardenberg, A., Festa-Bianchet, M., and Côté, S.D. (2007). Early onset of vegetation growth versus rapid green-up: impacts on juvenile mountain ungulates. *Ecology*, 88, 381–390.

Pettorelli, N., Bro-Jørgensen, J., Durant, S.M., Blackburn, T., and Carbone, C. (2009). Energy availability and density estimates in African ungulates. *American Naturalist*, 173, 698–704.

Pettorelli, N., Ryan, S., Mueller, T., *et al.* (2011). The Normalized Difference Vegetation Index (NDVI): unforeseen successes in animal ecology. *Climate Research*, 46, 15–27.

Pettorelli, N., Chauvenet, A.L.M., Duffy, J.P., Cornforth, W.A., Meillere, A., and Baillie, J.E.M. (2012). Tracking the effect of climate change on ecosystem functioning using protected areas: Africa as a case study. *Ecological Indicators*, 20, 269–276.

Peňuelas, J., Filella, I., Biel, C., Serrano, L., and Savé, R. (1993). The reflectance at the 950–970 nm region as an indicator of plant water status. *International Journal of Remote Sensing*, 14, 1887–1905.

Peňuelas, J., Gamon, J.A., Fredeen, A.L., Merino, J., and Field, C.B. (1994). Reflectance indices associated with physiological changes in nitrogen and water-limited sunflower leaves. *Remote Sensing of Environment*, 48, 135–146.

Peňuelas, J., Piňol, J., Ogaya, R., and Filella, I. (1997). Estimation of plant water concentration by the reflectance water index. *International Journal of Remote Sensing*, 18, 2869–2875.

Pfeifer, M., Disney, M., Quaife, T., and Marchant, R. (2011). Terrestrial ecosystems from space: a review of earth observation products for macroecology applications. *Global Ecology and Biogeography*, 21, 603–624.

Philippon, N., Jarlan, L., Martiny, N., Camberlin, P., and Mougin, E. (2007). Characterization of the interannual and intraseasonal variability of West African vegetation between 1982 and 2002 by means of NOAA AVHRR NDVI data. *Journal of Climate*, 20, 1202–1218.

Phillips, S., Anderson, R., and Schapire, R. (2006). Maximum entropy modeling of species geographic distributions. *Ecological Modelling*, 190, 231–259.

Phillips, L.B., Hansen, A.J., and Flather, C.H. (2008). Evaluating the species energy relationship with the newest measures of ecosystem energy: NDVI versus MODIS primary production. *Remote Sensing of Environment*, 112, 4381–4392.

Pianka, E.R. (1966). Latitudinal gradients in species diversity: a review of concepts. *American Naturalist*, 100, 33–46.

Pielke, R.A., Avissar, R., Raupach, M., Dolman, A.J., Zeng, X., and Denning, A.S. (1998). Interactions between the atmosphere and terrestrial ecosystems: influence on weather and climate. *Global Change Biology*, 4, 461–475.

Pimm, S.L., Jones, H.L., and Diamond, J. (1988). On the risk of extinction. *American Naturalist*, 132, 757–785.

Pinter Jr, P.J., Jackson, R.D., Ezra, C.E., and Gausman, H.W. (1985). Sun angle and canopy architecture effects on the reflectance of six wheat cultivars. *International Journal of Remote Sensing*, 6, 1813–1825.

Pinty, B. and Verstraete, M.M. (1992). GEMI: a non-linear index to monitor global vegetation from satellites. *Vegetatio*, 101, 15–20.

Pinzon, J.E., Brown, M.E., and Tucker, C.J. (2004). *Monitoring Seasonal and Interannual Variations in Land-surface Vegetation from 1981–2003 Using GIMMS NDVI*. <ftp://landval.gsfc.nasa.gov/Documentation/GIMMS_NDVI_8km_doc.pdf>

Pittiglio, C., Skidmore, A.K., van Gils, H.A.M.J., and Prins, H.H.T. (2012). Identifying transit corridors for elephant using a long time-series. *International Journal of Applied Earth Observation and Geoinformation*, 14, 61–72.

Plummer, S.E., North, P.R., and Briggs, S.A. (1994). The angular vegetation index: an atmospherically resistant index for the second along track scanning radiometer (ATSR-2). In: *Proceedings of the Sixth International Symposium Physical Measurements and Signatures in Remote Sensing, Val d'Isere, France*, 717–722.

Portier, C., Duncan, P., Gaillard, J.M., Guillon, N., and Sempéré, A. (2000). Growth of European roe deer: patterns and rates. *Acta Theriologica*, 45, 87–94.

Posse, G. and Cingolani, A.M. (2004). A test of the use of NDVI data to predict secondary productivity. *Applied Vegetation Science*, 7, 201–208.

Potapov, P., Hansen, M.C., and Pittman, S.V. (2008). Combining MODIS and Landsat imagery to estimate and map boreal forest cover loss. *Remote Sensing of Environment*, 112, 3708–3719.

Powell, R.A. (2012). Movements, home ranges, activity, and dispersal. In: Boitani, L. and Powell, R.A. (eds), *Carnivore Ecology and Conservation: A Handbook of Techniques*. Oxford University Press, London, pp. 188–217.

Pressey, R.L. (1996). Protected areas: where should they be and why should they be there? In: Spellerberg IF (ed.), *Conservation Biology*. Harlow: Longman, pp. 171–185.

Prihodko, L. and Goward, S.N. (1997). Estimation of air temperature from remotely sensed surface observations. *Remote Sensing of Environment*, 60, 335–346.

Prince, S.D. and Goward, S.N. (1995). Global primary production: a remote sensing approach. *Journal of Biogeography*, 22, 815–835.

Prince, S.D. and Tucker, C.J. (1986). Satellite remote sensing of rangelands in Botswana. II. NOAA AVHRR and herbaceous vegetation. *International Journal of Remote Sensing*, 7, 1555–1570.

Prince, S.D., Goetz, S.J., and Goward, S.N. (1995). Monitoring primary productivity from earth observing satellites. *Water, Air, and Soil Pollution*, 82, 509–522.

Prince, S.D., Becker-Reshef, I., and Rishmawi, K. (2009). Detection and mapping of long-term land degradation using local net production scaling: application to Zimbabwe. *Remote Sensing of Environment*, 113, 1046–1057.

Propastin, P., Fotso, L., and Kappas, M. (2010). Assessment of vegetation vulnerability to ENSO warm events over Africa. *International Journal of Applied Earth Observation and Geoinformation*, 12S, 83–89.

Proulx, R. and Fahrig, L. (2010). Detecting human-driven deviations from trajectories in landscape composition and configuration. *Landscape Ecology*, 25, 1479–1487.

Qi, J., Chehbouni, A., Huete, A.R., Kerr, Y.H., and Sorooshian, S. (1994). A modified soil adjusted vegetation index. *Remote Sensing of Environment*, 47, 1–25.

Qian, H., Wang, X., Wang, S., and Li, Y. (2007). Environmental determinants of amphibian and reptile species richness in China. *Ecography*, 30, 471–482.

R Development Core Team (2013). *R: a language and environment for statistical computing.* <http://www.r-project.org/>

Rabe, M.J., Rosenstock, S.S., and deVos, J.C. (2002). Review of big-game survey methods used by wildlife agencies of the western United States. *Wildlife Society Bulletin*, 30, 46–52.

Rahman, L.M. (2000). The Sundarbans: a unique wilderness of the world legal status. *USAD Forest Services Proceedings RMRS-P*-15, 2, 143–148.

Rasmussen, M.S. (1992). Assessment of millet yields and production in northern Burkina Faso using integrated NDVI from the AVHRR. *International Journal of Remote Sensing*, 13, 3431–3442.

Rasmussen, H.B., Wittemyer, G., and Douglas-Hamilton, I. (2006). Predicting time-specific changes in demographic processes using remote-sensing data. *Journal of Applied Ecology*, 43, 366–376.

Redford, K.H. (1992). The empty forest. *BioScience*, 42, 412–422.

Reed, B., Brown, J., Vanderzee, D., Loveland, T., Merchant, J., and Ohlen, D. (1994). Measuring phenological variability from satellite imagery. *Journal of Vegetation Sciences*, 5, 703–714.

Reyers, B., Pettorelli, N., Katzner, T., *et al*. (2010). Animal conservation and ecosystem services: garnering the support of mightier forces. *Animal Conservation*, 13, 523–525.

Reynolds, J.F. and Stafford Smith, M. (2002). *Global Desertification: Do Humans Create Deserts?* Dahlem University Press, Berlin.

Reynolds-Hogland, M.J. and Mitchell, M.S. (2007). Effects of roads on habitat quality for bears in the southern Appalachians: a long-term study. *Journal of Mammalogy*, 88, 1050–1061.

Richardson, A.D., Duigan, S.P., and Berlyn, G.P. (2002). An evaluation of noninvasive methods to estimate foliar chlorophyll content. *New Phytologist*, 153, 185–194.

Richardson, A.J. and Wiegand, C.L. (1977). Distinguishing vegetation from soil background information. *Photogrammetric Engineering and Remote Sensing*, 43, 1541–1552.

Ripple, W.J. (1985). Asymptotic relationship characteristics of grass vegetation. *Photogrammetric Engineering and Remote Sensing*, 51, 1915–1921.

Ripple, W.J. and Beschta, R.L. (2004). Wolves and the ecology of fear: can predation risk structure ecosystems. *BioScience*, 54, 755–766

Ripple, W.J. and Beschta, R.L. (2012a). Trophic cascades in Yellowstone: the first 15 years after wolf reintroduction. *Biological Conservation*, 145, 205–213.

Ripple, W.J. and Beschta, R.L. (2012b). Large predators limit herbivore densities in northern forest ecosystems. *European Journal of Wildlife Research*, 58, 733–742.

Ripple, W.J., Larsen, E.J., Renkin, R.A., and Smith, D.W. (2001). Trophic cascades among wolves, elk and aspen on Yellowstone National Park's northern range. *Biological Conservation*, 102, 227–234.

Roberts, J.P. and Schnell, G.D. (2006). Comparison of survey methods for wintering grassland birds. *Journal of Field Ornithology*, 77, 46–60.

Robinson, I.S. (2003). *Measuring the oceans from space*. Springer Verlag, Berlin.

Robinson, T., Rogers, D., and Williams, B. (1997). Mapping tsetse habitat suitability in the common fly belt of Southern Africa using multivariate analysis of climate and remotely sensed vegetation data. *Medical and Veterinary Entomology*, 11, 235–245.

Rodgers III, J.C., Murrah, A.W., and Cooke, W.H. (2009). The impact of Hurricane Katrina on the coastal vegetation of the Weeks Bay Reserve, Alabama from NDVI data. *Estuaries and Coasts*, 32, 496–507.

Rodó, X., Baert, E., and Comín, F.A. (1997). Variations in seasonal rainfall in Southern Europe during the present century: relationships with the North Atlantic Oscillation and the El Niño–Southern Oscillation. *Climate Dynamics*, 13, 275–284.

Rodwell, M.J. (2003). On the Predictability of North Atlantic Climate. In: Hurrell, J.W., Kushnir, Y., Ottersen, G., and Visbeck, M. (eds), *The North Atlantic Oscillation: Climatic Significance and Environmental Impact. Geophysical Monograph*, 134, 173–192.

Rogers, J.C. and McHugh, M.J. (2002). On the separability of the North Atlantic oscillation and Arctic oscillation. *Climate Dynamics*, 19, 599–608.

Ronchail, J., Cochonneau, G., Molinier, M., *et al*. (2002). Interannual rainfall variability in the Amazon basin and seasurface temperatures in the equatorial Pacific and the tropical Atlantic Oceans. *International Journal of Climatology*, 22, 1663–1686.

Rondeaux, G. and Baret, F. (1996). Optimization of soil-induced vegetation indices. *Remote Sensing of Environment*, 55, 95–107.

Ross, K.W., Brown, M.E., Verdin, J.P., and Underwood, L.W. (2009). Review of FEWS NET biophysical monitoring requirements. *Environmental Research Letters*, 4, 024009. doi:10.1088/1748-9326/4/2/024009.

Rossow, W.B. and Schiffer, R.A. (1999). Advances in understandingclouds from ISCCP. *Bulletin of the American Meteorological Society*, 80, 2261–2287.

Roughgarden, J., Running, S.W., and Matson, P.A. (1991). What does remote sensing do for ecology? Ecology, 72, 1918–1922.

Roura-Pascual, N., Suarez, A.V., Gomez, C., *et al*. (2004) Geographical potential of Argentine ants (*Linepithema humile Mayr*) in the face of global climate change. *Proceedings of the Royal Society of London, Series B*, 271, 2527–2535.

Rouse, J.W., Haas, R. H., Schell, J.A., and Deering, D.W. (1974). Monitoring vegetation systems in the Great Plains with ERTS. *Proceedings of the Third Earth Resources Technology Satellite-1 Symposium, December 10–15 1974, Greenbelt, MD*. NASA, Washington, DC, pp. 301–317.

Rowell, D.P., Folland, C.K., Maskell, K., and Ward, M.N. (1995). Variability of summer rainfall over tropical north Africa (1906–92), Observations and modelling. *Quarterly Journal of the Royal Meteorological Society*, 121, 669–674.

Roy, D.P., Kennedy, P., and Folving, S. (1997). Combination of the Normalized Difference Vegetation Index and surface temperature for regional scale European Forest cover mapping using AVHRR data. *International Journal of Remote Sensing*, 18, 1189–1195.

Running, S.W. (1990). Estimating primary productivity by combining remote sensing with ecosystem simulation. In: Hobbs R.J. and Mooney H.A. (eds), *Remote Sensing of Biosphere Functioning*. Springer-Verlag, New York, pp. 65–86.

Rutberg, A.T. (1987). Adaptive hypotheses of birth synchrony in ruminants - an interspecific test. *American Naturalist*, 130, 692–710.

Ryan, S.J. (2006). Spatial ecology of African buffalo and their resources in a savanna landscape. PhD dissertation, University of California at Berkeley, Berkeley, CA.

Ryan, S.J., Knechtel, C.U., and Getz, W.M. (2006). Range and habitat selection of African buffalo in South Africa. *Journal of Wildlife Management*, 70, 764–776.

Ryan, S.J., Knechtel, C.U., and Getz W.M. (2007). Ecological cues, gestation length, and birth timing in African buffalo. *Behavioural Ecology*, 18, 635–644.

Réale, D., McAdam, A.G., Boutin, S., and Berteaux, D. (2003). Genetic and plastic responses of a northern mammal to climate change. *Proceedings of the Royal Society of London, Series B*, 270, 591–596.

Sachs, J.D. and Reid, W.V. (2006). Environment: investments toward sustainable development. *Science*, 312, 1002.

Sadleir, R.M.F.S. (1969). *The Ecology of Reproduction in Wild and Domestic Animals*. Methuen, London.

Said, S., Gaillard, J.M., Duncan, P., *et al*. (2005). Ecological correlates of home range size in spring-summer for female roe deer (*Capreolus capreolus*) in a deciduous woodland. *Journal of Zoology*, 267, 301–308.

Saino, N., Szép T., Ambrosini, R., Romano, M., and Møller, A.P. (2004). Ecological conditions during winter affect sexual selection and breeding in a migratory bird. *Proceedings of the Royal Society of London, Series B*, 271, 681–686.

Saleska, S.R., Miller, S.D., Matross, D.M., *et al*. (2003). Carbon in Amazon forests: unexpected seasonal fluxes and disturbance-induced losses. *Science*, 302, 1554–1557.

Saleska, S.R., Didan, K., Huete, A.R., and Rocha, H.R.D. (2007). Amazon forests green-up during 2005 drought. *Science*, 318, 612.

Salvador, R., Valeriano, J., Pons, X., and Diaz-Delgado, R. (2000). A semi-automatic methodology to detect fire scars in shrubs and evergreen forests with Landsat MSS time series. *International Journal of Remote Sensing*, 21, 655–671.

Salzer, D. and Salafsky, N. (2003). *Allocating Resources Between Taking Action, Assessing Status, and Measuring Effectiveness*. The Nature Conservancy and Foundations for Success Working Paper, Draft version, 17 March 2003. TNC, Arlington, Virginia.

Samborski, S.M., Tremblay, N., and Fallon, E. (2009). Strategies to make use of plant sensors-based diagnostic information for nitrogen recommendations. *Agronomy Journal*, 101, 800–816.

Santin-Janin, H., Garel, M., Chapuis, J.L., and Pontier, D. (2009). Assessing the performance of NDVI as a proxy for plant biomass using non-linear models: a case study on the Kerguelen archipelago. *Polar Biology*, 32, 861–871.

Sanz, J.J., Potti, J., Moreno, J., Merino, S., and Frias, O. (2003). Climate change and fitness components of a migratory bird breeding in the Mediterranean region. *Global Change Biology*, 9, 461–472.

Sawyer, H. and Kauffman, M.J. (2011) Stopover ecology of a migratory ungulate. *Journal of Animal Ecology*, 80, 1078–1087.

Schaub, M., Kania, W., and Koppen, U. (2005). Variation in primary production during winter indices synchrony in survival rates in migratory white storks. *Journal of Animal Ecology*, 74, 656–666.

Schnur, M.T., Xie, H., and Wang, X. (2010). Estimating root zone soil moisture at distant sites using MODIS NDVI and EVI in a semi-arid region of southwestern USA. *Ecological Informatics*, 5, 400–409.

Scholes, R.J., Mace, G.M., Turner, W., *et al*. (2008). Toward a global biodiversity observing system. *Science*, 321, 1044–1045.

Schultz, P.A. and Halpert, M.S. (1993). Global correlation of temperature, NDVI and precipitation. *Advances in Space Research*, 13, 277–280.

Schwartz, M.D. and Reiter, B.E. (2000). Changes in North American spring. *International Journal of Climatology*, 20, 929–932.

Schwartz, M.D., Ahas, R., and Aasa, A. (2006). Onset of spring starting earlier across the Northern Hemisphere. *Global Change Biology*, 12, 343–351.

Schödelbauerová, I., Roberts, D.L., and Kindlmann, P. (2009). Size of protected areas is the main determinant of species diversity in orchids. *Biological Conservation*, 142, 2329–2334.

Seber, G.A.F. (1992). A review of estimating animal abundance II. *International Statistical Review*, 60, 129–166.

Secretariat of the Convention on Biological Diversity (2009). *Connecting Biodiversity and Climate Change Mitigation and Adaptation: Report of the Second Ad Hoc Technical*

Expert Group on Biodiversity and Climate Change. Montréal, Technical Series No. 41, 126 pp.

Secretariat of the Convention on Biological Diversity (2010). *Global Biodiversity Outlook 3*. Montréal, 94 pp. <http://gbo3.cbd.int>

Segah, H., Tani, H., and Hirano, T. (2010). Detection of fire impact and vegetation recovery over tropical peat swamp forest by satellite data and ground-based NDVI instrument. *International Journal of Remote Sensing*, 31, 5297–5314.

Seiferling, I.S., Proulx, R., Peres-Neto, P.R., Fahrig, L., and Messier, C. (2012). Measuring protected-area isolation and correlations of isolation with land-use intensity and protection status. *Conservation Biology*, 26, 610–618.

Sellers, P.J. (1985). Canopy reflectance, photosynthesis and transpiration. *International Journal of Remote Sensing*, 6, 1335–1372.

Sellers, P.J., Los, S.O., Tucker, C.J., *et al.* (1994). A global 1 by 1 degree NDVI data set for climate studies. Part 2: The generation of global fields of terrestrial biophysical parameters from the NDVI. *International Journal of Remote Sensing*, 15, 3519–3545.

Senay, G.B. and Elliott, R.L. (2002). Capability of AVHRR data in discriminating rangeland cover mixtures. *International Journal of Remote Sensing*, 23, 299–312.

Serrano, L., Ustin, S.L., Roberts, D.A., Gamon, J.A., and Penuelas, J. (2000). Deriving water content of chaparral vegetation from AVIRIS data. *Remote Sensing of Environment*, 74, 570–581.

Sesnie, S.E., Dickson, B.G., Rosenstock, S.S., and Rundall, J.M. (2012). A comparison of Landsat TM and MODIS vegetation indices for estimating forage phenology in desert bighorn sheep (*Ovis canadensis nelsoni*) habitat in the Sonoran Desert, USA. *International Journal of Remote Sensing*, 33, 276–286.

Seto, K.C. and Fragkias, M. (2007). Mangrove conversion and aquaculture development in Vietnam: a remote sensing-based approach for evaluating the Ramsar Convention on Wetlands. *Global Environmental Change*, 17, 486–500.

Seto, K.C., Fleishman, E., Fay, J.P., and Betrus, C.J. (2004). Linking spatial patterns of bird and butterfly species richness with Landsat TM derived NDVI. *International Journal of Remote Sensing*, 25, 4309–4324.

Shaman, J. and Tziperman, E. (2011). An atmospheric teleconnection linking ENSO and southwestern European precipitation. *Journal of Climate*, 24, 124–139.

Shekede, M.D., Kusangaya, S., and Schmidt, K. (2008). Spatio-temporal variations of aquatic weeds abundance and coverage in Lake Chivero, Zimbabwe. *Physics and Chemistry of the Earth*, 33, 714–721.

Shen, L., Xu, H., and Guo, X. (2012). Satellite remote sensing of Harmful Algal Blooms (HABs) and a potential synthesized framework. *Sensors*, 12, 7778–7803.

Shibayama, M., Salli, A., Hame, T., *et al.* (1999). Detecting phenophases of subarctic shrub canopies by using automated reflectance measurements. *Remote Sensing of Environment*, 67, 160–180.

Shililu, J., Ghebremeskel, T., Mengistu, S., *et al.* (2003). Distribution of anopheline mosquitoes in Eritrea. *American Journal of Tropical Medicine and Hygiene*, 69, 295–302.

Shishov, V.V., Vaganov, E.A., Hughes, M.K., and Koretz, M.A. (2002). Spatial variations in the annual tree-ring growth in Siberia in the past century. *Doklady Earth Sciences*, 387, 1088–1091.

Shochat, E., Stefanov, W.L., Whitehouse, M.E.A., and Faeth, S.H. (2004). Urbanization and spider diversity: influences of human modification of habitat structure and productivity. *Ecological Applications*, 14, 268–280.

Sims, D.A. and Gamon, J.A. (2002). Relationships between leaf pigment content and spectral reflectance across a wide range of species, leaf structures and developmental stages. *Remote Sensing of Environment*, 81, 337–354.

Sims, D.A., Rahman, A.F., Cordova, V.D., *et al.* (2006). On the use of MODIS EVI to assess gross primary productivity of North American ecosystems. *Journal of Geophysical Research*, 111, G04015.

Sims, N.C. and Thoms, M.C. (2002). What happens when flood plains wet themselves: vegetation response to inundation on the lower Balonne flood plain. In: *The Structure, Function and Management Implications of Fluvial Sedimentary Systems. Proceedings of an international symposium held at Alice Springs, Australia, September 2002*. IAHS Publ. no. 276, pp. 195–202.

Singh, N.J. and Milner-Gulland, E.J. (2011). Conserving a moving target: planning protection for a migratory species as its distribution changes. *Journal of Applied Ecology*, 48, 35–46.

Singh, N.J., Grachev, I.A., Bekenov, A.B., and Milner-Gulland, E.J. (2010). Tracking greenery in central Asia—the migration of the Saiga antelope. *Diversity and Distributions*, 16, 663–675.

Smith, D.W. and Bangs, E.E. (2009). Reintroduction of wolves to Yellowstone National Park: history, values and ecosystem restoration. In: Hayward, M.W. and Somers, M.J. (eds) *Reintroduction of Top-order Predators*. Wiley–Blackwell, pp. 92–125.

Smith, P., Gregory, P.J., van Vuuren, D., *et al.* (2010). Competition for land. *Philosophical Transactions of the Royal Society, Series B*, 365, 2941–2957.

Snyder, R.L., de Melo-Abreu, J.P., and Matulich, S. (2005). *Frost Protection: Fundamentals, Practice and Economics*. FAO Environment and Natural Resources Service Series, No. 10. FAO, Rome.

Soininen, J. and Luoto, M. (2012). Is catchment productivity a useful predictor of taxa richness in lake plankton communities? *Ecological Applications*, 22, 624–633.

Song, Y., Ma, M.G., and Veroustraete, F. (2010). Comparison and conversion of AVHRR GIMMS and SPOT VEGETATION NDVI data in China. *International Journal of Remote Sensing*, 31, 2377–2392.

Soriano, A. and Paruelo, J.M. (1992). Biozones: vegetation units defined by functional characters identifiable with the aid of satellite sensor images. *Global Ecology and Biogeography Letters*, 2, 82–89.

Soulé, M.E. (1985). What is conservation biology? A new synthetic discipline addresses the dynamics and problems of perturbed species, communities, and ecosystems. *BioScience*, 35, 727–734.

Spanner, M.A., Pierce, L.L., Running, S.W., and Peterson, D.L. (1990). The seasonality of AVHRR data of temperate coniferous forests: relationships with leaf area index. *Remote Sensing of Environment*, 33, 97–112.

Spruce, J.P., Smoot, J., and Graham, W. (2009). Developing new coastal forest restoration products based on Landsat, ASTER, and MODIS data. *OCEANS 2009, MTS/IEEE Biloxi—Marine Technology for Our Future:Global and Local Challenges; 26–29 October 2009, Biloxi, MS*.

Spruce, J.P., Sader, S., Ryan, R.E., *et al.* (2011). Assessment of MODIS NDVI time series data products for detecting forest defoliation by gypsy moth outbreaks. *Remote Sensing of Environment*, 115, 427–437.

Stamps, J.A. and Swaisgood, R.R. (2007). Some place like home: experience, habitat selection and conservation biology. *Applied Animal Behaviour Sciences*, 102, 392–409.

Stark, C.H. and Richards, K.G. (2008). The continuing challenge of agricultural nitrogen loss to the environment in the context of global change and advancing research. *Dynamic Soil, Dynamic Plant*, 2, 1–12.

Steinbauer, M.J. (2011). Relating rainfall and vegetation greenness to the biology of spur-throated and Australian plague locusts. *Agricultural and Forest Entomology*, 13, 205–218.

Steltzer, H. and Post, E. (2009). Seasons and life cycles. *Science*, 324, 886.

Stem, C., Margoluis, R., Salafsky, N., Brown, M. (2005). Monitoring and evaluation in conservation: a review of trends and approaches. *Conservation Biology*, 19, 295–309.

Stenseth, N.C., Ottersen, G., Hurrell, J.W., *et al.* (2003) Studying climate effects on ecology through the use of climate indices: the North Atlantic Oscillation, El Niño Southern Oscillation and beyond. *Proceedings of the Royal Society of London, Series B*, 270, 2087–2096.

Stephens, P.A., Sutherland, W.J., and Freckleton, R.P. (1999). What is the Allee effect? *Oikos*, 87, 185–190.

Sternberg, T., Tsolmon, R., Middleton, N., and Thomas, D. (2011). Tracking desertification on the Mongolian steppe through NDVI and field-survey data. *International Journal of Digital Earth*, 4, 50–64.

Stige, L.C., Stave, J., Chan, K.-S., *et al.* (2006). The effect of climate variation on agro-pastoral production in Africa. *Proceedings of the National Academy of Sciences of the USA*, 103, 3049–3053.

Stimson, H.C., Breshears, D.D., Ustin, S.L., and Kefauver, S.C. (2005). Spectral sensing of foliar water conditions in two co-occurring conifer species: *Pinus edulis* and *Juniperus monosperma*. *Remote Sensing of Environment*, 96, 108–118.

St-Louis, V., Pidgeon, A.M., Clayton, M.K., Locke, B.A., Bash, D., Radeloff, V.C. (2009) Satellite image texture and a vegetation index predict avian biodiversity in the Chihuahuan Desert of New Mexico. *Ecography* 32, 468–480.

Stone, M.L., Solie, J.B., Whitney, R.W., Raun, W.R., and Lees, H.L. (1996). Sensors for the detection of nitrogen in winter wheat. *Technical Paper Series* No. 961757. SAE, Warrendale, PA.

Stowe, L.L., McClain, E.P., Carey, R., *et al.* (1991). Global distribution of cloud cover derived from NOAA/AVHRR operational satellite data. *Advances in Space Research*, 3, 51–54.

Stowe, L.L., Davis, P.A., and McClain, E.P. (1999). Scientific basis and initial evaluation of the CLAVR-1 global clear/cloud classification algorithm for the advanced very high resolution radiometer. *Journal of Atmospheric and Oceanic Technology*, 16, 656–681.

Strand, H., Höft, R., Strittholt, J., Miles, L., Horning, N., Fosnight, E., and Turner, W. (2007). *Sourcebook on Remote Sensing and Biodiversity Indicators*. Secretariat of the Convention on Biological Diversity, Montréal, Technical Series no. 32, 203 pages.

Street, L.E., Shaver, G.R., Williams, M., and Van Wijk, M.T. (2007). What is the relationship between changes in canopy leaf area and changes in photosynthetic CO_2 flux in arctic ecosystems. *Journal of Ecology*, 95, 139–150.

Stöckli, R. and Vidale, P.L. (2004). European plant phenology and climate as seen in a 20-year AVHRR land-surface parameter dataset. *International Journal of Remote Sensing*, 25, 3303–3330.

Suarez-Seoane, S., Osborne, P.E., and Alonso, J.C. (2002). Large-scale habitat selection by agricultural steppe birds in Spain: identifying species–habitat responses using generalized additive models. *Journal of Applied Ecology*, 39, 755–771.

Sumfleth, K. and Duttmann, R. (2008). Prediction of soil property distribution in paddy soil landscapes using terrain data and satellite information as indicators. *Ecological Indicators*, 8, 485–501.

Sun, Z., Chang, N.-B., Opp, C., and Hennig, T. (2011). Evaluation of ecological restoration through vegetation patterns in the lower Tarim River, China with MODIS NDVI data. *Ecological Informatics*, 6, 156–163.

Swets, D.L., Reed, B.C., Rowland, J.D., and Marko, S.E. (1999). A weighted least-squares approach to temporal NDVI smoothing. In: *Proceedings of the 1999 American Society for Photogrammetry and Remote Sensing Annual Conference: From Image to Information, Portland, Oregon, May 17–21, 1999*.

Swihart, R.K., Slade, N.A., and Bergstrom, B.J. (1988). Relating body size to the rate of home range use in mammals. *Ecology*, 69, 393–399.

Swingland, I.R. and Greenwood, P.J. (1987). *The Ecology of Animal Movement*. Clarendon Press, Oxford.

Symeonakis, E. and Drake, N. (2004). Monitoring desertification and land degradation over sub-Saharan Africa. *International Journal of Remote Sensing*, 25, 573–592.

Szep, T. and Møller, A.P. (2004). Using remote sensing data to identify migration and wintering areas and to analyze effects of environmental conditions on migratory birds. In: Greenberg, R. and Marra P.P. (eds), *Birds in Two Worlds*. Smithsonian Institution, Washington, DC, pp. 390–400.

Taddei, R. (1997). Maximum Value Interpolated: a maximum value composite method improvement in vegetation index profiles analysis. *International Journal of Remote Sensing*, 18, 2365–2370.

Tait, A. and Zheng, X.G. (2003). Mapping frost occurrence using satellite data. *Journal of Applied Meteorology*, 42, 193–203.

Tan, S. and Narayanan, R.M. (2004). Design and performance of a multiwavelength airborne polarimetric lidar for vegetation remote sensing. *Applied Optics*, 43, 2360–2368.

Tanadini, M., Schmidt, B.R., Meier, P., Pellet, J., and Perrin, N. (2012). Maintenance of biodiversity in vineyard-dominated landscapes: a case study on larval salamanders. *Animal Conservation*, 15, 136–141.

Tang, Z., Fang, J., Sun, J., and Gaston, K.J. (2011). Effectiveness of protected areas in maintaining plant production. *PLoS ONE*, 6, e19116. doi:10.1371/journal.pone.0019116

Tanré, D., Holben, B.N., and Kaufman, Y.J. (1992). Atmospheric correction algorithm for NOAA-AVHRR products: theory and application. *IEEE Journal of Geosciences and Remote Sensing*, 30, 231–248.

Taylor, S., Kumar, L., and Reid, N. (2011). Accuracy comparison of Quickbird, Landsat TM and SPOT 5 imagery for Lantana camara mapping. *Journal of Spatial Science*, 56, 241–252.

Teal, R.K., Tubana, B., Girma, K., *et al.* (2006). In-season prediction of corn grain yield potential using normalized difference vegetation index. *Agronomy Journal*, 98, 1488–1494.

Texeira, M., Paruelo, J.M., and Jobbagy, E. (2008). How do forage availability and climate control sheep reproductive performance? An analysis based on artificial neural networks and remotely sensed data. *Ecological Modelling*, 217, 197–206.

Thiam, A.K. (1997). Geographic information systems and remote sensing methods for assessing and monitoring land degradation in the Sahel: the case of Southern Mauritania. Doctoral dissertation, Clark University, Worcester, MA.

Thomas, W. (1997). A three-dimensional model for calculating reflection functions of inhomogeneous and orographically structured natural landscapes. *Remote Sensing of Environment*, 59, 44–63.

Thomas, S. (2008). Urbanization as a driver of change. *Arup Journal*, 1, 58–67.

Thomas, D.L. and Taylor, E.J. (2006). Study designs and tests for comparing resource use and availability II. *Journal of Wildlife Management*, 70, 324–336.

Tierney, G.L., Faber-Langendoen, D., Mitchell, B.R., Shriver, W.G., and Gibbs, J.P. (2009). Monitoring and evaluating the ecological integrity of forest ecosystems. *Frontiers in Ecology and the Environment*, 7, 308–316.

Tilling, A.K., O'Leary, G.J., Ferwerda, J.G., *et al.* (2007). Remote sensing of nitrogen and water stress in wheat. *Field Crops Research*, 104, 77–85.

Tilman, D., Reich, P.B., and Isbell, F. (2012). Biodiversity impacts ecosystem productivity as much as resources, disturbance, or herbivory. *Proceedings of the National Academy of Sciences of the USA*, 109, 10394–10397.

Timko, J.A. and Innes, J.L. (2009). Evaluating ecological integrity in national parks: case studies from Canada and South Africa. *Biological Conservation*, 142, 676–688.

Tottrup, C. and Rasmussen, M.S. (2004). Mapping long-term changes in savannah crop productivity in Senegal through trend analysis of time series of remote sensing data. *Agriculture, Ecosystems and Environments*, 103, 545–560.

Tourre, Y.M., Jarlan, L., Lacaux, J.-P., Rotela, C.H., and Lafaye, M. (2008). Spatio-temporal variability of NDVI—precipitation over southernmost South America: possible linkages between climate signals and epidemics. *Environmental Research Letters*, 3, 044008 (9 pp.). doi:10.1088/1748-9326/3/4/044008.

Toutin, T. (2004). Review article: geometric processing of remote sensing images: models, algorithms and methods. *International Journal of Remote Sensing*, 25, 1893–1924.

Townshend, J.R.G. and Justice, C.O. (1986). Analysis of the dynamics of African vegetation using the normalized difference vegetation index. *International Journal of Remote Sensing*, 7, 1435–1446.

Townshend, J.R.G. and Justice, C.O. (1988). Selecting the spatial resolution of satellite sensors required for global monitoring of land transformations. *International Journal of Remote Sensing*, 9, 187–236.

Townshend, J.R.G., Justice, C.O., and Kalb, V.T. (1987). Characterization and classification of South American land cover types using satellite data. *International Journal of Remote Sensing*, 8, 1189–1207.

Trenberth, K.E. (1997). The definition of El Niño. *Bulletin of the American Meteorological Society*, 78, 2771–2777.

Trenberth, K.E. and Hurrell, J.W. (1994) Decadal atmosphere—ocean variations in the Pacific. *Climate Dynamics*, 9, 303–319.

Trigo, R.M., Osborn, T.J., and Corte-Real, J.M. (2002). The North Atlantic Oscillation influence on Europe: climate impacts and associated physical mechanisms. *Climate Research*, 20, 9–17.

Trimble, M.J., Ferreira, S.M., and van Aarde, R.J. (2009). Drivers of megaherbivore demographic fluctuations: inference from elephants. *Journal of Zoology (London)*, 279, 18–26.

Tucker, C.J. (1977). Spectral estimation of grass canopy variables. *Remote Sensing of Environment*, 6, 11–26.

Tucker, C.J. (1979). Red and photographic infrared linear combinations for monitoring vegetation. *Remote Sensing of Environment*, 8, 127–150.

Tucker, C.J. (1980). Remote sensing of leaf water content in the near infrared. *Remote Sensing of Environment*, 10, 23–32.

Tucker, C.J. and Yager, K. (2011). Ten Years of MODIS in space: lessons learned and future perspectives. *Italian Journal of Remote Sensing*, 43, 7–18.

Tucker, C.J., Holben, B.N., Elgin, J.H., and McMurtrey, J.E. (1981). Remote sensing of total dry-matter accumulation in winter wheat. *Remote Sensing of Environment*, 11, 171–189.

Tucker, C.J., Vanpraet, C.L., Boerwinkel, E., and Gaston, A. (1983). Satellite remote sensing of total dry matter production in the Senegalese Sahel. *Remote Sensing of Environment*, 13, 461–474.

Tucker, C.J., Townsend, J.R.G., and Goff, T.E. (1985a). African land-cover classification using satellite data. *Science*, 227, 369–375.

Tucker, C.J., Vanpraet, C.L., Sharman, M.J., and Van Ittersum, G. (1985b). Satellite remote sensing of total herbaceous biomass production in the Senegalese Sahel: 1980–1984. *Remote Sensing of Environment*, 17, 233–249.

Tucker, C., Justice, C., and Prince, S. (1986). Monitoring the grasslands of the Sahel 1984–1985. *International Journal of Remote Sensing*, 7, 1571–1581.

Tucker, C.J., Newcomb, W.W., Los, S.O., and Prince, S.D. (1991). Mean and inter-year variation of growing-season normalized difference vegetation index for the Sahel 1981–1989. *International Journal of Remote Sensing*, 12, 1113–1115.

Tucker, C.J., Grant, D.M., and Dykstra, J.D. (2004). NASA's global orthorectified Landsat data set. *Photogrammetric Engineering & Remote Sensing*, 70, 313–322.

Tucker, C.J., Pinzon, J.E., Brown, M.E., *et al.* (2005). An extended AVHRR 8-km NDVI data set compatible with MODIS and SPOT vegetation NDVI data. *International Journal of Remote Sensing*, 26, 4485–4498.

Tufto, J., Andersen, R., and Linnell, J.D.C. (1996). Habitat use and ecological correlates of home range size in a small cervid: the roe deer. *Journal of Animal Ecology*, 65, 715–724.

Tuomisto, H., Ruokolainen, K., Aguilar, M., and Sarmiento, A. (2003). Floristic patterns along a 43-km long transect in an Amazonian rain forest. *Journal of Ecology*, 91, 743–756.

Turner, D.P., Cohen, W.B., Kennedy, R.E., Fassnacht, K.S., and Briggs, J.M. (1999). Relationships between leaf area index and Landsat TM spectral vegetation indices across three temperate zone sites. *Remote Sensing of Environment*, 70, 52–68.

Turner, D.P., Ritts, W.D., Cohen W.B., *et al.* (2006). Evaluation of MODIS NPP and GPP products across multiple biomes. *Remote Sensing of Environment*, 102, 282–292.

Turner, W., Spector, S., Gardiner, N., Fladeland, M., Sterling, E., and Steininger, M. (2003) Remote sensing for biodiversity science and conservation. *Trends in Ecology and Evolution*, 18, 306–314.

Tuxen, K.A., Schile, L.M., Kelly, M., and Siegel, S.W. (2008). Vegetation colonization in a restoring tidal marsh: a remote sensing approach. *Restoration Ecology*, 16, 313–323.

Underwood, E., Ustin, S., and DiPietro, D. (2003). Mapping nonnative plants using hyperspectral imagery. *Remote Sensing of Environment*, 86, 150–161.

United Nations Economic and Social Commission for Asia Pacific (2005). *Statistical Indicators for Asia and the Pacific*. <http://www.unescap.org/STAT/data/statind/pdf/t2_dec04.pdf>.

United Nations International Strategy for Disaster Reduction (2009). *2009 UNISDR Terminology on Disaster Risk Reduction*. UN, Geneva.

Ünsalan, C. and Boyer, K.L. (2004). Linearized vegetation indices based on a formal statistical framework. *IEEE Transactions on Geoscience and Remote Sensing*, 42, 1575–1585.

Ustin, S.L. and Gamon, J.A. (2010). Remote sensing of plant functional types. *New Phytologist*, 186, 795–816.

Ustin, S.L., Roberts, D.A., and Gamon, J.A. (2004). Using imaging spectroscopy to study ecosystem processes and properties. *BioScience*, 53, 523–534.

Ustin, S.L., Gitelson, A.A., Jacquemoud, S., *et al.* (2009). Retrieval of foliar information about plant pigment systems from high resolution spectroscopy. *Remote Sensing of Environment*, 113, S67–S77.

Vaiopoulos, D., Skianis, G.A., and Nikolakopoulos, K. (2004). The contribution of probability theory in assessing the efficiency of two frequently used vegetation indices. *International Journal of Remote Sensing*, 25, 4219–4236.

van der Molen, M.K., Dolman, A.J., Ciais, P., et al. (2011). Drought and ecosystem carbon cycling. *Agricultural and Forest Meteorology*, 151, 765–773.

Van Dijk, A., Callis, S.L., Sakamoto, C.M., and Decker, W.L. (1987). Smoothing vegetation index profiles: an alternative method for reducing radiometric disturbance in NOAA/AVHRR data. *Photogrammetric Engineering Remote Sensing*, 53, 1059–1067.

van Dyke, F. (2008). *Conservation Biology: Foundations, Concepts, Applications*, 2nd edn. Springer-Verlag, Berlin.

Van Leeuwen, W.J.D., Orr, B.J., Marsh, S.E., and Hermann, S.M. (2006). Multi-sensor NDVI data continuity: Uncertainties and implications for vegetation monitoring applications. *Remote Sensing of Environment*, 100, 67–81.

Van Moorter, B., Bunnefeld, N., Panzacchi, M., Rolandsen, C.M., Solberg, E.J., and Saether, B.-E. (2013). Understanding scales of movement: animals ride waves and ripples of environmental change. *Journal of Animal Ecology*, 2013 15 Feb (Epub ahead of print).

Van Wagtendonk, J.W., and Root, R.R. (2003). The use of multi-temporal Landsat Normalized Difference Vegetation Index (NDVI) data for mapping fuel models in Yosemite National Park, USA. *International Journal of Remote Sensing*, 24, 1639–1651.

Van der Wal, D. and Herman, P.M.J. (2012). Ecosystem Engineering Effects of Aster tripolium and Salicornia procumbens Salt Marsh on Macrofaunal Community Structure. *Estuaries and Coasts*, 35, 714–726.

Venugopal, G. (1998). Monitoring the effects of biological control of water hyacinths using remotely sensed data: a case study of Bangalore, India. *Singapore Journal of Tropical Geography*, 19, 92–105.

Verbesselt, J., Robinson, A., Stone, C., and Culvenor, D. (2009). Forecasting tree mortality using change metrics derived from MODIS satellite data. *Forest Ecology and Management*, 258, 1166–1173.

Verbesselt, J., Hyndman, R., Newnham, G., and Culvenor, D. (2010). Detecting trend and seasonal changes in satellite image time series. *Remote Sensing of Environment*, 114, 106–115.

Verhoef, W., Menenti, M., and Azzali, S. (1996). A colour composite of NOAA-AVHRRNDVI based on time series (1981–1992). *International Journal of Remote Sensing*, 17, 231–235.

Verlinden, A. and Masogo, R. (1997). Satellite remote sensing of habitat suitability for ungulates and ostrich in the Kalahari of Botswana. *Journal of Arid Environments*, 35, 563–574.

Vicente-Serrano, S.M. and Heredia-Laclaustra, A. (2004). NAO influence on NDVI trends in the Iberian peninsula. *International Journal of Remote Sensing*, 25, 2871–2879.

Vierling, K.T., Bassler, C., Brandl, R., Vierling, L.A., Weiss, I., and Muller, J. (2011). Spinning a laser web: predicting spider distributions using LiDAR. *Ecological Applications*, 21, 577–588.

Viovy, N., Arino, O., and Belward, A.S. (1992). The Best Index Slope Extraction (BISE): a method for reducing noise in NDVI time-series. *International Journal of Remote Sensing*, 12, 1585–1590.

Virginia, R.A. and Wall, D.H. (2001). Ecosystem function, principles. In: Levin, S.A. (ed.), *Encyclopedia of Biodiversity*. Academic Press, San Diego.

Vitousek, P.M., D'Antonio, C.M., and Asner, G.P. (2011). Invasions and ecosystems: vulnerabilities and the contribution of new technologies. In Richardson D.M. (ed.). *Fifty Years of Invasion Ecology: The Legacy of Charles Elton*. Wiley–Blackwell, Oxford, pp. 277–288.

Vitousek, P.M., D'Antonio, C.M., Loope, L.L., and Westbrooks, R. (1996). Biological invasions as global environmental change. *American Scientist*, 84, 468–478.

Vogelman, T.C., Rock, B.N., and Moss, D.M. (1993). Red edge spectral measurements from sugar maple leaves. *International Journal of Remote Sensing*, 14, 1563–1575.

Vos, J. and Bom, M. (1993). Hand-held chlorophyll meter: a promising tool to assess the nitrogen status of potato foliage. *Potato Research*, 36, 301–308.

Wabnitz, C. C., Andrefouet, S., Torres-Pulliza, D., Muller-Karger, F.E., and Kramer, P.A. (2008). Regional-scale seagrass habitat mapping in the Wider Caribbean region using Landsat sensors: applications to conservation and ecology. *Remote Sensing of Environment*, 112, 3455–3467.

Wallace, J.M. and Gutzler, D.S. (1981). Teleconnections in the geopotential height field during the Northern Hemisphere winter. *Monthly Weather Review*, 109, 784–812.

Walter, M.J. and Hone, J. (2003). A comparison of 3 aerial survey techniques to estimate wild horse abundance in the Australian Alps. *Wildlife Society Bulletin*, 31, 1138–1149.

Walther, G.R., Post, E., Convey, P., et al. (2002). Ecological responses to recent climate change. *Nature*, 416, 389–395.

Wang, G. (2003). Reassessing the impact of North Atlantic Oscillation on the sub-Saharan vegetation productivity. *Global Change Biology*, 9, 493–499.

Wang, G., Hobbs, N.T., Boone, R.B., et al. (2006) Spatial and temporal variability modify density dependence in populations of large herbivores. *Ecology*, 87, 95–102.

Wang, J., Rich, P.M., and Price, K.P. (2003). Temporal responses of NDVI to precipitation and temperature in the central Great Plains, USA. *International Journal of Remote Sensing*, 24, 2345–2364.

Wang, J., Rich, P.M., Price, K.P., and Kettle, W.D. (2004). Relations between NDVI and tree productivity in the central Great Plains. *International Journal of Remote Sensing*, 25, 3127–3138.

Wang, Q., Watanabe, M., Hayashi, S., and Murakami, S. (2003). Using NOAA AVHRR data to assess flood damage in China. *Environmental Monitoring and Assessment*, 82, 119–148

Wang, Q., Tenhunen, J., Dinh, N.Q., Reichstein, M., Vesala, T., and Keronen, P. (2004). Similarities in ground- and satellite-based NDVI time series and their relationship to physiological activity of a Scots pine forest in Finland. *Remote Sensing of Environment*, 93, 225–237.

Wang, Q., Adiku, S., Tenhunen, J., and Granier, A. (2005). On the relationship of NDVI with leaf area index in a deciduous forest site. *Remote Sensing of Environment*, 94, 244–255.

Wang, W., Anderson, T.B., Phillips, N., Kaufman, K.R., Potter, C., and Myneni, B.R. (2006), Feedbacks of vegetation on summertime climate variability over the North American grasslands. *Earth Interactions*, 10, 1–27.

Wang, X., Piao, S., Ciais, P., et al. (2011). Spring temperature change and its implication in the change of vegetation growth in North America from 1982 to 2006. *Proceedings of the National Academy of Sciences of the USA*, 108, 1240–1245.

Webb, L.B., Whetton, P.H., and Barlow, E.W.R. (2011). Observed trends in winegrape maturity in Australia. *Global Change Biology*, 17, 2707–2719.

Wheeler, S.G., Misra, P.N., and Holmes, A.Q. (1976). *Linear dimensionality of Landsat agricultural data with implications for classifications*. Proceedings of the Symposium on Machine Processing of remotely sensed data, LARS, Purdue University, Indiana.

White, G.C. and Garrott, R.A. (1990). *Analysis of wildlife radio-tracking data*. Academic Press, New York, NY.

White, R.G. (1983). Foraging patterns and their multiplier effects on productivity of northern ungulates. *Oikos*, 40, 377–384.

White, M.A., de Beurs, K.M., Didan, K., et al. (2009). Intercomparison, interpretation, and assessment of spring phenology in North America estimated from remote sensing for 1982–2006. *Global Change Biology*, 15, 2335–2359.

Whittle, M., Quegan, S., Uryu, Y., Stuewe, M., and Yulianto, K. (2012). Detection of tropical deforestation using ALOS-PALSAR: a Sumatran case study. *Remote Sensing of Environment*, 124, 83–98.

Wiegand, T., Naves, J., Garbulsky, M.F., and Fernandez, N. (2008). Animal habitat quality and ecosystem functioning: exploring seasonal patterns using NDVI. *Ecological Monographs*, 78, 87–103.

Wiens, J.A. (1989). Spatial scaling in ecology. *Functional Ecology*, 3, 385–397.

Wiens, J., Sutter, R., Anderson, M., et al. (2009). Selecting and conserving lands for biodiversity: The role of remote sensing. *Remote Sensing of Environment*, 113, 1370–1381.

Wilfong, B.N., Gorchov, D.L., and Henry, M.C. (2009). Detecting an invasive shrub in deciduous forest understories using remote sensing. *Weed Science*, 57, 512–520.

Willems, E.P., Barton, R.A., and Hill, R.A. (2009). Remotely sensed productivity, regional home range selection, and local range use by an omnivorous primate. *Behavioural Ecology*, 20, 985–992.

Williams, B.K., Nichols, J.D., and Conroy, M.J. (2002). *Analysis and Management of Animal Populations*. Academic Press, San Diego.

Willis, C.G., Ruhfel, B., Primack, R.B., Miller-Rushing, A.J., and Davis, C.C. (2008). Phylogenetic patterns of species loss in Thoreau's woods are driven by climate change. *Proceedings of the National Academy of Sciences of the USA*, 105, 17029–17033.

Winnie, J.A., Cross, P., and Getz, W. (2008). Habitat quality and heterogeneity influence distribution and behaviour in African buffalo. *Ecology*, 89, 1457–1468.

Wise, R.M., van Wilgen, B.W., and Le Maitre, D.C. (2012). Costs, benefits and management options for an invasive alien tree species: the case of mesquite in the Northern Cape, South Africa. *Journal of Arid Environments*, 84, 80–90.

Wittemyer, G., Rasmussen, H.B., and Douglas-Hamilton, I. (2007). Breeding phenology in relation to NDVI variability in free-ranging African elephant. *Ecography*, 30, 42–50.

Woodcock, C.E., Allen, R., Anderson, M., et al. (2008). Free access to Landsat imagery. *Science*, 320, 1011.

Woodhouse, I.H., Nichol, C., Sinclair, P., et al. (2011). A multispectral canopy LiDAR demonstrator project. *IEE Geoscience and Remote Sensing Letters*, 8, 839–843.

Worm, B., Barbier, E.B., Beaumont, N., et al. (2006). Impacts of biodiversity loss on ocean ecosystem services. *Science*, 314, 787–790.

Wotton, B.M., Nock, C.A., and Flannigan, M.D. (2010). Forest fire occurrence and climate change in Canada. *International Journal of Wildland Fire*, 19, 253–271.

Wright, D.H. (1983). Species–energy theory: an extension of species–area theory. *Oikos*, 41, 496–506.

Wright, D.L., Rasmussen, V.P., Ramsey, R.D., and Baker, D.J. (2004) Canopy reflectance estimation of wheat nitrogen content for grain protein management. *GIScience and Remote Sensing*, 41, 287–300.

Wu, Z., Dijkstra, P., Koch, G.W., Penuelas, J., and Hungate, B.A. (2011). Responses of terrestrial ecosystems to temperature and precipitation change: a meta-analysis of experimental manipulation. *Global Change Biology*, 17, 927–942.

Wulder, M.A., Dymond, C.C., White, J.C., Leckie, D.G., and Carroll, A.L. (2006). Surveying mountain pine beetle damage of forests: a review of remote sensing opportunities. *Forest Ecology and Management*, 221, 27–41.

Wulder, A., White, J.C., Coops, N.C., and Butson, C.R. (2008). Multi-temporal analysis of high spatial resolution imagery for disturbance monitoring. *Remote Sensing of Environment*, 112, 2729–2740.

Wulder, M.A., Masek, J.G., Cohen, W.B., Loveland, T.R., and Woodcock, C.E. (2012a). Opening the archive: how free data has enabled the science and monitoring promise of Landsat. *Remote Sensing of Environment*, 122, 2–10.

Wulder, M.A., White, J.C., Nelson, R.F., *et al*. (2012b). Lidar sampling for large-area forest characterization: a review. *Remote Sensing of Environment*, 121, 196–209.

Yamasaki, H. (1997). A function of colour. *Trends in Plant Science*, 2, 7–8.

Yang, C.M. and Cheng, C.H. (2001). Spectral characteristics of rice plants infested by brown planthoppers. *Proceedings of the National Scientific Council of the Republic of China, Part B*, 253, 180–186.

Yang, Y., Yang, L., and Merchant, J.W. (1997). An assessment of AVHRR/NDVI—ecoclimatological relations in Nebraska, USA. *International Journal of Remote Sensing*, 18, 2161–2180.

Yazdani, R., Ryerson, A.R., and Derenyi, E. (1981). Vegetation change detection in an area—a simple approach for use with geo-data base. *Proceedings of the 7th Canadian Symposium on Remote Sensing, Winnipeg, Manitoba, Canada*, pp. 88–92.

Yoder, B.J. and Pettigrew-Crosby, R.E. (1995). Predicting nitrogen and chlorophyll content and concentration from reflectance spectra (400–2500 nm) at leaf and canopy scales. *Remote Sensing of Environment*, 53, 199–211.

Yoder, B.J. and Waring, R.H. (1994). The normalized difference vegetation index of small Douglas-fir canopies with varying chlorophyll concentrations. *Remote Sensing of Environment*, 49, 81–91.

Young, K.D., Ferreira, S.M., and van Aarde, R.J. (2009). Elephant spatial use in wet and dry savannas of southern Africa. *Journal of Zoology (London)*, 278, 189–205.

Yu, F., Price, K.P., Ellis, J., and Kastens, D. (2004). Satellite observations of the seasonal vegetation growth in Central Asia: 1982–1990. *Photogrammetric Engineering and Remote Sensing*, 70, 461–469.

Zeng, N., Yoon, J., Marengo, J., *et al*. (2008). Causes and impact of the 2005 Amazon drought. *Environmental Research Letters*, 3, 014002. doi:10.1088/1748-9326/3/1/014002.

Zha, Y., Gao, J., Ni, S., and Shen, N. (2005). Temporal filtering of successive MODIS data in monitoring a locust outbreak. *International Journal of Remote Sensing*, 26, 5665–5674.

Zhang, J.-H., Wang, K., Bailey, J.S., and Wang, R.-C. (2006). Predicting nitrogen status of rice using multispectral data at canopy scale. *Pedosphere*, 16, 108–117.

Zhang, J., Wu, L., Huang, G., Zhu, W., and Zhang, Y. (2011). The role of May vegetation greenness on the southeastern Tibetan Plateau for East Asian summer monsoon prediction. *Journal of Geophysical Research*, 116, D05106, doi:10.1029/2010JD015095

Zhang, X., Friedl, M.A., Schaaf, C.B., *et al*. (2003). Monitoring vegetation phenology using MODIS. *Remote Sensing of Environment*, 84, 471–475.

Zhang, W., Wang, K., Chen, H., He, X., and Zhang, J. (2012). Ancillary information improves kriging on soil organic carbon data for a typical karst peak cluster depression landscape. *Journal of the Science of Food and Agriculture*, 92, 1094–1102.

Zhao, M. and Running, S.W. (2010). Drought-induced reduction in global terrestrial net primary production from 2000 through 2009. *Science*, 329, 940–943.

Zhou, L., Tucker, C.J., Kaufmann, R.K., Slayback, D., Shabanov, N.V., and Myneni, R.B. (2001). Variations in northern vegetation activity inferred from satellite data of vegetation index during 1981 to 1999. *Journal of Geophysical Research*, 106, 20 069–20 083.

Zhou, Y.-B., Newman, C., Xu, W.-T., *et al*. (2011). Biogeographical variation in the diet of Holarctic martens (genus Martes, Mammalia: Carnivora: Mustelidae): adaptive foraging in generalists. *Journal of Biogeography*, 38, 137–147.

Zinner, D., Pelaez, F., and Torkler, F. (2002). Distribution and habitat of grivet monkeys (*Cercopithecus aethiops aethiops*) in eastern and central Eritrea. *African Journal of Ecology*, 40, 151–158.

Index

A
Above-ground biomass 11, 21, 60, 139
Absorption 6, 18, 30
Abundance 95–97, 120–121
African-Eurasian waterbird agreement 129
African convention on the conservation of nature and natural resources 135
Aichi targets 129–130
Albedo 152
Annual relative range 34, 64–65, 107
Anthocyanin 76
Assisted colonization 105
Arctic Oscillation 48, 52
Atmospheric correction 22, 26–27, 38, 41–42
Atmospherically resistant vegetation index 23, 26
Advanced Very High Resolution Radiometer 20, 33, 36, 40, 48, 58

B
Band 7, 20, 28
Biodiversity 5, 97, 143
 Biodiversity indicator 142–143
Bidirectional reflectance distribution function 32
Body mass 90–92

C
Carnivore 82–83, 85, 118
Carotene 18, 76
Carotenoid *see* Carotene
Change detection 13
Chlorophyll 6, 18, 76–78
Conference of Parties 129, 131
Conservation introduction 105
Conservation translocation *see* Translocation
Constancy 65–66
Contingency 65–66
Convention on Biological Diversity 101, 128–129, 131, 143
Convention on International Trade in Endangered Species 131–133, 135
Convention on Migratory Species 129, 131, 141
Corrected transformed vegetation index 23–24
Corridors 107–109
Crop yield 79

D
Deforestation 11, 60, 62, 67–68, 139
Densely vegetated areas 22, 26, 124
Difference vegetation index 23, 26
Digital elevation model 11, 41
Digital observatory on protected areas 135–136, 145
Disease 62–63, 86, 119
Dispersal 87–88
Distribution 72–73, 85, 111–112, 122–123
Drought 58–59

E
Ecological indicator 128
Ecological integrity 103
Ecological replacement 105
Ecosystem 5, 143
 Ecosystem distribution 12, 56–58
 Ecosystem functioning 5
 Ecosystem services 5, 140
 Integrated ecosystem measurements 13
Electromagnetic spectrum 6–8
El Nino Southern Oscillation 46–50, 52–53, 55
Enhanced vegetation index 27, 150–151
Equivalent water thickness 77
Essential biodiversity variable 142–144
Eutrophication 149
Evapotranspiration 21, 32, 44
Extreme winter warming event 63

F
Famine early warning system 59
Flood 58, 61
Fraction of the Photosynthetically Active Radiation 21, 151
Frost 61
Functional attribute 63

G
GeoEye 10
Geographic information system 14–16, 41
Global inventory modelling and mapping studies 33, 37, 106
Green NDVI 76
Greenhouse gas 2, 3, 44, 60
Greenness above bare soil 23, 26
Greenness vegetation index 23, 26
Ground-truthing 16, 42–43

H
Habitat 85–86
 Habitat use 83–84
 Habitat restoration 136
 Habitat selection 83–84
 Habitat suitability 86–87, 109
Home range 84
Hyperspectral 19–20, 152–153

I
IKONOS 9, 14, 78
Infestation 58, 62–63
Infrared 7, 18
Integrated NDVI 33–34, 64–65, 74, 79, 107
Inter-calibration 28
International Panel on Biodiversity and Ecosystem Services 133
International Union for the Conservation of Nature 101, 106

193

International Panel on Climate Change 3, 52, 65, 128
Introduced species 110
Invasive species 110–113, 149, 152
Irradiance 6, 18, 22, 24

K
Key period 90–92

L
Large Area Crop Inventory Experiment 22, 79
Landcover classification 13, 16, 56
Landsat 10, 12, 19, 22, 26, 30, 33, 37, 60, 146–147
Land degradation 67–69
Leaf Area Index 22, 32, 40–41, 44, 151
Light detection and ranging 11, 68, 146, 153
Lusaka agreement 135

M
Mangrove 10, 139
Match-mismatch hypothesis 93
Maximum value composite 40
Mesophyll 19, 25
Mid-infrared 18, 27
Migration 87–89
Misregistration 41
Moderate resolution imaging spectroradiometer 27, 33, 37, 67, 69, 77
Multispectral 7, 19–20

N
National Aeronautics and Space Administration 10, 20, 79
Natural capital 140
Near-infrared 19, 27, 30
Non-such index 23, 26
Normalized burn ratio 60
Normalized difference water index 77
North Atlantic Oscillation 47, 49–52
National Oceanic and Atmospheric Administration 19, 33, 36, 79

O
Omnivore 117
Orbital degradation 38
Orthorectification 14

P
Pacific Decadal Oscillation 49–50
Perpendicular vegetation index 23, 26

Photosynthetic capacity 32, 44
Photosynthetically active radiation 22, 32
Photosynthesis 18, 27, 30, 32
Pigment 18, 76
Pixel 8, 19
Plant community 73
Plant growth 74
Plant mortality 75
Plant nitrogen 77–78
Plant water content 76–77
Predictability 65–66
Primary production 18, 21–22, 32, 40, 44
    Gross primary production 18, 151
    Net primary production 18, 32, 40, 151
Protected area 64–65, 101–104, 128, 135, 138
    Protected area effectiveness 102–104

Q
Quickbird 9, 19, 75, 78

R
Radio detection and ranging 7, 11, 68, 153
Radiation 6–7, 63
Radiometer 7
Radiometric resolution 8, 19–20
Ramsar 131–132, 136–138
Ratio vegetation index 22–23, 26
Red edge 76
Reducing emissions from deforestation and forest degradation 139
Reflectance 13, 18, 22, 24–25, 30, 32, 78
Reflection 6, 18, 30
Reinforcement 105
Reintroduction 81, 104–105
Remote sensing 6, 12
Reproductive success 92–94
Respiration 18

S
Satellite Pour l'Observation de la Terre 9, 19, 79
Scattering 6, 30
Seagrass meadow 10
Sea surface temperature 46, 54–55
Sensor 7, 9, 19
    Optical sensor 7
    Passive sensor 7–8

Active sensor 7–8
Sensor calibration 27, 38
Sensor zenith angle 24–25
Sensor drift 38
Simple ratio 21–23
Smoothing techniques 38–40
Software 15–16, 145
Soil 26–27, 32, 38, 78–79
Soil-adjusted vegetation index 23, 26
Soil brightness index 23, 26
Solar zenith angle 24–25
Spatial resolution 8, 13, 19–20, 119–122
Species-area relationship 72
Species distribution model 73, 86, 113
Species richness 70, 72, 80, 97–98, 126, 149
Spectral resolution 8, 19–20, 58
Spectral signature 5, 13, 19
Structural attribute 63

T
Temporal resolution 8, 13, 19–20, 81, 119–122
Texture 35–36, 98
Transformed vegetation index 22–23
Translocation 105
Transmittance 6, 24–25, 30
Transmission error 38

U
United States Geological Survey 10
United Nations Environmental Programme 128
United Nations Framework Convention on Climate Change 131, 133–134, 139, 143

V
Vegetation index number 22–23
Visible spectrum 6, 18

W
Water availability 66
Wavelength 6, 20, 30
Wetland 136–139, 148–149
Wildlife management 81
Wild fire 58, 60
World heritage convention 132

X
Xanthophyll 18

Y
Yellow vegetation index 23, 26